Lecture Notes in Artificial Intelligence   9150

Subseries of Lecture Notes in Computer Science

LNAI Series Editors

Randy Goebel
  University of Alberta, Edmonton, Canada
Yuzuru Tanaka
  Hokkaido University, Sapporo, Japan
Wolfgang Wahlster
  DFKI and Saarland University, Saarbrücken, Germany

LNAI Founding Series Editor

Joerg Siekmann
  DFKI and Saarland University, Saarbrücken, Germany

T0220841

More information about this series at http://www.springer.com/series/1244

Manfred Kerber · Jacques Carette
Cezary Kaliszyk · Florian Rabe
Volker Sorge (Eds.)

Intelligent Computer Mathematics

International Conference, CICM 2015
Washington, DC, USA, July 13–17, 2015
Proceedings

 Springer

Editors
Manfred Kerber
University of Birmingham
Birmingham
UK

Florian Rabe
Jacobs University Bremen
Bremen
Germany

Jacques Carette
McMaster University
Hamilton, ON
Canada

Volker Sorge
University of Birmingham
Birmingham
UK

Cezary Kaliszyk
University of Innsbruck
Innsbruck
Austria

ISSN 0302-9743   ISSN 1611-3349   (electronic)
Lecture Notes in Artificial Intelligence
ISBN 978-3-319-20614-1   ISBN 978-3-319-20615-8   (eBook)
DOI 10.1007/978-3-319-20615-8

Library of Congress Control Number: 2015942531

LNCS Sublibrary: SL7 – Artificial Intelligence

Springer Cham Heidelberg New York Dordrecht London

Printed on acid-free paper

Springer International Publishing AG Switzerland is part of Springer Science+Business Media
(www.springer.com)

Preface

Digital and computational solutions are becoming the prevalent means for the generation, communication, processing, storage, and curation of mathematical information. Separate communities have been developed to investigate and build computer-based systems for computer algebra, automated deduction, and mathematical publishing as well as novel user interfaces. While all of these systems excel in their own right, their integration can lead to synergies offering significant added value. The Conference on Intelligent Computer Mathematics (CICM) offers a venue for discussing and developing solutions to the great challenges posed by the integration of these diverse areas.

CICM has been held annually as a joint meeting since 2008, co-locating related conferences and workshops to advance work in these subjects. Previous meetings have been held in Birmingham (UK 2008), Grand Bend (Canada 2009), Paris (France 2010), Bertinoro (Italy 2011), Bremen (Germany 2012), Bath (UK 2013), and Coimbra (Portugal 2014).

CICM 2015 was held in Washington, DC, during July 13–17, 2015, with four invited presentations, five main tracks, a number of workshops, a doctoral mentoring program, and an informal track to share work in progress. The program of the meeting, as well as additional materials, are available at http://cicm-conference.org/2015/cicm.php.

We were pleased to have four distinguished invited speakers to make presentations on a set of subjects, each touching on several CICM topics. This volume includes one abstract of a presentation by Leonardo de Moura (Microsoft Research) on "Formalizing Mathematics with The Lean Theorem Prover" and two full papers that were each written with co-authors by Tobias Nipkow (Technische Universität München) on "Mining the Archive of Formal Proofs" and Richard Zanibbi (Rochester Institute of Technology) on "Math Search for the Masses: Multimodal Search Interfaces and Appearance-Based Retrieval." Furthermore, we had a talk by Jim Pitman (University of California, Berkeley) about GDML, the Global Digital Math Library project.

The abstract follows this preface, the two full papers are the first two in these proceedings.

As in previous years, we had several tracks, five altogether: Calculemus, Digital Mathematics Libraries (DML), and Mathematical Knowledge Management (MKM), which mirror the three main communities that form CICM, and two tracks on Systems and Data and Projects and Surveys.

The Calculemus track of CICM examines the integration of symbolic computation and mechanized reasoning, the Digital Mathematics Libraries track deals with math-aware technologies, standards, algorithms and processes, the Mathematical Knowledge Management track is concerned with all aspects of managing mathematical knowledge in informal, semi-formal, and formal settings.

The Systems and Projects track of previous years was split into two different tracks, first Systems and Data to explicitly stress the importance of data collections, and second Projects and Surveys to also encourage survey papers, which may become a starting point for people interested in a particular subject.

Prior to the creation of CICM, two of the present tracks already had a significant history: there had been 15 previous Calculemus meetings and six MKM conferences. In 2007, when Calculemus and MKM were held together in Hagenberg, Austria, as part of the RISC Summer, it was decided to continue to hold these meetings together. This led to the first CICM in 2008. The DML track has been present since that first CICM, at first as a workshop. The other tracks were added in the form of a Systems and Projects track in 2011.

This year, CICM had 43 submissions. Of these, there were 27 full papers and 16 shorter descriptions (on systems, data, or projects). A small number of papers was moved between tracks when it was felt there would be a more natural fit. Each submission received at least three reviews, some four or even five. The review phase included a response period, in which authors could clarify points raised by the reviewers. This made for a highly productive round of deliberations before the final decisions were taken. In the end, the track Program Committees decided to accept 16 full papers (two of them surveys) and nine in Systems and Data or Projects for these proceedings.

The Program Committee work for the tracks was managed using the EasyChair system. This allowed committee members to declare actual or potential conflicts of interest, and thereby be excluded from any deliberations on those papers. Submissions on which track chairs had conflicts were handled by the general program chair. In this way, committee members could (and did) submit papers for consideration without compromising the peer-review process.

As in previous years, several workshops and informal programs were organized in conjunction with CICM 2015. This year these were:

- CICM Doctoral Program, providing a dedicated forum for PhD students to present their on-going or planned research and receive feedback, advice and suggestions from a dedicated research advisory board.
- CICM Work-in-Progress Session, a forum for the presentation of original work not yet in a suitable form for communication as a formal paper.
- ThEdu 2015 – Theorem Provers Components for Educational Software, with the goal of combining and focus systems from theorem proving, computer algebra, and dynamic geometry to enhance existing educational software and the design of the next generation of mechanized mathematics assistants. ThEdu was organized by Walther Neuper, Graz University of Technology, Austria, and Pedro Quaresma, University of Coimbra, Portugal.
- MathUI 2015 – 10th Workshop on Mathematical User Interfaces, an international workshop to discuss how users can be best supported when doing/learning/searching for/interacting with mathematics using a computer. MathUI was organized this year by Andrea Kohlhase, University of Applied Sciences Neu-Ulm, and Paul Libbrecht, University of Education of Weingarten, Germany.

- Formal Mathematics for Mathematicians, a workshop dealing with developing large repositories of advanced mathematics. It was organized by Adam Grabowski, Artur Kornilowicz, University of Białystok, Poland, Krystyna Kuperberg, Auburn University, USA, Adam Naumowicz, University of Białystok, Poland, and Josef Urban, Radboud University, The Netherlands.

We thank all those who contributed to this meeting. We are grateful for the support by The George Washington University, the National Institute of Standards and Technology, the Sloan Foundation, MapleSoft, Wolfram Inc., and Microsoft. We would like to thank the EasyChair team (Andrei Voronkov et al.) for the EasyChair system, which we found indispensable. We would like also to thank the invited speakers, the contributing authors, the reviewers, the members of the Program Committee and the local organizers, all of whose efforts contributed to the practical and scientific success of the meeting.

May 2015

Manfred Kerber
Jacques Carette
Cezary Kaliszyk
Florian Rabe
Volker Sorge

Organization

CICM Steering Committee

Serge Autexier (Publicity/Workshop Officer)
Thierry Bouche (DML Delegate)
Bill Farmer (Treasurer)
Manfred Kerber (CICM PC Chair 2015)
Michael Kohlhase (Secretary)
Florian Rabe (MKM Delegate)
Renaud Rioboo (CALCULEMUS delegate)
Stephen Watt (CICM PC Chair 2014)

CICM 2015 Organizing Committee

General Program Chair, Projects and Surveys Track Chair

Manfred Kerber   University of Birmingham, UK

Local Arrangements Chairs

Bruce R. Miller   NIST, USA
Abdou Youssef   George Washington University, USA

Calculemus Track Chair

Jacques Carette   McMaster University, Canada

DML Track Chair

Volker Sorge   University of Birmingham, UK

MKM Track Chair

Cezary Kaliszyk   University of Innsbruck, Austria

Systems and Data Track Chair

Florian Rabe   Jacobs University Bremen, Germany

Doctoral Program Chair

Umair Siddique   Concordia University, Canada

Publicity and Workshops Chair

Serge Autexier   DFKI, Germany

Program Committee

Calculemus Track Program Committee

Jacques Carette   McMaster University, Canada
Frédéric Chyzak   Inria, France
James H. Davenport   University of Bath, UK
Madalina Erascu   Institute e-Austria and West University of Timisoara,
                          Romania
Michal Konecny   Aston University, UK
Anders Mörtberg   University of Gothenburg, Sweden
François Pessaux   ENSTA ParisTech, France
Renaud Rioboo   ENSIIE, France
Bas Spitters   The Netherlands
Makarius Wenzel   Germany
David Wilson   University of Bath, UK
Wolfgang Windsteiger   RISC Institute, JKU Linz, Austria

DML Track Program Committee

Volker Sorge   University of Birmingham, UK
Thiery Bouche   University of Grenoble, France
Joe Cornelli   Planet Math, USA
Thomas Fischer   University Library Göttingen, Germany
Toshihiro Kanahori   Tsukuba University, Japan
Peter Krautzberger   MathJax Consortium, USA
Ross Moore   University of Queensland, Australia
Jiri Rakosník   Czech Academy of Sciences, Czech Republic
David Ruddy   Cornell University Library, USA
Noureddin Sadawi   Brunel University, UK
Petr Sojka   Masaryk University, Czech Republic

MKM Track Program Committee

Cezary Kaliszyk   University of Innsbruck, Austria
Andrea Asperti   University of Bologna, Italy
David Aspinall   University of Edinburgh, UK
Pierre Corbineau   Verimag
Marcos Cramer   University of Luxembourg, Luxembourg
Oleg Golubitsky   Google, Inc.
Gudmund Grov   Heriot-Watt University, UK
Predrag Janičić   University of Belgrade, Serbia
Andrea Kohlhase   University of Applied Sciences Neu-Ulm, Germany
George Labahn   University of Waterloo, Canada

Bruce Miller   NIST, USA
Grant Passmore   University of Cambridge and University of Edinburgh, UK
Erik Postma   Maplesoft
Aleksy Schubert   University of Warsaw, Poland
Christoph Schwarzweller   Gdansk University, Poland
Alan Sexton   University of Birmingham, UK
Elena Smirnova   Texas Instruments, USA
Sofiène Tahar   Concordia University, Canada
Christian Urban   King's College London, UK
Josef Urban   Radboud University, The Netherlands

Systems and Data Track Program Committee

Florian Rabe   Jacobs University Bremen, Germany
Serge Autexier   DFKI, Germany
Paul-Olivier Dehaye   University of Zurich, Switzerland
Matthew England   University of Bath, UK
Yannis Haralambous   Institut Mines-Télécom, France
Moa Johansson   Chalmers University, Sweden
Christoph Lange   University of Bonn, Germany
Adam Naumowicz   University of Białystok, Poland
Pedro Quaresma   University of Coimbra, Portugal
Alan Sexton   University of Birmingham, UK
Geoff Sutcliffe   University of Miami, USA
Frank Tompa   University of Waterloo, Canada
Josef Urban   Radboud University, The Netherlands

Papers in the Projects and Surveys track were reviewed by Program Committee members from the other tracks.

Additional Reviewers

Caminati, Marco B.
Chrząszcz, Jacek
Hardin, Thérèse
Hölzl, Johannes
Jackson, Paul
Kohlhase, Michael
Kumar, Ramana
Lüth, Christoph
Maggesi, Marco

Mahmoud, Mohamed Yousri
Pałka, Michał
Rosén, Dan
Sacerdoti Coen, Claudio
Siddique, Umair
Smallbone, Nicholas
Staton, Sam
Youssef, Abdou
Zeilberger, Noam

Abstracts of Invited Talks

Formalizing Mathematics with the Lean Theorem Prover

Leonardo de Moura

Microsoft Research
leonardo@microsoft.com

Abstract. Lean is a new open source theorem prover being developed at Microsoft Research and Carnegie Mellon University, with a small trusted kernel based on dependent type theory. It aims to bridge the gap between interactive and automated theorem proving, by situating automated tools and methods in a framework that supports user interaction and the construction of fully specified axiomatic proofs. The goal is to support both mathematical reasoning and reasoning about complex systems, and to verify claims in both domains. Lean is an ongoing and long-term effort, and much of the potential for automation will be realized only gradually over time, but it already provides many useful components, integrated development environments, and a rich API which can be used to embed it into other systems. It is currently being used to formalize basic datatypes and algebraic structures, a library for homotopy type theory, rudimentary category theory, and elements of non-abelian topology. The core parts of the Lean standard library have been developed constructively, but we also provide a smooth transition to classical logic. If users want to work classically, they just have to load the classical axioms and/or files built on them.

In this talk, we provide a short introduction to the Lean theorem prover, describe how mathematical structures (e.g., groups, rings and fields) are encoded in the system, quotient types, the type class mechanism, and the main ideas behind the novel elaboration algorithm implemented in Lean. More information about Lean can be found at http://leanprover.github.io. The interactive book "Theorem Proving in Lean"[1] is the standard reference for Lean. The book is available in PDF and HTML formats. In the HTML version, all examples and exercises can be executed in the reader's web browser. This book is part of the course material for the interactive theorem proving course[2] offered in the spring of 2015 at Carnegie Mellon University.

---

[1] http://leanprover.github.io/tutorial
[2] http://www.cs.cmu.edu/~emc/15815-s15

Mining the Archive of Formal Proofs

Jasmin Christian Blanchette[1,2], Maximilian Haslbeck[3], Daniel Matichuk[4,5], and Tobias Nipkow[3]

[1] Inria Nancy & LORIA, Villers-lès-Nancy, France
[2] Max-Planck-Institut für Informatik, Saarbrücken, Germany
[3] Fakultät für Informatik, Technische Universität München, Munich, Germany
nipkow@in.tum.de
[4] NICTA, Sydney, Australia
[5] University of New South Wales, Sydney, Australia

Abstract. The Archive of Formal Proofs is a vast collection of computer-checked proofs developed using the proof assistant Isabelle. We perform an in-depth analysis of the archive, looking at various properties of the proof developments, including size, dependencies, and proof style. This gives some insights into the nature of formal proofs.

Math Search for the Masses: Multimodal Search Interfaces and Appearance-Based Retrieval

Richard Zanibbi and Awelemdy Orakwue

Document and Pattern Recognition Lab, Department of Computer Science,
Rochester Institute of Technology, Rochester, NY 14623, USA
`rlaz@cs.rit.edu`

Abstract. We summarize math search engines and search interfaces produced by the Document and Pattern Recognition Lab in recent years, and in particular the m_{in} math search interface and the *Tangent* search engine. Source code for both systems are publicly available. "The Masses" refers to our emphasis on creating systems for mathematical non-experts, who may be looking to define unfamiliar notation, or browse documents based on the visual appearance of formulae rather than their mathematical semantics.

Contents

Projects and Surveys

Systems and Data

Invited Talks

Mining the Archive of Formal Proofs

Jasmin Christian Blanchette[1,2], Maximilian Haslbeck[3], Daniel Matichuk[4,5],
and Tobias Nipkow[3](✉)

[1] Inria Nancy and LORIA, Villers-lès-Nancy, France
[2] Max-Planck-Institut für Informatik, Saarbrücken, Germany
[3] Fakultät für Informatik, Technische Universität München, Munich, Germany
nipkow@in.tum.de
[4] NICTA, Sydney, Australia
[5] University of New South Wales, Sydney, Australia

Abstract. The Archive of Formal Proofs is a vast collection of computer-checked proofs developed using the proof assistant Isabelle. We perform an in-depth analysis of the archive, looking at various properties of the proof developments, including size, dependencies, and proof style. This gives some insights into the nature of formal proofs.

1 Introduction

The *Archive of Formal Proofs* (*AFP*, http://afp.sf.org) is an online library of proof developments for the proof assistant Isabelle [21] contributed by its users. The AFP is organized like a scientific journal. Each contribution is called an *article* and is a collection of Isabelle *theories*, i.e., files with definitions, lemmas, and proofs in Isabelle's input language Isar [20,29]. A few articles are ML programs that realize specialized definition or proof facilities. The AFP was started in 2004. This paper refers to the AFP snapshot of 16 April 2015, which contains a total of 64,497 lemmas. The term *lemmas* subsumes theorems and corollaries throughout the paper.

The purpose of this paper is to analyze the following properties of the AFP: general size statistics, dependency graph, proof style and proof size, and performance of fully automatic proof. We attempt to answer a number of questions:

- To what extent are AFP articles reused as the basis of other articles?
- How large are articles?
- What percentage of text is taken up by definitions, lemmas, and proofs?
- How many contributors are behind the AFP, and how large are their contributions?
- Does the dependency graph share characteristics with citation graphs in the scientific literature?
- How did the AFP evolve over time?
- Can we estimate the size of a proof from the statement to be proved?
- What is the typical structure of lemma statements? Are they mostly equalities, Horn clauses, or more complex formulas?

© Springer International Publishing Switzerland 2015
M. Kerber et al. (Eds.): CICM 2015, LNAI 9150, pp. 3–17, 2015.
DOI: 10.1007/978-3-319-20615-8_1

- How successful is the automatic proof tool Sledgehammer [4,22] at discharging various goals from the AFP, as opposed to the smaller benchmark sets used in earlier evaluations?

This appears to be the first in-depth analysis of a large collection of computer-checked formal proofs, with the partial exception of Josef Urban's work on the Mizar Problems for Theorem Proving [26], a library of problems for first-order automatic theorem proving generated from the Mizar Mathematical Library [1].

The AFP is heavily biased towards computer science: of the 215 articles, 146 are indexed under computer science and only 82 articles under either mathematics or logic. (Entries may occur in multiple categories.) In contrast, the Mizar Mathematical Library is heavily biased towards mathematics.

Although Isabelle is a generic proof assistant supporting several object logics, all AFP articles use higher-order logic (HOL) as their object logic. Isabelle's version of HOL corresponds to Church's simple type theory [8] extended with polymorphism and Haskell-style type classes. HOL allows nested function types and quantification over functions. Predicates are simply functions to the Boolean type. Named functions are called *constants* in HOL terminology, even if they take arguments. Thus, in the formula $x + 0 = x$, both 0 and + are constants, whereas x is a variable. Otherwise, HOL conventions are a mixture of mathematics and functional programming.

2   Sizes

In its 11 years, the AFP has grown to one million lines of "code" (LOC)— 1,018,800 LOC to be precise—where "code" refers to definitions and proofs (including comments but not empty lines). The growth of the AFP over time is shown in Figs. 1 and 2. The growth in Fig. 1 looks roughly linear. If one examines the growth rate, it fluctuates but has an upward trend. The development of the total number of authors that have ever contributed to the AFP is shown in Fig. 3 and it is similar to the size graph. The number of authors active each year, shown in short lighter (in colour: pink) bars in Fig. 3, has only been growing slowly.

The distribution of sizes of the articles is shown in Fig. 4. Half the articles have up to 2,000 LOC, beyond that the number of articles decays sharply, but with a long tail, not all of which is visible in the figure: 9 AFP articles are larger than 20,000 LOC; the largest AFP article (77,100 LOC) is `JinjaThreads`, a formalization of a dialect of Java by Andreas Lochbihler [16]. At first sight it is not clear what distribution Fig. 4 follows. Initially, it shows an exponential decay, but a better fit is a power law: a log-log plot shows something close to a straight line. Many similar phenomena, e.g., file sizes on the Internet [9], also follow a power law.

2.1   Definitions vs. Lemmas vs. Proofs

The three largest categories of text in the AFP are proofs, lemma statements, and definitions (in that order):

Proofs:	593,828 LOC (58 %)
Lemma statements:	192,576 LOC (19 %)
Definitions:	85,808 LOC (8 %)

Above we have only counted proofs associated with lemmas.

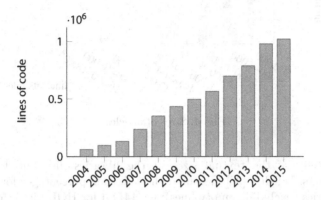

Fig. 1. Size of AFP over time (cumulative)

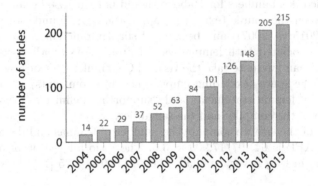

Fig. 2. Number of AFP articles over time (cumulative)

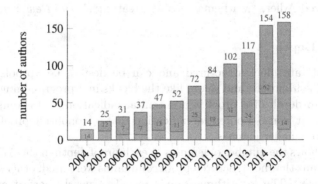

Fig. 3. Number of AFP authors over time (cumulative)

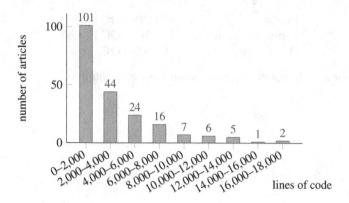

Fig. 4. Sizes of articles

In his study of the textual sizes of the libraries distributed with four proof assistants, Wiedijk [30] measured the following percentages for the above three categories (excluding empty lines): 62/14/1.4 for HOL Light, 50/21/8 for Isabelle, 60/12/10 for Coq, and 84/9/3 for Mizar. The discrepancy between the AFP and Wiedijk's numbers for Isabelle should not surprise because of the differences between the source texts (e.g., applications vs. foundations, degree of polish, age (2015 vs. 2007)) and because of slightly different counting schemes (e.g., proofs associated with lemmas vs. all proofs). As another example that these numbers can fluctuate take the verified C compiler (in Coq) by Leroy [15]: only 44 % of the space (excluding empty lines and comments) is taken up by proofs, 21 % by lemma statements and "supporting definitions", and 24 % by the definition of the compiler and the semantics.

If instead of the size we compare the number of lemmas and definitions, the ratio for the AFP is $64{,}497/17{,}909 \approx 3.6$. The ratio for the proof of the odd-order theorem in Coq is very similar: $13{,}000/4{,}000 \approx 3.25$ [11]. This echoes an old adage:

> One good definition is worth three theorems.
> — Alfred Adler, "Mathematics and Creativity," *The New Yorker* (1972)

2.2 Proof Depth

Isabelle proofs are block-structured and can be nested, i.e., proofs can have subproofs, sub-subproofs, and so on, like the blocks in a programming language. The maximum depth of a proof is a potential indication of its complexity. At the same time, it is also a potential indication that a monolithic proof should be refactored into smaller lemmas.

Figure 5 shows the number of lemmas at each proof depth in the AFP. A proof depth of 1 means that no structured proof commands were used, otherwise known as a *proof script*. The logarithmic y-axis reveals a nearly perfect exponential distribution. The vast majority of proof goals never exceeds a depth of 1. But

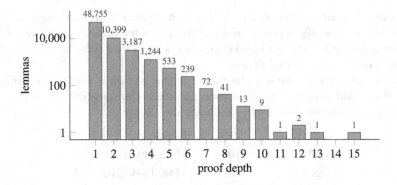

Fig. 5. Number of lemmas at each proof depth

still more than 1 % have a depth of 5 or more, and there is even one proof of depth 15.

It would be interesting to compare this with other block-structured proof languages, such as Mizar [18] and the TLA$^+$ Proof System [7]. Proofs written in the latter system tend to have a richer hierarchic structure. It is not clear to us whether this is due to the application area (the verification of concurrent and distributed algorithms), to specific features of the proof language, or simply to the personal preferences of its users.

3   The Imports Graph

Figure 6 shows the AFP *imports graph*. The nodes of the graph are the AFP articles. We say that an AFP article E_2 *imports* an article E_1 if some theory

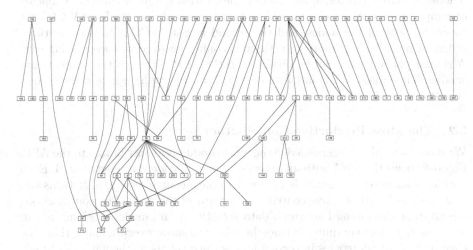

Fig. 6. The AFP imports graph

of E_2 imports some theory of E_1. In this case there is a directed edge from E_1 to E_2. We say that E_2 *depends on* E_1 if there is a non-empty path from E_2 to E_1. The graph is a dag; in Fig. 6 edges always go downwards. We do not show transitive edges and isolated nodes.

One of the questions we want to answer in this section is whether the imports graph share characteristics with citation graphs in the scientific literature.

The key characteristics of the graph are as follows:

Nodes:	215	Depth:	6
Isolated nodes:	106	Max. out-degree:	9
Edges:	97	Max. in-degree:	4

The top three articles with respect to their out-degree, i.e., the three most popular articles, are shown in Fig. 7. The top article, `Collections`, is a framework for efficient implementations of common collection types like sets and maps [14]. All three articles are of a generic nature and designed for reuse.

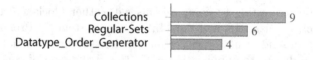

Fig. 7. The most popular AFP articles

3.1   Weakly Connected Components

Now we consider the weakly connected components (WCC) in the graph, i.e., the maximal subgraphs such that from any node there is a path to any other node where edges may be followed either forwards or backwards. Each WCC is a group of loosely related articles. When mining the citation graphs of different subfields of computer science, it was observed [2] that there is always one large WCC that covers 80–90 % of the nodes, and the second largest WCC is smaller by three orders of magnitude: almost everything is connected to almost everything else. A similar phenomenon can be observed in the AFP, although the numbers are smaller. The largest WCC has 70 nodes (1/3 of the graph), whereas the second largest one has 5 nodes.

3.2   The Most Productive Contributors

We should also like to acknowledge the most productive contributors to the AFP. Figure 8 shows the top 5 authors in terms of lines of code they contributed. Each author of an n-author article is assumed to have contributed $1/n$. It turns out that the top 5 authors have contributed a third of the AFP. The contributions are all in programming languages, data structures, and model checking, not in mathematics. The computer science bias is even more overwhelming than the classification of all articles into computer science versus mathematics and logic mentioned in the introduction would have lead us to expect.

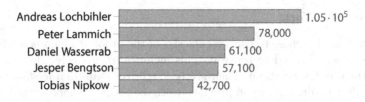

Fig. 8. The most productive authors

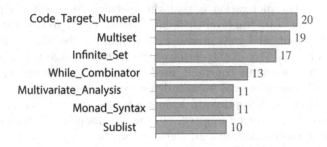

Fig. 9. The most popular library theories

3.3 Library Theories

Isabelle/HOL provides the theories Main and Complex_Main that contain many standard theories such as sets, relations, lists, natural numbers, integers, and real numbers. Many applications build on either of those. In addition, there is a large number of library theories that are included with Isabelle/HOL and can be imported selectively. The difference to the AFP is that these library theories are originated by the Isabelle developers and are specifically designed for reuse. Figure 9 shows the most popular library theories and how often they are used. The top theories are a mixture of specific theories (e.g., Multiset) and substantial mathematical developments (e.g., Multivariate_Analysis [12]).

Comparing Fig. 9 with Fig. 7, we find that the top library theories are imported twice as often as the top AFP articles, although both are designed for reuse. There are a number of explanations: library theories are developed by the Isabelle developers who concentrate on the most fundamental theories; adding a theory to the library is a more lightweight process than adding a new article to the AFP; the library was started at least four years before the AFP.

4 Lemma Statement Size vs. Proof Size

In addition to complete proof developments, the size of individual lemma proofs can also be considered. A naive measurement could consider the number of lines between the lemma and qed keywords to be the size of a proof. However, the intent of measuring the proof size is generally to estimate the effort required to produce that proof. Such a naive measurement will fail to capture the cumulative nature of proofs; for example, a simple corollary will seem to have a trivial proof despite depending on some large result.

4.1   Previous Work

Matichuk *et al.* [17] chose to consider the size of the proof of a lemma as the total number of lines required to prove it. This includes (recursively) the sizes of all lemmas the proof depends on. Their study sought to establish a *leading measure* of proof effort by building upon previous work [25], which demonstrated a linear relationship between proof effort (measured in person-weeks) and proof size. Matichuk *et al.* investigated the relationship between proof size and statement size of lemmas. The motivation is that the specification size for a proof (i.e. the statement size of its top-level theorem) is much easier to calculate early in a proof development, and thus it would be valuable to be able to use this to estimate the eventual size of the entire proof. Statement size was measured as the total number of unique constants used in the lemma statement, including (recursively) all constants used in their definitions.

This proved to be susceptible to lemma *over-specification*, where constants were mentioned but never interpreted, i.e., the lemma could instead have been abstracted over those constants. In these cases, the size of the lemma statement was much larger than its proof would indicate. This prompted an *idealized* measure, which discounts constants whose definitions are never unfolded in the entire proof.

The study examined six software verification proof developments: four proofs from the L4.verified project as well as JinjaThreads and SATSolver Verification from the AFP. They compared the raw statement and idealized statement sizes to the proof sizes for all lemmas in each proof and found a consistent quadratic relationship, which was strengthened by the idealized measure. However, the exact nature of the relationship was different between each proof: a model built against one proof does not necessarily fit others.

4.2   Analysis Against the AFP

We performed the same analysis against three of the largest AFP articles, shown in Table 1 and Fig. 10. Here R^2 is the usual coefficient of determination for statistical models, where an R^2 of 1 indicates that the model fits the data perfectly. We see that Group-Ring-Module partially fits a quadratic model, which can be explained by the hierarchical nature of the proofs. Although JinjaThreads and Group-Ring-Module both fit a quadratic model, the quadratic coefficient on their regression lines differ by an order of magnitude. Psi_Calculi, does not fit the same relationship. There is a column of data points at a statement size of about 100 (or about 60 idealized), indicating that most lemmas in the development actually have the same statement size. This can be explained by the fact that Psi_Calculi is a language formalization. Each lemma mentions the inductive set which defines the language semantics, and the size of that constant dominates the statement size of the lemma. This indicates that this measure of lemma statement size is too coarse for Psi_Calculi: no model built against this data will be able to discriminate between long and short proofs.

Table 1. R^2 and coefficients a, b, c for quadratic regression with equation $ax^2 + bx + c$
statement size versus proof size

AFP article	Measure	R^2	a	b	c
JinjaThreads	Raw	0.346	0.04	10.04	287.22
	Idealized	0.712	0.12	16.48	283.49
Group-Ring-Module	Raw	0.487	1.29	29.20	154.81
	Idealized	0.622	2.26	20.56	58.25
Psi_Calculi	Raw	0.349	0.85	69.21	609.56
	Idealized	0.431	4.87	198.49	798.34

The results are similarly diverse when performing this analysis against the
entire AFP. Using the idealized measure, approximately half of the articles have
an R^2 of less than 0.5 (or have too few data points to build a model), 50 have
an R^2 between 0.5, and 0.7 and 50 have an R^2 greater than 0.7. The best fitting
articles are primarily those related to software verification, e.g. JinjaThreads,
SATSolverVerification, DiskPaxos. Among these, the quadratic coefficients
span two orders of magnitude, from 0.07 to 21.29. The variation in the consistency of this relationship, and the differences between the models, demonstrates
proof size cannot be estimated based solely on this coarse measure of lemma
statement size. It is, however, an indication that it could be used as part of a
more sophisticated measure, which considers the particular domain of the proof
as well as additional measures of lemma statement complexity, such as those
discussed in the following section.

5   Lemma Statement Complexity

We are interested in the complexity of the 64,497 lemma statements, using more
traditional metrics than in the previous section, such as clause and literal counts.
We used Isabelle's clausifier to rewrite formulas into conjunctive normal form
(CNF), i.e. as a conjunction of clauses, each of which is a disjunction of literals.
Literals of the form $\neg\, a$ are negative; otherwise, they are positive.

Out of the AFP's 64,497 lemmas, the clausifier times out or fails on 171
of them. These are completely excluded from the statistics below. A manual
inspection reveals that these are typically highly complex formulas, such as custom induction schemas.

As measures of formula complexity, Fig. 11 gives the number of clauses per
lemma, and Fig. 12 gives the number of literals per lemma.

A few lemmas give rise to zero clauses: these are typically simple tautologies
identified as such by the clausifier. Most formulas give rise to exactly one clause.
These are further classified as follows:

- 11,444 formulas correspond to unit equality clauses, i.e., simple equations of
  the form $t = u$ for two terms t, u.
- 22,475 formulas correspond to conditional equality clauses, i.e., clauses that
  contain at least one positive literal of the form $t = u$.

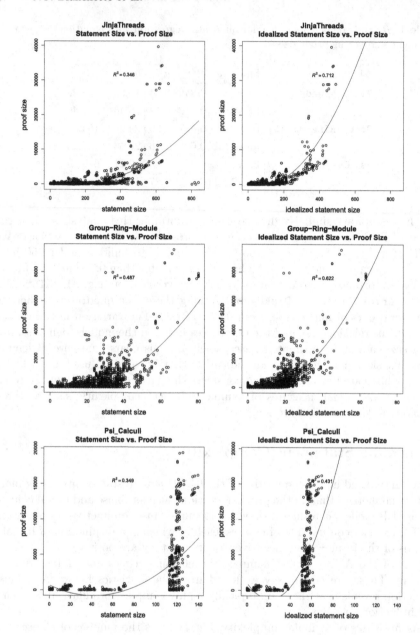

Fig. 10. Relation between statement size and proof size

– 46,186 formulas correspond to Horn clauses, i.e., clauses that contain at most one positive literal. These can be seen as implications $a_1 \wedge \cdots \wedge a_n \implies a$ or $a_1 \wedge \cdots \wedge a_n \implies \texttt{False}$ and are relatively easy to reason automatically about.

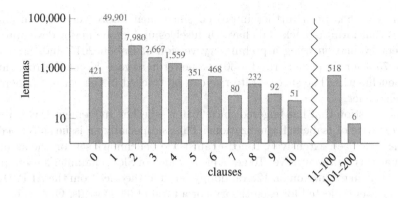

Fig. 11. Number of lemmas with a corresponding number of clauses

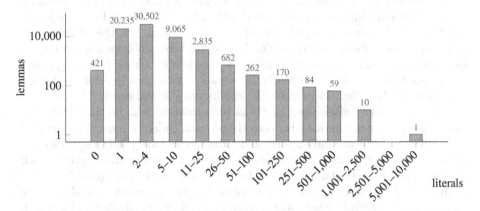

Fig. 12. Number of lemmas with a corresponding number of literals

6 Proof Automation with Sledgehammer

Sledgehammer is a proof tool for Isabelle that exploits powerful first-order automatic theorem provers, notably E [24], SPASS [28], Vampire [23], and Z3 [10]. Given a proof goal, it heuristically selects a few hundred potentially relevant lemmas, invokes the external provers, and upon success produces a proof snippet that can be inserted in the user's formalization to discharge the goal. Similar tools are available for other proof assistants, notably MizAR [27] for Mizar and HOLyHammer [13] for HOL Light.

Sledgehammer was introduced in the 2007 edition of Isabelle and started to be used seriously in 2009. In their "Judgement Day" study from 2010, Böhme and Nipkow [6] evaluated Sledgehammer on 1240 subgoals emerging from seven theory representing various applications of Isabelle to computer science and mathematics. They reported a success rate of 46 % for three provers (E, SPASS, and Vampire) run in parallel for 30 s, meaning that nearly half of the goals in these seven theories could be discharged automatically, with no user guidance.

The tool has been further improved since then. Moreover, the automatic provers that form its back-ends have themselves undergone major development. A recent evaluation using a preliminary version of Isabelle2015 finds a success rate of 75 % for six provers run for 30 s on newer hardware [5], for the Judgement Day benchmarks. The success rate rise can be tracked in the various papers on Sledgehammer.

One question that has lingered since 2010 is whether Judgement Day is really representative of Isabelle formalizations. This is especially an issue since Sledgehammer has been extensively tuned against the benchmark set, on the assumption that it is representative. To get a clearer idea of Sledgehammer's usefulness, we now ran an evaluation on 128 randomly selected theories from the AFP. Up to 100 goals were selected for each theory, for a total of 6,934 goals. Our evaluation data is publicly available.[1]

The evaluation harness invokes Sledgehammer on each goal. The hardware setup consists of Linux servers equipped with Intel Core2 Duo CPUs running at 2.40 GHz. Each prover was given 30 s to solve each goal, but the 30 s slot was split into several slices, each corresponding to different problems and options to the prover. Lemmas were selected using the static MePo filter [19], as opposed to the machine learning based MaSh [5], whose development has fully stabilized only after the Isabelle2014 release. The results are summarized in Fig. 13.

As we remarked elsewhere [3], "It is important to bear in mind that the evaluation is not a competition between the provers. Different provers are invoked with different problems and options, and although we have tried to optimize the setup for each, we might have missed an important configuration option. Each number must be seen as a lower bound on the potential of the prover."

In case of success, the search is followed by reconstruction in Isabelle. For most goals, the reconstructed proof is a one-line call to an Isabelle proof method, such as *simp* (term rewriting), *metis* (a built-in resolution prover), or *blast* (a tableau prover). This call can then be inserted in the Isabelle formalization to discharge the goal. Reconstruction is a success if at least one of the attempted proof methods succeeds within 2 s. The percentage of goals with successful one-line proofs is given in the "One-line" column of Fig. 13. A few goals require a more detailed proof, expressed in the Isar format. The goal is considered solved if the Isar proof is successfully generated and replayed. This is reflected in the "+ Isar" column of Fig. 13.

	One-line	+ Isar	+ Oracle
E	49.7	51.4	52.5
SPASS	49.4	50.5	52.0
Vampire	49.5	51.0	51.8
Z3	49.6	50.0	53.7

Fig. 13. Success rate of Sledgehammer invocations per automatic prover (%)

---

[1] http://www21.in.tum.de/~blanchet/afp_mining_data.tgz.

When both reconstruction approaches fail, the user can still trust the external prover (and Sledgehammer's translation from HOL to the prover's formalism) as an oracle. In practice, most Isabelle users would prefer to work on a manual proof instead. These reconstruction failures are recorded in the "+ Oracle" column of Fig. 13.

The provers are neck and neck. Trusted as oracles, they prove 60.7 % of the goals when used in combination. This is significantly lower than the most recent evaluations based on Judgement Day. We offer the following possible explanations:

- Our evaluation uses the official release (Isabelle2014), instead of a preliminary version of Isabelle2015. It misses out on MaSh [5], on the improved Isar proof generation module [3], and on modern versions of provers. Recently, the SMT solver CVC4 has been integrated with Isabelle and is now, by a clear margin, the most successful prover [5].
- Sledgehammer's development since 2010 has been guided by experimental results on the Judgement Day suite, under the assumption that it is representative of Isabelle. Hence, it is not surprising that Sledgehammer should perform particularly well on these benchmarks.
- There is a lot of variation between theories. For theories with at least ten goals, our evaluation found success rates varying between 10 % and 100 %. We cannot exclude that the seven Judgement Day theories are particularly easy for Sledgehammer. Indeed, one third of Judgement Day consists of a large mathematical theory (`Fundamental_Theorem_Algebra`) whose goals are particularly easy.

7 Conclusion

We can summarize our findings by answering the questions raised in the introduction:

- There is too little reuse to our taste: the top 3 articles are reused 9, 6, and 4 times.
- There is some similarity to citation graphs in the computer science literature: the largest weakly connected component (WCC) in the AFP imports graph is 10 times larger than the next smaller WCC.
- The growth of the AFP appears to be only slightly better than linear although we hope it is only early days.
- The sizes of articles seem to follow a power law with a long tail. Two thirds are less than 4,000 lines long but 9 are longer than 20,000 lines.
- Over the whole AFP, 58 % of the text is taken up by proofs, 19 % by lemma statements, and 8 % by definitions.
- Lemma statement size is quadratically related to lemma proof size in approximately half of the AFP articles (with $R^2 > 0.5$). The exact nature of the relationship is not consistent, however. A more sophisticated measure for statement size/complexity is required to build a predictive model.

- The syntactic nesting depth of proofs follows an exponential decay. Almost 99 % of all proofs have a depth of 4 or less, but there is one proof of depth 15.
- One third of all lemma statements are equations, two thirds are Horn clauses.
- Sledgehammer can automate the proof of about 60 % of the goals that arise, which is respectable but less than on the Judgement Day benchmark suite.

Acknowledgement. Our colleague Johannes Hölzl suggested the proof depth diagram (Fig. 5), which we found insightful. Matichuk is partially supported by NICTA. NICTA is funded by the Australian Government through the Department of Communications and the Australian Research Council through the ICT Centre of Excellence Program. Nipkow is supported by DFG grant NI 491/16-1.

References

1. The Mizar mathematical library. http://mizar.org
2. An, Y., Janssen, J., Milios, E.: Characterizing and mining citation graphs of the computer science literature. Knowl. Inf. Syst. **6**, 664–678 (2004)
3. Blanchette, J.C., Böhme, S., Fleury, M., Smolka, S.J., Steckermeier, A.: Semi-intelligible Isar proofs from machine-generated proofs. Accepted in J. Autom. Reason. http://www21.in.tum.de/~blanchet/isar2.pdf
4. Blanchette, J.C., Böhme, S., Paulson, L.C.: Extending Sledgehammer with SMT solvers. J. Autom. Reason. **51**(1), 109–128 (2013)
5. Blanchette, J.C., Greenaway, D., Kaliszyk, C., Kühlwein, D., Urban, J.: A learning-based relevance filter for Isabelle/HOL (2015) (Submitted). http://www21.in.tum. de/~blanchet/mash2.pdf
6. Böhme, S., Nipkow, T.: Sledgehammer: judgement day. In: Giesl, J., Hähnle, R. (eds.) IJCAR 2010. LNCS, vol. 6173, pp. 107–121. Springer, Heidelberg (2010)
7. Chaudhuri, K., Doligez, D., Lamport, L., Merz, S.: The TLA$^+$ proof system: building a heterogeneous verification platform. In: Cavalcanti, A., Deharbe, D., Gaudel, M.-C., Woodcock, J. (eds.) ICTAC 2010. LNCS, vol. 6255, p. 44. Springer, Heidelberg (2010)
8. Church, A.: A formulation of the simple theory of types. J. Symb. Logic **5**(2), 56–68 (1940)
9. Crovella, M.E., Bestavros, A.: Self-similarity in world wide web traffic: evidence and possible causes. IEEE/ACM Trans. Network. **5**(6), 835–846 (1997)
10. de Moura, L., Bjørner, N.S.: Z3: an efficient SMT solver. In: Ramakrishnan, C.R., Rehof, J. (eds.) TACAS 2008. LNCS, vol. 4963, pp. 337–340. Springer, Heidelberg (2008)
11. Gonthier, G., et al.: A machine-checked proof of the odd order theorem. In: Blazy, S., Paulin-Mohring, C., Pichardie, D. (eds.) ITP 2013. LNCS, vol. 7998, pp. 163–179. Springer, Heidelberg (2013)
12. Hölzl, J., Immler, F., Huffman, B.: Type classes and filters for mathematical analysis in Isabelle/HOL. In: Blazy, S., Paulin-Mohring, C., Pichardie, D. (eds.) ITP 2013. LNCS, vol. 7998, pp. 279–294. Springer, Heidelberg (2013)
13. Kaliszyk, C., Urban, J.: HOL(y)Hammer: online ATP service for HOL light. Math. Comput. Sci. **9**(1), 5–22 (2015)
14. Lammich, P., Lochbihler, A.: The Isabelle collections framework. In: Kaufmann, M., Paulson, L.C. (eds.) ITP 2010. LNCS, vol. 6172, pp. 339–354. Springer, Heidelberg (2010)

15. Leroy, X.: A formally verified compiler back-end. J. Autom. Reason. **43**, 363–446 (2009)
16. Lochbihler, A.: Java and the Java memory model — a unified, machine-checked formalisation. In: Seidl, H. (ed.) ESOP 2012. LNCS, vol. 7211, pp. 497–517. Springer, Heidelberg (2012)
17. Matichuk, D., Murray, T., Andronick, J., Jeffery, R., Klein, G., Staples, M.: Empirical study towards a leading indicator for cost of formal software verification. In: Canfora, G., Elbaum, S. (eds.) International Conference on Software Engineering (ICSE 2015). ACM (2015)
18. Matuszewski, R., Rudnicki, P.: Mizar: the first 30 years. Mech. Math. Appl. **4**(1), 3–24 (2005)
19. Meng, J., Paulson, L.C.: Lightweight relevance filtering for machine-generated resolution problems. J. Appl. Logic **7**(1), 41–57 (2009)
20. Nipkow, T., Klein, G.: Concrete Semantics with Isabelle/HOL. Springer, Heidelberg (2014). http://concrete-semantics.org
21. Nipkow, T., Paulson, L.C., Wenzel, M. (eds.): Isabelle/HOL. LNCS, vol. 2283. Springer, Heidelberg (2002)
22. Paulson, L.C., Blanchette, J.C.: Three years of experience with Sledgehammer, a practical link between automatic and interactive theorem provers. In: Sutcliffe, G., Schulz, S., Ternovska, E. (eds.) International Workshop on the Implementation of Logics (IWIL 2010). EPiC Series, vol. 2, pp. 1–11. EasyChair (2012)
23. Riazanov, A., Voronkov, A.: The design and implementation of Vampire. AI Commun. **15**(2–3), 91–110 (2002)
24. Schulz, S.: System description: E 1.8. In: McMillan, K., Middeldorp, A., Voronkov, A. (eds.) LPAR-19 2013. LNCS, vol. 8312, pp. 735–743. Springer, Heidelberg (2013)
25. Staples, M., Jeffery, R., Andronick, J., Murray, T., Klein, G., Kolanski, R.: Productivity for proof engineering. In: Morisio, M., Dybå, T., Torchiano, M. (eds.) Empirical Software Engineering and Measurement (ESEM 2014), pp. 15:1–15:4. ACM, New York (2014)
26. Urban, J.: MPTP 0.2: design, implementation, and initial experiments. J. Autom. Reason. **37**(1–2), 21–43 (2006)
27. Urban, J., Rudnicki, P., Sutcliffe, G.: ATP and presentation service for Mizar formalizations. J. Autom. Reason. **50**(2), 229–241 (2013)
28. Weidenbach, C., Dimova, D., Fietzke, A., Kumar, R., Suda, M., Wischnewski, P.: SPASS version 3.5. In: Schmidt, R.A. (ed.) CADE-22. LNCS, vol. 5663, pp. 140–145. Springer, Heidelberg (2009)
29. Wenzel, M.: Isabelle/Isar—a versatile environment for human-readable formal proof documents. Ph.D. thesis, Institut für Informatik, Technische Universität München (2002). http://tumb1.biblio.tu-muenchen.de/publ/diss/in/2002/wenzel.html
30. Wiedijk, F.: Statistics on digital libraries of mathematics. Stud. Logic, Gramm. Rhetor. **18**(31), 137–151 (2009)

Math Search for the Masses: Multimodal Search Interfaces and Appearance-Based Retrieval

Richard Zanibbi[✉] and Awelemdy Orakwue

Document and Pattern Recognition Laboratory, Department of Computer Science,
Rochester Institute of Technology, Rochester, NY 14623, USA
rlaz@cs.rit.edu

Abstract. We summarize math search engines and search interfaces produced by the Document and Pattern Recognition Lab in recent years, and in particular the m_{in} math search interface and the *Tangent* search engine. Source code for both systems are publicly available. "The Masses" refers to our emphasis on creating systems for mathematical non-experts, who may be looking to define unfamiliar notation, or browse documents based on the visual appearance of formulae rather than their mathematical semantics.

Keywords: Mathematical Information Retrieval (MIR) · User interface design · Handwriting recognition · Character recognition

1 Introduction: Why Math Search Pertains to the Masses

Mathematical notation is a *natural* language used to define the models, metrics and analytical tools of modern societies. It is natural in the sense that the notation is re-purposed and adapted for different mathematical concepts, problems, and communities, leading to various dialects. The influence of mathematical notation, while quiet, is pervasive. Whether it is choosing foods to purchase based on their cost and quantified nutritional information, using demographic and usage statistics to determine which forms of entertainment to attract and promote, determining where to build manufacturing sites, or how to represent a problem and its potential solutions in science and technology, math notation is an essential tool that shapes both our personal lives and environment. Given this, math literacy is critical for participating fully in the modern world, and considerable attention continues to be focused on strengthening mathematics education.

However, for many persons of all ages, mathematical notation is a source of significant frustration or anxiety at times due to real or perceived difficulties with interpreting unfamiliar notation. This is particularly true when the text accompanying mathematical notation is found to be confusing. To search the internet for alternative sources about the notation, users must formulate their query in text, even if they are unclear what about what the represented concept is. Mathematical experts might search the internet using LaTeX for an expression,

© Springer International Publishing Switzerland 2015
M. Kerber et al. (Eds.): CICM 2015, LNAI 9150, pp. 18–36, 2015.
DOI: 10.1007/978-3-319-20615-8_2

or using the (often, already known) name for the concept [42]. Even experts relate to the odd experience of revisiting concepts expressed in a notation distinct from that used when they originally learned a concept, and the difficulties this introduces in interpreting the notation.

While mathematical concepts can often be notated various ways, some psychological studies suggests that the appearance of math notation affects our reasoning about it [15], and that individuals will often space subexpressions systematically when entering math, even if mathematically unnecessary [14]. The studies suggest that our perception of math notation may be grounded in visual structure, i.e. how it looks.

An important related concern is hit content summarization, i.e. how search hits are presented to the user [36]. In a recent study it was confirmed that as one expects, the formatting of math expressions significantly affects relevance assessment of search hits [25]. The normal hit format provided by Google (e.g. as raw LATEX or Presentation MathML) was compared with the same hits with formulae rendered, and on average participants had a 17 % higher relevance assessment accuracy in the rendered condition.

We propose that if it is natural to use words from unclear texts in queries, it is also natural to use mathematical notation from unclear texts *directly* in queries. A recent study illustrates the benefit of this approach [35]. When undergraduate students were asked to learn about the binomial coefficient, and shown the expression $\binom{4}{2}$, many did not know what the notation represented. When given an interface in which they could draw, recognize the spatial layout of symbols and then search for the expression, most participants found this to be both intuitive and helpful.

In the remainder of our paper, we summarize research carried out to realize the entry and retrieval of math based on its appearance, as this is what one works from when notation is unfamiliar or difficult to interpret, or when trying to locate similarly structured expressions (e.g. when browsing formulae in a collection).

2   Math Encodings: Symbol Layout Trees and Operator Trees

In practice, math encodings are most commonly used to represent the appearance and mathematical content of formulae. Appearance-based encodings such as LATEX [13] and Presentation MathML[1] are used to display mathematical expressions. A number of web browsers support Presentation MathML directly (e.g. Firefox), and online tools such as MathJax[2] may also be used to render LATEX and MathML contained in HTML pages. Mathematical content encodings such as Content MathML can be used for evaluation and symbolic manipulation by Computer Algebra Systems (e.g. Mathematica, Maple).

---

[1] http://www.w3.org/Math/.

[2] https://www.mathjax.org/.

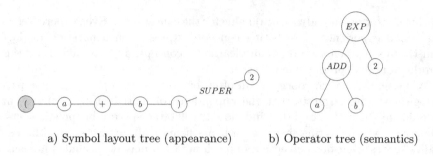

a) Symbol layout tree (appearance)   b) Operator tree (semantics)

Fig. 1. Symbol layout tree and operator tree for $(a+b)^2$

As illustrated in Fig. 1, the appearance and content of mathematical expressions are hierarchical. As a result, both appearance and content-based encodings define trees, which are annotated in various ways to support applications. Encodings for appearance represent a spatial arrangement of symbols on baselines (writing lines), which we term a *Symbol Layout Tree*. In Fig. 1a the symbol layout tree is rooted at the leftmost parenthesis ('('), and there are two writing lines present: the main baseline, and a superscripted baseline.

The *Operator Tree* in Fig. 1b represents a hierarchical application of operations to operands, from the leaves to the root of the tree. Relative to the layout tree in Fig. 1a, in the operator tree operator symbols are replaced by their associated operation (e.g. '+' becomes ADD), implicit operations are made explicit (e.g. superscript becomes EXP (exponent)), and parentheses used for grouping in the expression appearance are removed. Groupings are redundant in an operator tree, where ordering constraints are explicit.

Due to symbols and spatial relationships in formulae being frequently redefined by authors, it is impossible to define a mapping from formula appearance to formula semantics. To create this mapping the domain of discourse (e.g. algebra vs. calculus vs. logic, etc.) along with the specific environment defining constants, variables, and operations in an expression are required. The mapping from operator trees to layout trees is also one-to-many, as a single operator tree may be written various ways. For example, 'x^2' and '$(x)^2$' can represent the same operation, and operations may be associated with different symbols (e.g. '÷' vs. '/' for division) and symbol arrangements (e.g. $1/2$ vs. $\frac{1}{2}$).

The flexibility of mathematical notation benefits both authors and their technical communities. However, this flexibility and dependency on context for interpretation poses substantial challenges for automated recognition and retrieval of mathematical notation [3, 37].

3 m_{in}: A Multimodal Math Search Interface

Figure 2 shows the m_{in} search interface which runs in standard web browsers on desktop and tablet computers (e.g. iPads [26]). m_{in} is implemented in Javascript and HTML, with symbol recognition and parsing performed by external web services.[3] In Fig. 2 we see a text box for keywords at top-right, while the large white canvas at bottom is used to enter formulae. A list of formula are stored in the 'deck,' the wide rectangular panel at the top of the interface. The deck has operations to add, remove, and switch between formulae.

m_{in}'s design seeks to allow mathematical non-experts to easily 'draw' math expressions for queries by switching seamlessly between typing, freehand drawing, and importing formula images. Another design goal is providing clear, intuitive feedback for the recognized structure of an expression (i.e. the symbol layout tree). Our design was influenced by earlier math editing and recognition prototypes, particularly the pen-based Freehand Formula Entry System (FFES [29,41]), the vector graphics-based XPress system [23], and the *infty* math OCR system [33].

Figure 7 shows the entry of the formula in Fig. 2. A combination of typed LATEX (e.g. '$x_i - x$' in the top-left panel) and handwriting (shown as red lines) are used in queries. As handwritten symbols are recognized, they gradually fade and are replaced by the recognized symbol. LATEX strings are replaced by Math-Jax renderings. In the final step the symbol layout is parsed, and symbols are gradually moved in an animation to ideal positions. The fonts and locations for recognized symbols are obtained from Support Vector Graphics (SVG) MathJax output produced using the LATEX string for recognized symbol layout. Handwriting is visible in the Editing mode, where pen/touch strokes appear above recognition results (see Fig. 2).

Figures 4 and 5 illustrate additional operations for image input, using the deck to store and combine formulae, matrix entry, and correcting symbol recognition errors.

3.1   Human Studies: Formula Entry Operations and Recognition Visualization

A pair of human studies have influenced the design of m_{in}. The first study compared visualizations of recognition results for handwritten formulae. In the first condition, results were shown separately from user input in a rendered LATEX image, and in the second condition handwritten symbols were gradually rescaled and moved to ideal positions using a *style-preserving morph* [41]. Overall, participants found results from the rendered image clearer, but surprisingly there was no significant difference in entry time for users using the image or morphing feedback, despite symbol recognition results not being visible in the morphing condition unless individual symbols were selected. Also, in the image condition some participants became stuck, as they were unable to find where their expression was recognized incorrectly (this finding has been replicated in other studies since; see [37]). This did not happen when the participant's symbols were 'morphed' in-place.

---

[3] Source code: http://www.cs.rit.edu/~dprl/Software.html.

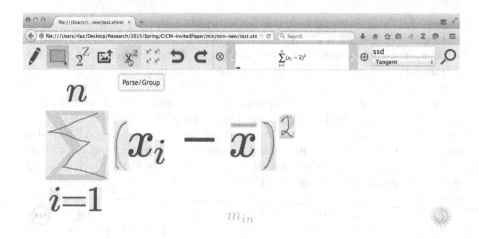

Fig. 2. Combined Keyword ('ssd') and Formula Query in m_{in} (Editing Mode). At top, from left-to-right buttons are provided for symbol entry, selection and correction, image import, parsing/grouping symbols, creating an expression grid (matrix), and undo/redo. The 'deck' stores a list of entered formulae, which may be combined (see Fig. 4). On pressing the search button (the magnifying glass), LaTeX for the formula shown in the 'deck' is concatenated with keywords and sent to a selected search engine (Color figure online)

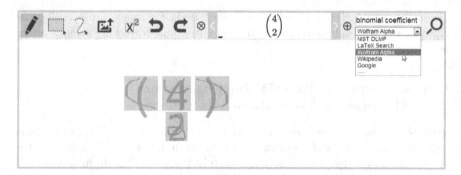

Fig. 3. m_{in} circa Spring 2013 (Drawing Mode [35]). The drop-down list of search engines is visible. The third button from the left at top corrects stroke groupings, and was later replaced by the symbol correction button. Browser fonts drawn above handwritten strokes and simple symbol repositioning visualize recognition. In the new m_{in} strokes gradually fade and are replaced by recognized symbols in draw mode

In m_{in} a style-preserving morph is performed when a button to recognize expression structure is pressed (the 'Parse/Group' button). This 'morph' is actually a modified version, described below.

The second human study evaluated an earlier version of m_{in} (see Fig. 3), and identified opportunities for improvement [35]. In particular, the undergraduate college students that participated in the study found that while symbol

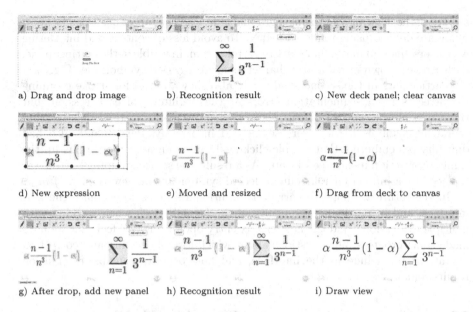

a) Drag and drop image

b) Recognition result

c) New deck panel; clear canvas

d) New expression

e) Moved and resized

f) Drag from deck to canvas

g) After drop, add new panel

h) Recognition result

i) Draw view

Fig. 4. Image Input and the Formula 'Deck.' The panel (deck) showing images at top of the interface stores formulae. An image creates the first expression (a,b), which is then added to a second expression (c,d,e) by dragging its panel from the deck to the canvas (f,g), and then storing the combined result in a third slider panel (g,h,i)

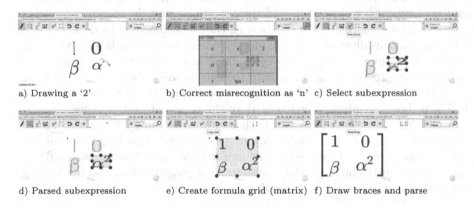

a) Drawing a '2'

b) Correct misrecognition as 'n' c) Select subexpression

d) Parsed subexpression

e) Create formula grid (matrix) f) Draw braces and parse

Fig. 5. Matrix Entry and Symbol Correction. As shown in this example, m_{in} allows subexpressions to be grouped separately (c,d) or as a grid of expressions (e). Symbol recognition errors are corrected using a transparent pop-up window (b). The window appears after selecting a symbol and then pressing the 'relabel' button at top

recognition results were now visible when drawing (they would appear above user strokes, see Fig. 3), this cluttered the canvas, making it difficult to see errors. The symbol placement from the original style preserving morph is also coarse and sometimes confusing (e.g. making adjacent symbols appear subscripted [41]). To address these issues, in the new m_{in} recognized symbols replace handwritten strokes in the drawing view using a gradual fade, and the target positions for morphing are defined using rendered LATEX. To avoid loss of context and interfering with users' 'mental maps,' handwritten strokes remain visible in the editing mode, with strokes shown in red above characters for recognized symbols (see Fig. 2).

Participants in the second study were also shown a tool for stroke grouping to correct symbol segmentation errors at the beginning of each session, but there was almost no use of this tool, with participants instead deleting and redrawing symbols if they were segmented incorrectly. Participants also had difficulty remembering that double-clicking/tapping on a symbol brings up the symbol correction menu (see Fig. 5b). As a result, the stroke grouping button was replaced by a button for relabeling selected symbols in the new version of m_{in}.

In the future we hope to carry out additional studies to evaluate the new interface, and in particular the utility of the formula deck when working with multiple expressions, and the new matrix entry operations. We feel that we have made some progress, but questions about which formula editing and correction operations to include in the interface, and how best to visualize recognition results remain.

3.2 m_{in} System Architecture and Recognition Modules

Figure 6 presents a global view of m_{in}'s architecture. Users input keywords as text, and math using a combination of LATEX, handwritten symbols and images. There are two primary data structures that define the interpretation of a formula on the canvas: a list of symbols and their locations, and the recognized symbol layout tree for the formula. For clarity, we do not show the formula 'deck' in Fig. 6 (see Fig. 4 for an illustration of the deck). When the user clicks on the search button, keywords and the current expression shown in the deck are concatenated in a query string, which is then sent to the currently selected search engine.

Both automatic (solid arrows) and user operations (dotted arrows) are shown in Fig. 6. Users manipulate symbols through entry, deletion, moving, resizing and merging (e.g. to combine two dashes into an '=' sign). Users also invoke the parser to update the symbol layout tree and produce LATEX. When the 'Parse/Group' button is pushed by the user, formula structure is visualized two ways, by moving symbols on the canvas to ideal positions, and also by showing the rendered LATEX in the formula 'deck.'

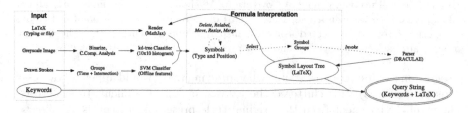

Fig. 6. m_{in} Architecture. User interactions are shown using dotted arrows, and solid arrows represent automated processing.

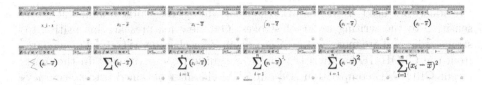

Fig. 7. Entering Formula in Fig. 2 (from top-left, left-to-right). Symbols are entered through typing and drawing. At bottom-right the 'Parse/Group' button is pressed, symbol layout is recognized and symbols are gradually moved ('morphed') to ideal positions (Color figure online)

Currently, operations for entering symbols and recognizing symbol layout are independent of one another. In particular, layout analysis is not performed until explicitly requested by the user. While it is beneficial to integrate classification, segmentation and parsing for automatic recognition [37], this type of integration may be unhelpful in an interactive system, such as when decisions previously accepted by the user are modified (e.g. if handwritten symbols are re-segmented and re-classified). Our parser only revises symbols when a compound token is detected, such as replacing a horizontal line with '+' above by '±.' Our hope is that this behavior is both convenient and predictable.

In the remainder of this section we discuss the recognition modules used in m_{in}. The current recognition modules were designed with accuracy, speed, and simplicity in mind. The symbol recognition and parsing modules run externally to m_{in} on servers, with requests made and results transmitted back using simple XML encodings. This allows recognition modules to be easily replaced by services provided on other servers that accept and produce the same encodings. Despite this network overhead, recognition is fast and most first-time users are unaware that recognition is performed remotely.

3.3 Symbol Entry and Correction

For formula entry, the grouping (segmentation) of handwritten strokes and symbols in images is performed using simple methods. It is assumed that handwritten symbols are entered one-at-a-time, and recognition is invoked after a short delay (e.g. 1–2 s), or when a drawn stroke intersects other strokes (in which case the strokes are merged into a single symbol). For typeset symbols in images, each separate region of connected black pixels (*connected component*) is treated as a separate symbol. This results in many symbols being over-segmented initially (e.g. '='), but in many cases the DRACULAE parser can locate and correct this by matching rewriting local structures in the symbol layout tree [40]. A pair of parameters are used to control the location of the centroid used to represent symbol locations, and thresholds to define vertical spatial regions around symbols (above, below, superscipt, subscript).

Handwritten symbol classification is performed by a Support Vector Machine with a Gaussian kernel applied to modified off-line (i.e. image-based) features [7].

Previously we used Hidden Markov Models [9]. These worked well, but were sensitive to the writing order of strokes. Our new features are insensitive to stroke order and are more accurate as a result. Our classifier is trained using data from the CROHME handwritten math recognition competitions.[4] In the most recent CROHME competition [19] our SVM classifier obtained a test accuracy of 88.7 % for 101 symbol classes, and 83.6 % when invalid symbols are included (102 classes). These rates are within 2–3% of those obtained by the winning system from MyScript Corporation.[5]

Typeset symbols in images (especially digitally-born) tend to be clean and regular, and so we use a simple nearest neighbor classifier. Connected components are assigned to classes using a 10×10 histogram of pixel counts. We currently use a kd-tree implementation from the Python-based scikit-learn library[6] for fast approximate nearest-neighbor classification. The classifier is trained using the Infty data sets [34]. An earlier version obtained recognition rates over 97 % for 190 classes on 70,637 test samples in the *Infty* data set using pixel histograms [43]. We do not yet support .pdf input [2,16], but hope to in the future.

In the current version of m_{in}, mis-segmented symbols from handwriting or images are deleted and re-entered by users, for example using handwriting or typing. Both of the handwritten and image-based symbol recognizers return a ranked list of classes that can be selected from the symbol correction menu (see Fig. 5b). This menu also includes a list of symbols organized by type (e.g. digits, latin letters, greek letters, operators, etc.).

3.4   Parsing Symbol Layout and Generating LaTeX

Symbol layout in a formula written on the m_{in} canvas is parsed using DRAC-ULAE [40], implemented in the TXL tree rewriting language [5]. DRACULAE employs a compiler design, performing a series of tree rewriting passes that: (1) produce an initial symbol layout tree, (2) replace compound tokens (e.g. replacing two vertically adjacent dashes by '='), (3) rewrite structures such as fractions, and (4) translate the resulting tree to LaTeX. DRACULAE also produces operator trees where possible (i.e. a 'semantic' encoding), but this is unused in m_{in}. In the initial layout analysis step, DRACULAE uses a fast greedy algorithm to recursively locate symbols on the main baseline, and then assigns remaining symbols to regions around baseline symbols (e.g. above, below, superscript, subscripts, within for roots, etc.).

As shown in Fig. 5, users can invoke DRACULAE to parse a subexpression which is then grouped into a unit, 'locking' its interpretation [17] and preventing modification by subsequent parses. Symbols and grouped subexpressions may also be combined in a grid, e.g. to enter matrices. This operation uses simple horizontal and vertical bounding box projections to identify gaps for rows and

---

[4] http://www.isical.ac.in/~crohme/.

[5] http://www.myscript.com/.

[6] http://scikit-learn.org/.

columns - DRACULAE does not recognize matrix structure. Instead, we have DRACULAE treat grouped subexpressions as individual symbols during parsing. Matrix recognition remains a difficult open problem [19,37], but if accuracy can be increased, in the future it may be beneficial to recognize grid cells in addition to rows and columns of predefined cells.

As described earlier, MathJax is used to visualize recognized symbols, and define the ideal locations to which symbols on the canvas are repositioned (morphed) after parsing.

Parsing errors (e.g. detecting an adjacent symbol as subscripted) are corrected by some combination of moving symbols, undoing the previous parse operation (which 'morphs' symbols back to their previous positions), and pressing 'Parse/Group' again.

4   Appearance-Based Math Retrieval

In this section we summarize a number of different search engines and models designed to support math search using formula appearance. In particular, we describe the *Tangent* search engine and its integration with the m_{in} math search interface, along with methods for visual search of document images and videos.

4.1   Query-by-Expression for Symbolic Encodings (LATEX, MathML)

Approaches to query-by-expression may be categorized as *text-based* or *tree-based*, as determined by the structure used to represent and retrieve expressions. In text-based approaches, math expressions are linearized before indexing and retrieval. These linearizations are normalized to reduce variability in representation. Common normalizations include defining synonyms for symbols (e.g. function names), using canonical orderings for spatial relationships and commutative operators (e.g. to group 'a + b' with 'b + a'), enumerating variables, and replacing symbols by their mathematical type.

Linearized math expressions are often handled by term frequency-inverse document frequency-based (TF-IDF) techniques from text retrieval [18,30,39]. While linearization loses some formula structure information, it allows text and math retrieval to be carried out in a single framework (usually Lucene[7]). In a different approach, the largest common string subsequence is used to retrieve LATEX strings [31].

Tree-based approaches represent layout or operator trees for formulae directly. Methods have been developed that compress tree indices by storing identical subtrees in expressions uniquely [10], with exact matching and tree-edit distances used for retrieval [11]. *Substitution trees* designed for unification have been used to create tree-structured indices [12,27]. Descendants of an index tree node contain expressions that unify with the parameterized expression stored at the node (e.g. '$f($ $\boxed{1}$ $)$' unifies with '$f(a)$,' with substitution $\boxed{1} \rightarrow a$). A

---

[7] https://lucene.apache.org/.

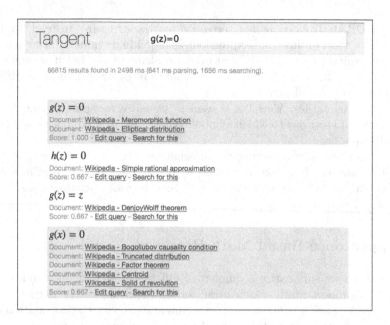

Fig. 8. Original Tangent Formula Search Engine [32]. The 'Edit query' links send a search hit to the m_{in} search interface for editing and re-submission to Tangent or other search engines. The 'Search for this' link supports browsing by allowing hits to be submitted as new queries. Queries may be typed in LaTeX into the text box shown at top, or submitted from m_{in} (see Fig. 2, where Tangent is the selected search engine)

more recent technique adapts TF-IDF retrieval for vectors of subexpressions and 'generalized' subexpressions where constants and variables are represented by a single symbol [16]. Subtrees are normalized for commutative operators and operator precedence, converting symbol layout trees to pseudo-operator trees.

An emerging class of 'spectral' tree-based approaches use sets of local structural representations rather then complete subtrees for retrieval. One system converts sub-expressions in operator trees to words representing individual arguments and operator-argument pairs [21]. A lattice over the sets of generated words are used to define similarity, and a breadth-first search constructs a neighbor graph traversed during retrieval. Another system employs an inverted index over paths in operator trees from the root to each operator and operand, using exact matching of paths for retrieval [8].

Over a number of years our group has developed a novel 'spectral' retrieval model, and a search engine implementing the model [22, 27, 32]. We discuss these in the next section.

4.2   The Tangent Math Search Engine

A screenshot of the *Tangent* search engine[8] is shown in Fig. 8. The query '$g(z) = 0$' is shown along with the top four matched expressions and their associated Wikipedia articles. The goal with this interface design was to make it easy to use retrieved expressions for editing and search. At the bottom of each hit is a rank score, along with a link to send the hit to m_{in} for editing, and a second link for using a hit to re-query the collection. This integration of m_{in} and Tangent allows for both visual and textual editing of formula queries.

In Fig. 8 we see that Tangent retrieves formulae with structure similar to the query, even when different symbols are used (e.g. 'g' replaced by 'h,' 'z' by 'x', and '0' by 'z'). This is interesting, because here only *exact* matching is used for retrieval [32]. Search results from this first version of Tangent often appear to have performed unification of symbols, but no unification is carried out. This is because the *relative positions of symbols* are used for matching.[9]

In Fig. 8, all four hits contain parentheses that are one symbol apart, with an equals sign at right. Matching additional symbol pairs lead to a higher rank. In this example, the first hit is an exact match, with score 1.0 (all symbol pairs are matched), while the remaining three hits have the same rank score, differing by the identity of exactly one symbol relative to the query. This causes relationships with the non-matching symbol to be treated as unmatched. In each case, the five pairs associated with the unmatched symbol are 'misses,' out of fifteen total symbol pairs (for six symbols, $\binom{6}{2} = 15$ symbol pairs, $10/15 = 0.667$).

More concretely, the spectral model used in Tangent represents symbol layout trees by the relative position of each symbol with its descendants in the tree. This is a 'bag-of-words' model, with 'words' representing relative symbol positions. Note that tuples are not generated for all pairs of symbols when there is branching in the layout tree, unlike the query and hits shown in Fig. 8 which lie on a single baseline. Tuple generation is illustrated in Fig. 10a–c. The fraction line has a relationship with every other symbol in the tree, each defined by a pair of integers giving the path length from the line to the symbol in the tree (shown as *Dist.* in Fig. 10c), and change in baseline position. The baseline position change is initially 0, increasing by one for each superscript/above relationship and decreasing by one for each subscript/below relationship along the path between two symbols (shown as *Vert.* in Fig. 10c [32]).

Tuples in Tangent Version 2 [22]. Later, changes were made to the tuple generation model, adding tuples for symbols in the leaves of layout trees. In Fig. 10c, we see that three tuples are defined for the symbols '2,' 'y' and 'z' at the leaves of the tree shown in Fig. 10b. This addition was made to allow single-symbol queries to be represented in the index, particularly to allow matching for matrix subexpressions comprised of a single symbol such as shown in Fig. 10d, where three of the matrix entries are a single digit.

---

[8] http://saskatoon.cs.rit.edu/tangent.
[9] This approach was motivated by a ranking function that used sets of matching symbols and symbol pairs to greatly improve initial retrieval results [27].

A representation for matrices and grid/array structures was also added, such as for expressions shown in Figs. 9a and 10d. Each grid is represented by a symbol named 'matrix' with its dimensions concatenated on the end (e.g. **matrix2x2** in Fig. 10f). This 'matrix' symbol is then used to represent the entire matrix contents. In Fig. 10f the structure of the expression treating the matrix as a unit is contained in rows 6–12 of the table.

Cells/subexpressions in a matrix or grid are represented as independent expressions; in Fig. 10f these are the last five rows of the table, representing 'x^2,' '0,' '0,' and '1.' The subexpression at each matrix location is represented by a tuple giving a row and column location, with the subexpression represented by its LATEX string (as shown in the top five rows of Fig. 10f). The idea in this case was to be able to detect when a particular subexpression is present, and also whether the subexpression is located at the correct location in the matrix.

Finally, to support participation in the NTCIR-11 math retrieval tasks [1,28], the Tangent inverted index for tuples was expanded to include entries where one of the two symbols are undefined (e.g. 'x^2' would be represented concretely, and by '$?^2$' and '$x^?$', where '?' represents a wildcard). Figure 9a shows an example of a query containing wildcards. In both tasks, symbols could be replaced by wildcard symbols, which our group interpreted as being any individual symbol. Relationships between two wildcard symbols are not indexed, as in some cases this will match a vast number of entries in the index (for example, consider 'something next to something').

Retrieval. Formula retrieval is performed using an inverted index over symbol pair relationship tuples, mapping tuples to the expressions that contain them. Expressions are represented uniquely, with a separate table recording which documents contain an expression [22].

Queries are first converted to a set of unique tuples with associated counts. Unique tuples are then used to locate matching expressions from the inverted index, and determine the number of tuples from the query matched in each retrieved expression. Matched expressions are ranked by the harmonic mean of the percentage of pairs matched in the query, and the percentage of pairs matched in the candidate. This may be understood as the f-measure for *recall* of query tuples in the candidate, and *precision* of tuples in the candidate. This ranking metric prefers larger query set tuple matches, while penalizing unmatched tuples.

In the second version of Tangent, the engine was modified to support both text and multiple formulae in queries. Lucene was used for text retrieval, and formulae were retrieved using Tangent's formula search engine. The formula match score for a document was computed as the sum of the highest formula match scores located for each query expression in a document, each weighted by the relative size of each expression [22]. The final rank score for a document was a linear combination of the Lucene-based keyword score and the formula match score. With this, Tangent was now able to handle combined text and formula queries.

Formula Query: $\mathbb{P}\big[\,\boxed{X}\geq\boxed{t}\,\big]\leq\dfrac{\mathbf{E}\big[\,\boxed{X}\,\big]}{\boxed{t}}$ $\qquad \mu(A)=\begin{cases}1 & \text{if } 0\in A\\ 0 & \text{if } 0\notin A.\end{cases}$

Keyword: Markov inequality

a) Math-2 #39 $\qquad\qquad$ b) Wikipedia #49

Fig. 9. Sample Queries from NTCIR-11. Query (a) contains four wildcard symbols (shown in boxes), and two keywords. Queries for the Wikipedia subtask were single expressions. Query (b) has no wildcards and includes a tabular/matrix layout

Results. A human evaluation compared search results returned by the original Tangent and a Lucene (text-based) retrieval model [32]. *Precision-at-k* is the percentage of hits in the top-k results deemed 'relevant' to a query. In this case, participants were asked to rate hits by their similarity to the query using a 5-point Likert scale from 'Very dissimilar' to 'Very similar,' with ratings of 'Similar' or 'Very Similar' treated as relevant. The average precision-at-1 and precision-at-10 values for Tangent were 99 % and 60 %, and 60 % and 39 % for the text-based system. This confirms that using more tree structure produces search results that are perceived as more similar, a result also confirmed in the recent NTCIR math retrieval tasks [1].

The second version Tangent was entered in the NTCIR-11 Math Main Task [1] and the NTCIR-11 Wikipedia formula retrieval subtask [28]. Queries from each task are shown in Fig. 9. The Main task had 50 combined formula and text queries, for a subset of the arXiv containing 100,000 technical papers with substantial mathematics broken up at paragraphs into 8.3 million segments, treated as the documents for the task. Two human evaluators judged hits for the main task to produce precision and related metrics. Tangent produced the highest precision-at-5 measure (92 %), using a 50 % weighting for combining the text and formula match scores.

The Wikipedia subtask was a query-by-expression task with 100 queries for 35,000 articles from Wikipedia [28]. This task used an automated evaluation protocol, ranking system by specific-item retrieval measures (e.g. the rank at which the article from which query expressions are taken were located, and the number of exact matches returned in the top-k hits), without measures for relevance or similarity. For the Wikipedia task, the formula retrieval engine matched the highest top-1 score (68 %, obtained by three systems), and overall was amongst the best performing systems in the competition, hampered primarily by queries that contained a large fraction of wildcard symbols (e.g. $\frac{?}{?}$). Considering the manner in which keyword searches are often carried out using a small number of concrete terms, to us it is unclear how frequently queries with a large number of wildcards would be used in a practical setting vs. copying or creating concrete expressions for inclusion in queries. That said, we believe that re-ranking initial results so that variable-variable relationships are not ignored can be used to mitigate this issue.

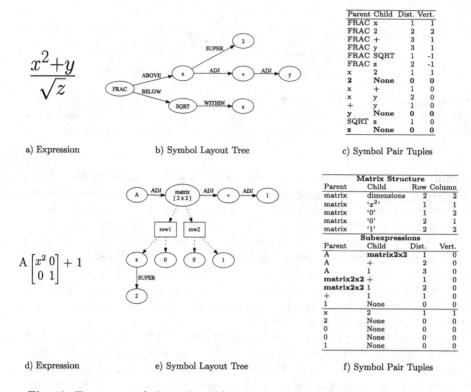

Parent	Child	Dist.	Vert.
FRAC	x	1	1
FRAC	2	2	2
FRAC	+	3	1
FRAC	y	3	1
FRAC	SQRT	1	-1
FRAC	z	2	-1
x	2	1	1
2	None	0	0
x	+	1	0
x	y	2	0
+	y	1	0
y	None	0	0
SQRT	z	1	0
z	None	0	0

a) Expression   b) Symbol Layout Tree   c) Symbol Pair Tuples

Matrix Structure		
Parent	Child	Row Column
matrix	dimensions	2 2
matrix	'x^2'	1 1
matrix	'0'	1 2
matrix	'0'	2 1
matrix	'1'	2 2

Subexpressions		
Parent	Child	Dist. Vert.
A	matrix2x2	1 0
A	+	2 0
A	1	3 0
matrix2x2	+	1 0
matrix2x2	1	2 0
+	1	1 0
1	None	0 0
x	2	1 1
2	None	0 0
0	None	0 0
0	None	0 0
1	None	0 0

d) Expression   e) Symbol Layout Tree   f) Symbol Pair Tuples

Fig. 10. Tangent: symbol pair-based layout representation in for two expressions

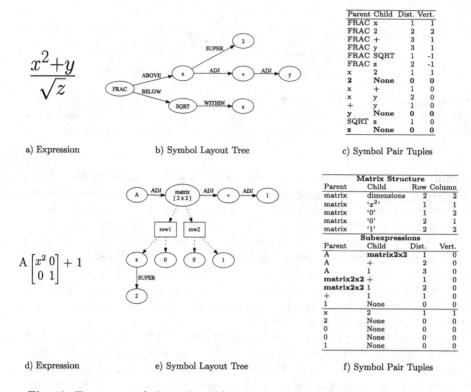

Fig. 11. User interface for evaluating image-based query-by-expression using handwritten queries [38]

4.3   Image-Based Formula Retrieval

For space we will cover this topic just briefly, but we believe that this is an important future direction for research. Figure 11 shows an evaluation interface for the first image-based handwritten math retrieval system [38].

In this system layout and contour features measured from an image of a handwritten mathematical expression are used to search document images for similar expressions. Formulae are indexed using X-Y cutting trees [20], with Dynamic Time Warping of upper and lower image contours used to produce the final ranking (adapting an earlier handwritten word spotting technique [24]). We were very surprised that our first prototype allowed 10 participants to locate the page from which handwritten queries were taken 63 % of the time in the top 10 on average (20 queries). If the original query images were used, then 90 % of the original queries could be located in the top 10 results.

Related work is currently underway, using image-based retrieval of math in lecture videos using snapshots [6] and handwritten queries [4].

5   Conclusion: Text + Diagram Search for the Masses

We have summarized our work on creating interfaces and search engines that support math retrieval using the appearance of mathematical expressions. Our aim in doing this is to help all persons, mathematical non-experts and experts, to retrieve mathematical information naturally using the appearance of expressions, in combination with keywords when appropriate.

A key direction for future research is the creation of intuitive, fast interfaces for diagram copying, editing and inclusion in search queries. m_{in} has made a start in this direction, but much remains to be done. Related to this, we believe that an important future line of research is redefining the conventional text 'search box' to include formulae directly.

Currently, spectral approaches to matching structure in trees appear to be the most promising for appearance-based formula retrieval, such as that used in Tangent. In addition to opportunities defined earlier, identifying ways to reduce index sizes and accelerate retrieval will be important for producing engines that will scale to very large collections, and ideally, internet search engine-scale collections.

In closing, there have been many advances in Mathematical Information Retrieval in recent years, and we believe that progress in searching for diagrammatic notations will dramatically alter the way in which people search for technical information. It will allow queries to move from "documents with words similar to *these*" to also include "documents with diagrams similar to *these*."

Acknowledgements. We thank George Nagy, Maria Zemankova, Christian Viard-Gaudin, Harold Mouchère, Frank Tompa and Andrew Kane for helpful discussions. This material is based upon work supported by the National Science Foundation (USA) under Grant Numbers IIS-1016815 and HCC-1218801.

References

1. Aizawa, A., Kohlhase, M., Ounis, I., Schubotz, M.: NTCIR-11 Math-2 task overview. In: Proceedings of the 11th NTCIR Conference, Tokyo, Japan, pp. 99–102, December 2014

2. Baker, J.B., Sexton, A.P., Sorge, V.: A linear grammar approach to mathematical formula recognition from PDF. In: Carette, J., Dixon, L., Coen, C.S., Watt, S.M. (eds.) Calculemus/MKM 2009. LNCS, vol. 5625, pp. 201–16. Springer, Heidelberg (2009)

3. Blostein, D., Zanibbi, R.: Processing mathematical notation. In: Doermann, D., Tombre, K. (eds.) Handbook of Document Image Processing and Recognition, pp. 679–702. Springer, London (2014)

4. Chatbri, H., Kwan, P.W., Kameyama, K.: A modular approach for query spotting in document images and its optimization using genetic algorithms. In: Proceedings of the IEEE Congress on Evolutionary Computation, Beijing, China, pp. 2085–2092, July 2014

5. Cordy, J.R.: The TXL source transformation language. Sci. Comput. Program. **61**(3), 190–10 (2006)

6. Davila, K., Agarwal, A., Gaborski, R., Zanibbi, R., Ludi, S.: AccessMath: Indexing and retrieving video segments containing math expressions based on visual similarity. In: Proceedings of the IEEE Western New York Image Processing Workshop, Rochester, NY, pp. 14–17 (2013)

7. Davila, K., Ludi, S., Zanibbi, R.: Using off-line features and synthetic data for on-line handwritten math symbol recognition. In: Proceedings of the International Conference Frontiers in Handwriting Recognition, Crete, Greece, pp. 323–328 (2014)

8. Hiroya, H., Saito, H.: Partial-match retrieval with structure-reflected indices at the NTCIR-10 math task. In: Proceedings of the NII Testbeds and Community for Information Access Research, Tokyo, Japan, pp. 692–695, June 2013

9. Hu, L., Zanibbi, R.: HMM-based recognition of online handwritten mathematical symbols using segmental k-means initialization and a modified pen-up/down feature. In: Proceedings of the International Conference Document Analysis and Recognition, pp. 457–462 (2011)

10. Kamali, S., Tompa, F.W.: A new mathematics retrieval system. In: Proceedings of the 19th ACM International Conference on Information and Knowledge Management, CIKM 2010, pp. 1413–1416. ACM, New York (2010)

11. Kamali, S., Tompa, F.W.: Structural similarity search for mathematics retrieval. In: Carette, J., Aspinall, D., Lange, C., Sojka, P., Windsteiger, W. (eds.) CICM 2013. LNCS, vol. 7961, pp. 246–62. Springer, Heidelberg (2013)

12. Kohlhase, M., Sucan, I.: A search engine for mathematical formulae. In: Calmet, J., Ida, T., Wang, D. (eds.) AISC 2006. LNCS (LNAI), vol. 4120, pp. 241–53. Springer, Heidelberg (2006)

13. Lamport, L.: LATEX: A Document Preparation System. Addison-Wesley Reading, MA (1986)

14. Landy, D., Goldstone, R.: Formal notations are diagrams: Evidence from a production task. Mem. Cogn. **35**(8), 2033–40 (2007)

15. Landy, D., Goldstone, R.: How abstract is symbolic thought? J. Exp. Psychol. Learn. Mem. Cogn. **35**(8), 720–33 (2007)

16. Lin, X., Gao, L., Hu, X., Tang, Z., Xiao, Y., Liu, X.: A mathematics retrieval system for formulae in layout presentations. In: Proceedings of the ACM SIGIR, pp. 697–706 (2014)

17. MacLean, S., Labahn, G.: A new approach for recognizing handwritten mathematics using relational grammars and fuzzy sets. Int. J. Doc. Anal. Recogn. (IJDAR) **16**(2), 1–25 (2012)

18. Miller, B.R., Youssef, A.: Technical aspects of the digital library of mathematical functions. Ann. Math. Artif. Intell. **38**, 121–36 (2003)

19. Mouchére, H., Viard-Gaudin, C., Zanibbi, R., Garain, U.: ICFHR 2014 competition on recognition of on-line handwritten mathematical expressions (CROHME 2014). In: Proceedings of the International Conference Frontiers in Handwriting Recognition, Crete, Greece, pp. 791–796 (2014)

20. Nagy, G., Seth, S.: Hierarchical representation of optically scanned documents. In: Proceedings of the Seventh International Conference on Pattern Recognition, Montreal, Canada, pp. 347–349 (1984)

21. Nguyen, T.T., Hui, S.C., Chang, K.: A lattice-based approach for mathematical search using formal concept analysis. Expert Syst. Appl. **39**(5), 5820–8 (2012)

22. Pattaniyil, N., Zanibbi, R.: Combining TF-IDF text retrieval with an inverted index over symbol pairs in math expressions: The Tangent math search engine at NTCIR 2014. In: Proceedings of the 1st NII Testbeds and Community for Information access Research (NTCIR), Tokyo, Japan (2014) (online, p. 8)

23. Pollanen, M., Wisniewski, T., Yu, X.: Xpress: A novice interface for the real-time communication of mathematical expressions. In: Proceedings of the Workshop Mathematical User-Interfaces, Linz, Austria, June 2007

24. Rath, T., Manmatha, R.: Word spotting for historical documents. Int. J. Doc. Anal. Recogn. **9**(2–4), 139–52 (2007)

25. Reichenbach, M., Agarwal, A., Zanibbi, R.: Rendering expressions to improve accuracy of relevance assessment for math search. In: Proceedings of the ACM SIGIR, Gold Coast, Australia, pp. 851–854 (2014)

26. Sasarak, C., Hart, K., Pospesel, R., Stalnaker, D., Hu, L., LiVolsi, R., Zhu, S., Zanibbi, R.: m_{in}: a multimodal web interface for web search. In: Symp. Human-Computer Interaction and Information Retrieval, Cambridge, MA, pp. (online, p. 4), Oct 2012

27. Schellenberg, T., Yuan, B., Zanibbi, R.: Layout-based substitution tree indexing and retrieval for mathematical expressions. In: Proceedings of the Document Recognition and Retrieval XVIII, pp. OI:1–8 (2012)

28. Schubotz, M.: Challenges of mathematical information retrieval in the NTCIR-11 Math Wikipedia Task. In: Proceedings of the SIGIR (2015, to appear)

29. Smithies, S., Novins, K., Arvo, J.: A handwriting-based equation editor. In: Proceedings of the Graphics Interface, Kingston, ON, June 1999

30. Sojka, P., Líška, M.: Indexing and searching mathematics in digital libraries. In: Davenport, J.H., Farmer, W.M., Urban, J., Rabe, F. (eds.) Calculemus/MKM 2011. LNCS, vol. 6824, pp. 228–43. Springer, Heidelberg (2011)

31. Pavan Kumar, P., Agarwal, A., Bhagvati, C.: A structure based approach for mathematical expression retrieval. In: Sombattheera, C., Loi, N.K., Wankar, R., Quan, T. (eds.) MIWAI 2012. LNCS, vol. 7694, pp. 23–34. Springer, Heidelberg (2012)

32. Stalnaker, D., Zanibbi, R.: Math expression retrieval using an inverted index over symbol pairs. In: Proceedings of the Document Recognition and Retrieval XXII. Proc. SPIE, San Francisco, USA, vol. 9402, pp. 940207–1:12, Feb 2015

33. Suzuki, M., Kanahori, T., Ohtake, N., Yamaguchi, K.: An integrated OCR software for mathematical documents and its output with accessibility. In: Miesenberger, K., Klaus, J., Zagler, W.L., Burger, D. (eds.) ICCHP 2004. LNCS, vol. 3118, pp. 648–55. Springer, Heidelberg (2004)

34. Uchida, S., Nomura, A., Suzuki, M.: Quantitative analysis of mathematical documents. Int. J. Doc. Anal. Recogn. **7**(4), 211–8 (2005)

35. Wangari, K., Zanibbi, R., Agarwal, A.: Discovering real-world use cases for a multimodal math search interface. In: Proceedings of the ACM SIGIR, Gold Coast, Australia, pp. 947–950, July 2014
36. Youssef, A.S.: Methods of relevance ranking and hit-content generation in math search. In: Kauers, M., Kerber, M., Miner, R., Windsteiger, W. (eds.) MKM/CALCULEMUS 2007. LNCS (LNAI), vol. 4573, pp. 393–406. Springer, Heidelberg (2007)
37. Zanibbi, R., Blostein, D.: Recognition and retrieval of mathematical expressions. Int. J. Doc. Anal. Recogn. (IJDAR) 15(4), 331–57 (2012)
38. Zanibbi, R., Yu, L.: Math spotting: Retrieving math in technical documents using handwritten query images. In: Proceedings of the International Conference on Document Analysis and Recognition, Beijing, China, pp. 446–451, September 2011
39. Zanibbi, R., Yuan, B.: Keyword and image-based retrieval of mathematical expressions. In: Proceedings of the Document Recognition and Retrieval XVIII, pp. 78740I–78740I (2011)
40. Zanibbi, R., Blostein, D., Cordy, J.R.: Recognizing mathematical expressions using tree transformation. IEEE Trans. Pattern Anal. Mach. Intell. 24(11), 1455–1467 (2002)
41. Zanibbi, R., Novins, K., Arvo, J., Zanibbi, K.: Aiding manipulation of handwritten mathematical expressions through style-preserving morphs. In: Proceedings of the Graphics Interface, Ottawa, ON, June 2001
42. Zhao, J., Kan, M.Y., Theng, Y.L.: Math information retrieval: user requirements and prototype implementation. In: JCDL 2008: Proceedings of the 8th ACM/IEEE-CS Joint Conference on Digital Libraries, pp. 187–196. ACM, New York (2008)
43. Zhu, S., Hu, L., Zanibbi, R.: Rotation-robust math symbol recognition and retrieval using outer contours and image subsampling. In: Proceedings of the Document Recognition and Retrieval XX, San Francisco, CA, pp. 5:1–5:12, Feb 2013

Calculemus

Towards Formal Fault Tree Analysis
Using Theorem Proving

Waqar Ahmed$^{(\boxtimes)}$ and Osman Hasan

School of Electrical Engineering and Computer Science (SEECS),
National University of Sciences and Technology (NUST), Islamabad, Pakistan
{waqar.ahmad,osman.hasan}@seecs.nust.edu.pk

Abstract. Fault Tree Analysis (FTA) is a dependability analysis technique that has been widely used to predict reliability, availability and safety of many complex engineering systems. Traditionally, these FTA-based analyses are done using paper-and-pencil proof methods or computer simulations, which cannot ascertain absolute correctness due to their inherent limitations. As a complementary approach, we propose to use the higher-order-logic theorem prover HOL4 to conduct the FTA-based analysis of safety-critical systems where accuracy of failure analysis is a dire need. In particular, the paper presents a higher-order-logic formalization of generic Fault Tree gates, i.e., AND, OR, NAND, NOR, XOR and NOT and the formal verification of their failure probability expressions. Moreover, we have formally verified the generic probabilistic inclusion-exclusion principle, which is one of the foremost requirements for conducting the FTA-based failure analysis of any given system. For illustration purposes, we conduct the FTA-based failure analysis of a solar array that is used as the main source of power for the Dong Fang Hong-3 (DFH-3) satellite.

Keywords: Higher-Order logic · Probabilistic analysis · Theorem proving · Satellite's solar arrays

1 Introduction

With the increasing usage of engineering systems in safety-critical domains, their dependability and failure analysis [1] has become a dire need to predict their reliability, availability and safety. One of the most widely used dependability and failure analysis techniques is the Fault Tree Analysis (FTA) method [2]. It is a graphical technique consisting of internal nodes, which are represented by gates like OR, AND and XOR, and the external nodes, that model the events which are associated with the occurrence of faults in sub-systems or components of the given system. The generic nature of these gates and events allows us to construct an efficient and accurate fault tree (FT) model for any given system. This FT can in turn be used to investigate the potential causes of a fault occurrence in a system and the calculation of minimal number of events that contribute towards the occurrence of a *top event*, i.e., a critical event, which can cause the

© Springer International Publishing Switzerland 2015
M. Kerber et al. (Eds.): CICM 2015, LNAI 9150, pp. 39–54, 2015.
DOI: 10.1007/978-3-319-20615-8_3

whole system failure upon its occurrence. Some noteworthy applications of FTA include the failure analysis of transportation systems [3], healthcare systems [4] and aerospace systems [5].

Traditionally, FTA is carried out by using paper-and-pencil proof methods, computer simulations and computer algebra systems. The first step in the paper-and-pencil proof methods is the construction of the FT of the given system on a paper. This is followed by the identification of the Minimal Cut Set (MCS) failure events, which contribute in the occurrence of the top event. These MCS failure events are generally modeled in terms of the exponential or weibull random variables and the Probabilistic Inclusion-Exclusion (PIE) principle [6] is then used to evaluate the exact probability of failure of the given system. However, this method is prone to human errors when it comes to the MCS and failure probability assessment of large safety-critical systems. For instance, in nuclear plants, where a fault tree model involves 50 to 130 levels of logic gates between the top event and the lowest basic events that are contributing to the top event [7]. So, there is a possibility, that many of these basic failure events may be overlooked while calculating MCS and thus not further incorporated in the FTA, which may lead to erroneous designs.

The FTA-based computer simulators, such as Relia-Soft [8] and ASENT Reliability analysis tools [9], provide graphical editors for the construction of FTs and the analysis is carried out by generating samples from the exponential and Weibull random variables that are associated with the events of the FT. These samples are then processed to evaluate the reliability and the failure probability of the complete system using computer arithmetic and numerical techniques. Although, these tools provide a more scalable alternative to the paper-and-pencil proof methods but the computational requirement increases drastically as the size of the FT increases. For example, if there are q terms involves in the MCS of a given FT then the total number of terms in the corresponding PIE principle will be $2q - 1$. In addition, these tools cannot ascertain absolute correctness or error-free analysis because of the involvement of pseudo random numbers and numerical methods and the inherent sampling-based nature of simulation.

Similarly, computer algebra systems (CAS), such as Mathematica [10], provide extensive features for FT-based failure analysis. For instance, the MCS expressions for any given system can be validated with failure distributions, such as Exponential or Weibull, by using symbolic and numerical algorithms. However, due to the presence of these unverified simplification algorithms, the analysis provided by CAS cannot be termed as sound and accurate.

Formal methods can overcome the above-mentioned inaccuracy limitations of the traditional techniques and thus have been used for FTA. The Interval Temporal Logic (ITS), i.e., a temporal logic that supports first-order logic, has been used, along with the Karlsruhe Interactive Verifier (KIV), for formal FTA of a rail-road crossing [11]. The work presented in [12] describes a deductive method for FT construction, in contrast to the intuitive approach followed in [11], by using the Observational Transition Systems (OTS) [12] and then the formal analysis of this FT is carried out using CafeOBJ [13], which is a formal specification language with interactive verification support. One of the main limitations of all the above-mentioned formal methods based works is the inability to conduct a probability theoretic FTA. The COMPASS tool-set [14], which is developed at RWTH Achen University in collaboration with the European Space Agency (ESA), caters for this problem and supports the formal FTA specifically for aerospace systems using the NuSMV and MRMC model checkers. However, the scope of these tools is somewhat limited in terms of handling failure analysis of large FTs, due to the inherent state-space explosion problem of model checking, and the fact that the computation of probabilities in these methods involve numerical methods, which compromises the accuracy of the results.

An accurate MCS calculation and exact failure probability assessment in the FTA is very important specially while dealing with safety-critical systems used in domains like transportation, aerospace or medicine. In order to achieve an accurate and precise FTA, we propose to conduct the formal FTA within the sound core of a higher-order-logic theorem prover [15]. Higher-order logic provides a precise deductive mechanism that can be used to model any mathematically expressive behavior including recursive definitions, random variables, fault tree events, which are the foremost building blocks for modeling FTs. Once the FTs are modeled in higher-order logic, we can deduce an accurate MCS by using formal reasoning based on the set-theoretic foundations. Moreover, FT properties, such as the probability of failure, can be formally verified using interactive theorem provers based on the PIE principles.

The foremost requirement for reasoning about reliability and failure related properties of a system in a theorem prover is the availability of the higher-order-logic formalization of probability theory. Hurd's formalization of measure and probability theories [16] is a pioneering work in this regard. Building upon this formalization most of the commonly-used continuous random variables [17] and some reliability theory fundamentals [18] have been formalized using the HOL theorem prover. However, Hurd's formalization of probability theory [16] only supports the whole universe as the probability space. This feature limits its scope in many aspects [19] and one of the main limitations, related to FTA-based analysis, is the nonability to reason about multiple continuous random variables [17]. Some recent probability theory formalizations [19,20] allow using any arbitrary probability space that is a subset of the universe and thus are more flexible than Hurd's formalization of probability theory. Particularly, Mhamdi's probability theory formalization [19], which is based on extended-real numbers (real numbers including $\pm\infty$), has been recently used to reason about the Reliability Block Diagram (RBD)-based analysis of a series pipelines structure [21], which involves multiple exponential random variables. The current paper is mainly inspired from this development as we use Mhamdi's formalized probability theory [19] for the formalization of all the commonly used FTA gates and the formal verification of their probabilistic properties. Moreover, we have also formally verified the PIE principle, which provides the foremost foundation for formal reasoning about the accurate failure analysis of any FT.

In order to illustrate the effectiveness of the proposed FTA approach, the paper presents a formal failure analysis, by taking a FT model, of a solar array that has been used in the DFH-3 Satellite, which was launched by the People's Republic of China on May 12, 1997 [5]. Solar arrays are one of the most vital components of the satellites because the mission success heavily depends upon the continuous reliable source of power [22]. Over the last ten years, 12 out of the 117 satellite's solar array anomalies, documented by the Airclaims Ascend SpaceTrak database, led to the total satellite failure [22,23]. Thus the absolute accuracy of the failure analysis of a solar array is a dire need in satellite missions and, to the best of our knowledge, it is the novelty of the proposed technique to meet this requirement. The satellite's solar array is a mechanical system, which mainly consists of various mechanisms, including: deployable, synchronization, locking and orientation. The FT of the solar array contains the failure events of these mechanisms and their interrelationships regarding the overall system failure. The paper presents the higher-order-logic modeling of this FT and the formal verification of the probability of failure of satellite's solar array system based on the probability of occurrence of the above-mentioned mechanism faults.

2   Probability Theory in HOL

In this section, we provide a brief overview of the HOL4 formalization of the probability theory [19], which we build upon in this paper. Based on the measure theoretic foundations, a probability space is defined as a triple (Ω, Σ, Pr), where Ω is a set, called the sample space, Σ represents a σ-algebra of subsets of Ω, where the subsets are usually referred to as measurable sets, and Pr is a measure with domain Σ and is 1 for the whole sample space. In the HOL4 probability theory formalization [19], given a probability space p, the functions space and subsets return the corresponding Ω and Σ, respectively. Based on this definition, all the basic probability axioms have been verified. Now, a random variable is a measurable function between a probability space and a measurable space, which essentially is a pair $(S, \mathcal{A})$, where S denotes a set and $\mathcal{A}$ represents a nonempty collection of sub-sets of S. A random variable is termed as discrete if S is a set with finite elements and continuous otherwise.

The cumulative distribution function (CDF) is defined as the probability of the event where a random variable X has a value less than or equal to some value x, i.e., $Pr(X \leq x)$. This definition characterizes the distribution of both discrete and continuous random variables and has been formalized [21] as follows:

⊢ ∀ p X x. CDF p X x = distribution p X {y | y ≤ Normal x}

The function Normal takes a *real* number as its input and converts it to its corresponding value in the *extended-real* data-type, i.e., it is the *real* data-type with the inclusion of positive and negative infinity. The function distribution takes three parameters: a probability space p, a random variable X and a set of *extended-real* numbers and returns the probability of the given random variable X acquiring all the values of the given set in probability space p.

The unreliability or the probability of failure $F(t)$ is defined as the probability that a system or component will fail by the time t. It can be described in terms of CDF, known as the failure distribution function, if the random variable X represent a time-to-failure of the component. This time-to-failure random variable X usually exhibits the exponential or weibull distribution.

The notion of mutual independence of n random variables is a major requirement for reasoning about the failure analysis of most of the FT gates. According to this notion, if we have N mutually independent failure events then

$$Pr(\bigcap_{i=1}^{N} L_i) = \prod_{i=1}^{N} Pr(L_i) \tag{1}$$

This concept has been formalized as follows [21]:

```
⊢ ∀ p L. mutual_indep p L = ∀ L1 n. PERM L L1 ∧
    1 ≤ n ∧ n ≤ LENGTH L ⇒
    prob p (inter_list p (TAKE n L1)) =
    list_prod (list_prob p (TAKE n L1))
```

The function `mutual_indep` accepts a list of events L and probability space p and returns $True$ if the events in the given list are mutually independent in the probability space p. The predicate `PERM` ensures that its two list arguments form a permutation of one another. The function `LENGTH` returns the length of the given list. The function `TAKE` returns the first n elements of its argument list as a list. The function `inter_list` performs the intersection of all the sets in its argument list of sets and returns the probability space if the given list of sets is empty. The function `list_prob` takes a list of events and returns a list of probabilities associated with the events in the given list of events in the given probability space. Finally, the function `list_prod` recursively multiplies all the elements in the given list of real numbers. Using these functions, the function `mutual_indep` models the mutual independence condition such that for any 1 or more events n taken from any permutation of the given list L, the property $Pr(\bigcap_{i=1}^{N} L_i) = \prod_{i=1}^{N} Pr(L_i)$ holds.

3   Formalization of Fault Tree Gates

In this section, we describe a generic formalization of commonly used FT gates given in Table 1. Our formalizations are generic in terms of the number of inputs n, i.e., our definitions can be used to model arbitrary-input FT gates.

3.1   Formal Definitions of Fault Tree Gates

If the occurrence of the output failure event is caused by the occurrence of all the input failure events then this kind of behavior can be modeled by using the AND FT gate. The function `AND_FT_gate`, given in Table 1, models this behavior as

Table 1. HOL4 formalization of fault tree gates

Fault Tree Gates	HOL Formalization
	⊢ ∀ p L. AND_FT_gate p L = inter_list p L
	⊢ ∀ L. OR_FT_gate L = union_list L
	⊢ ∀ p L1 L2. NAND_FT_gate p L1 L2 = inter_list p (compl_list p L1) ∩ inter_list p L2
	⊢ ∀ p L. NOR_FT_gate p L = p_space p DIFF (OR_gate L)
	⊢ ∀ p A B. XOR_FT_gate p A B = ((p_space p DIFF A ∩ B) ∪ (A ∩ p_space p DIFF B))
	⊢ ∀ p A. NOT_FT_gate p A = (p_space p DIFF A)

it accepts an arbitrary probability space p and returns the intersection of input failure events, given in the list L, by using the recursive function inter_list.

In the OR FT gate, the occurrence of the output failure event depends upon the occurrence of any one of its input failure event. The function OR_FT_gate, given in Table 1, models this behavior as it returns the union of the input failure list L by using the recursive function union_list. The NOR FT gate can be viewed as the complement of the OR FT gate and its output failure event occurs if none of the input failure event occurs.

The NAND FT gate models the behavior of the occurrence of an output failure event when at least one of the failure events at its input does not occur. This type of gate is used in FTs when the non-occurrence of the failure event in conjunction with the other failure events causes the top failure event to occur. This behavior can be expressed as the intersection of complementary and normal events [1], where the complementary events model the non-occurring failure events and the normal events model occurring failure events. It is important to note that the behavior of the NAND FT gate is usually not captured by the complement of the AND FT gate in the FTA literature [1]. The function NAND_FT_gate accepts a probability space p and two list of failure events $L1$ and $L2$. The function returns the intersection of non-occurring failure events, which in turn is modeled by passing the list of failure events $L1$ to the recursive function compl_list, and occurring failure events, which are given in the list $L2$, by utilizing the recursive function inter_list. The function compl_list returns a list of events such that each element of this list is the difference between the probability space p and the corresponding element of the given list.

The output failure event occurs in the 2-input XOR FT gate if only one, and not both, of its input failure events occur. The HOL representation of the behaviour of the XOR FT gate is presented in Table 1. The function NOT_FT_gate accepts an arbitrary failure event A along with probability space p and returns the complement to the probability space p of the given input failure event A.

3.2   Formal Verification of Failure Probability of Fault Tree Gates

The function AND_FT_gate, given in Table 1, can be used to evaluate the failure probability of the output failure event of the AND FT gate. If A_i represents the ith failure event with failure probability F_i at time t among the n mutually independent failure events of the AND FT gate then the generic mathematical expression for the failure probability of a n-input AND FT gate is as follows:

$$F_{AND\_gate}(t) = Pr(\bigcap_{i=2}^{N} A_i(t)) = \prod_{i=2}^{N} F_i(t) \qquad (2)$$

We formally verified this expression as the following theorem in HOL4:

Theorem 1. ⊢ ∀ p L. prob_space p ∧
   2 ≤ LENGTH L ∧ mutual_indep p L ⇒
   (prob p (AND_gate p L) = list_prod (list_prob p L))

The first assumption ensure that p is a valid probability space based on the probability theory in HOL4 [19]. The next two assumptions guarantee that the list of failure events must have at least two failure event and the failure events are mutually independent, respectively. The conclusion of the theorem represents Eq. (2). The proof of Theorem 1 is primarily based on some probability theory axioms and the mutual independence definition.

Similarly, if A_i represents the ith with failure event failure probability F_i at time t among the n mutually independent failure events of an OR FT gate then its failure probability expression is as follows:

$$F_{OR\_gate}(t) = Pr(\bigcup_{i=2}^{N} A_i(t)) = 1 - \prod_{i=2}^{N}(1 - F_i(t)) \qquad (3)$$

In order to formally verify the above equation, we first formally verify the following lemma that provides an alternate expression for the failure probability of an OR FT gate in terms of the failure probability of an AND FT gate:

Lemma 1. ⊢ ∀ L p. (prob_space p) ∧
(∀ x'. MEM x' L ⇒ x' ∈ events p) ⇒
  (prob p (OR_gate L) =
  1 - prob p (AND_gate p (compl_list p L))

Now, we can formally verify Eq. (3) in HOL4 as follows:

Theorem 2. ⊢ ∀ p L. (prob_space p) ∧
(2 ≤ LENGTH L) ∧ (mutual_indep p L) ∧
(∀ x'. MEM x' L ⇒ x' ∈ events p) ⇒
  (prob p (OR_gate L) =
  1 - list_prod (one_minus_list (list_prob p L)))

Where the function one_minus_list accepts a list of *real* numbers $[x_1, x_2, \cdots, x_n]$ and returns the list of *real* numbers such that each element of this list is 1 minus the corresponding element of the given list, i.e., $[1 - x_1, 1 - x_2, \cdots, 1 - x_n]$. The proof of Theorem 2 is primarily based on Lemma 1 and Theorem 1 along with the fact that given the list of n mutually independent events, the complement of these n events are also mutually independent.

Similarly, we also verified the failure probability theorems for other FT gates, given in Table 1, and the corresponding mathematical expressions and theorems are given in Table 2. All these results are verified under the same assumptions as the ones used in Theorems 1 and 2.

Table 2. Probability of failure of fault tree gates

Fault Tree Gates	Theorem's Conclusion
$F_{NOR}(t) = 1 - F_{OR}(t)$ $= \prod_{i=2}^{N}(1 - F_i(t))$	(prob p (NOR_FT_gate p L) = list_prod (one_minus_list (list_prob p L)))
$F_{NAND}(t) = Pr(\bigcap_{i=2}^{k} \overline{A_i}(t) \cap \bigcap_{j=k}^{N} A_i(t))$ $= \prod_{i=2}^{k}(1 - F_i(t)) * \prod_{j=k}^{N}(F_j(t))$	(prob p (NAND_FT_gate p L1 L2) = list_prod ((list_prob p (compl_list p L1))) * list_prod (list_prob p L2))
$F_{XOR}(t) = Pr(\bar{A}(t)B(t) \cup A(t)\bar{B}(t))$ $= (1 - F_A(t))F_B(t) +$ $F_A(t)(1 - F_B(t))$	(prob p (XOR_FT_gate p A B) = (1- prob p A)*prob p B + prob p A*(1 - prob p B)
$F_{NOT}(t) = Pr(A(t))$ $= (1 - F_A(t))$	prob p (NOT_FT_gate p A) = (1 - prob p A)

The proof script [24] of the above-mentioned formalization is composed of 4000 lines of HOL script and took about 200 man-hours. The main outcome of this exercise is that the definitions, given in Table 1, can be used to capture the behavior of most of the FTs in higher-order logic and the Theorems of Table 2 can then be used in conjunction with the formalization of the PIE principle, explained next, to formally verify the corresponding failure probabilities.

4   Formalization of Probabilistic Inclusion-Exclusion Principle

The probabilistic inclusion-exclusion principle (PIE) forms an integral part of the reasoning involved in verifying the failure probability of a FT. In FTA, firstly all the basic fault events are identified that can cause the occurrence of the system failure event. These fault events are then combined to model the overall fault behavior of the given system by using the fault gates. These combinations of basic failure events, called cut sets, are then reduced to minimal cut sets (MCS) by using some set-theory rules, such as idempotent, associative and commutative [25]. At this point, the PIE principle is used to evaluate the overall failure probability of the given system based on the MCS events.

If A_i represent the ith basic failure event or a combination of failure event then the failure probability of the given system can be expressed in terms of the probabilistic inclusion-exclusion principle as follows:

$$\mathbb{P}(\bigcup_{i=1}^{n} A_i) = \sum_{t \neq \{\}, t \subseteq \{1,2,\ldots,n\}} (-1)^{|t|+1} \mathbb{P}(\bigcap_{j \in t} A_j) \tag{4}$$

The above equation can be formalized in HOL4 is as follows:

Theorem 3. ⊢ ∀ p L1 L2. prob_space p ∧
(∀ x. MEM x L ⇒ x ∈ events p) ⇒
   (prob p (union_list L) =
   sum_set {t | t ⊆ set L ∧ t ≠ {} }
     (λt. -1 pow (CARD t + 1) * prob p (BIGINTER t)))

The assumptions of the above theorem are the same as the ones used in Theorem 1. The function sum_set takes an arbitrary set s with element of type α and a real-valued function f. It recursively sums the return value of the function f, which is applied on each element of the given set s. In the above theorem, the set s is represented by the term $\{x|C(x)\}$ that contains all the values of x, which satisfy condition C. Whereas, the λ abstraction function (λ t. -1 pow (CARD t + 1) * prob p (BIGINTER t)) models $(-1)^{|t|+1}\mathbb{P}(\bigcap_{j \in t} A_j)$, such that the functions CARD and BIGINTER return the number of elements and the intersection of all the elements of the given set, respectively. Thus, the conclusion of the theorem represents Eq. (4).

The formal reasoning about Theorem 3 is based upon the following lemma:

Lemma 2. ⊢ ∀ P. (∀ n. (∀ m. m < n ⇒ P m) ⇒ P n) ⇒ ∀ n. P n

Where n in our case is the length of the list L and m represent another list whose length is less then the length of the list L. The predicate P represents the conclusion of Theorem 3. The above property brings an important hypothesis in the assumption list, which has the same form as that of the conclusion of Theorem 3. Then, by utilizing induction and some properties of the function sum_set along with some fundamental axioms of probability, we can verify Theorem 3.

The proof script [24] for Theorem 3 is composed of 1000 lines of HOL code and involved 50 man-hours of proof effort. To the best of our knowledge, this is the first formal verification of the probabilistic inclusion exclusion principle, which, besides being used in FTA, is a widely used mathematical result in analyzing various bio-informatics [26] and telecommunication [27] systems.

5   Application: Satellite's Solar Array

The solar arrays used in satellite missions are usually in a folded position during the launch phase [5]. Once the satellite is deployed in the corresponding orbit then the solar arrays are unfolded and the goal is to keep them oriented towards the sun all the time to maximize the power generation for the satellite [5]. The faults in the solar array are mainly caused by the mechanical components that drive these mechanisms associated with the driving, deployment, synchronization, locking and orientation. For example, the solar array is usually driven by using a torsion spring [5]. Whereas, the closed cable loop (CCL) and the stepping or servo motors are used during the synchronization and orientation phases [5]. A FT can thus be constructed by considering the faults in these mechanical components, which are the fundamental causes of satellite' solar array mechanisms failure The FT for the solar array of the DFH-3 Satellite that was launched by the People's Republic of China on May 12, 1997 [28] is depicted in Fig. 1 and we formally analyze this FT in this paper.

The failure events, A, B, C, D in Fig. 1, represent the failures in the unlock mechanism, deployment process, locking process and orientation process, respectively. Whereas, the failure event E represents the failures in the corresponding mechanical parts of the system. These failure events are combined either by using the OR or AND FT gates by considering the behavior of the faults.

In order to formalize the solar array FT of Fig. 1, we first present the formal modeling of list of failure events that are associated with each corresponding fault of the solar array FT.

Definition 1. ⊢ ∀ p x. fail_event_list p [] x = [] ∧
　∀ p x h t. fail_event_list p (h::t) x =
　　PREIMAGE h {y | y ≤ Normal x } ∩ p_space p ::
　　fail_event_list p t x

The function fail_event_list accepts a probability space p, a list of random variables, representing the failure time of individual components, and a real number x, which represents the time index at which the failure of the component occurs. It returns a list of events, representing the failure of all the individual components at time x. The formal definitions of FT gates, given in Sect. 3, along with Definition 1 can be utilized to formally represent the FT of satellite's solar array in terms of its cut-set failure events. The HOL4 formalization of satellite's solar array FT is as follows:

Definition 2. ⊢ ∀ p x1 x2 x3 x4 x5 x6 x7 x8 x9 x10 x11
　　　　　　　x12 x13 x14 t.
Solar_FT p x1 x2 x3 x4 x5 x6 x7 x8 x9 x10 x11 x12 x13 x14 t =
OR_FT_gate [OR_FT_gate (fail_event_list p [x1; x2] t);
　　　　　OR_FT_gate [OR_FT_gate (fail_event_list p [x3; x4] t);
　　　　　　AND_FT_gate p (fail_event_list p [x5; x6] t); OR_FT_gate
(fail_event_list p [x3; x7; x8] t)];
　　　　　OR_FT_gate (fail_event_list p [x3; x9] t);
　　　　　OR_FT_gate (fail_event_list p [x10; x11] t);

```
OR_FT_gate [PREIMAGE x12 {y | y ≤ Normal t };
            PREIMAGE x13 {y | y ≤ Normal t };
            OR_FT_gate (fail_event_list p[x3; x14]t)]]
```

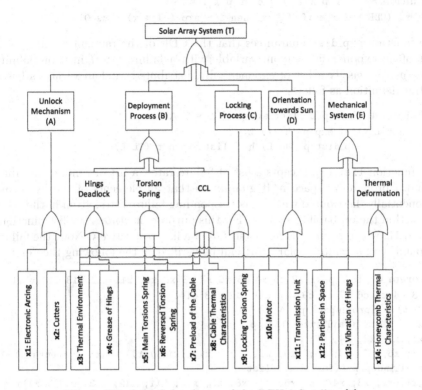

Fig. 1. FT of the Solar Array of the DFH-3 Satellite [5]

Where the random variables $x1 - x14$ model the time-to-failure of the solar array processes and components as depicted in Fig. 1. However, the cut-set failure events in the above definition is not minimal [5], i.e., there are some redundant failure events. For example, $x3$ is part of more than one OR FT gates. These kind of redundant failure events can be removed by verifying an accurate equivalent but reduced representation, i.e., the MCS, by using set theory laws, like idempotent, commutative and associative, as follows:

Lemma 2. ⊢ ∀ p x1 x2 x3 x4 x5 x6 x7 x8 x9 x10 x11 x12 x13 x14 t.
```
      prob_space p ⇒
(Solar_FT p x1 x2 x3 x4 x5 x6 x7 x8 x9 x10 x11 x12 x13 x14 t =
OR_FT_gate [OR_FT_gate (fail_event_list p [x1; x2; x3; x4] t);
           AND_FT_gate p (fail_event_list p [x5; x6] t);
           OR_FT_gate
             (fail_event_list p [x7; x8; x9; x10; x11; x12; x13; x14]t)])
```

We consider that random variables, associated with the failure events of the solar array FT, exhibit the exponential distribution, which can be formalized in HOL4 as follows:

Definition 3. ⊢ ∀ p X l. exp_dist p X l =
  ∀ x. (CDF p X x = if 0 ≤ x then 1 - exp (-l * x) else 0)

The function exp_dist guarantees that the CDF of the random variable X is that of an exponential random variable with a failure rate l in a probability space p. We classify a list of exponentially distributed random variables based on this definition as follows:

Definition 4. ⊢ ∀ p L. list_exp p [] L = T ∧
  ∀ p h t L. list_exp p (h::t) L =
        exp_dist p (HD L) h ∧ list_exp p t (TL L)

The function list_exp accepts a list of failure rates, a list of random variables L and a probability space p. It guarantees that all elements of the list L are exponentially distributed with the corresponding failure rates, given in the other list, within the probability space p. For this purpose, it utilizes the list functions HD and TL, which return the *head* and *tail* of a list, respectively. Now, the failure probability of satellite's solar array can be verified as the following theorem:

Theorem 4. ⊢ ∀ p x1 x2 x3 x4 x5 x6 x7 x8 x9 x10 x11 x12 x13 x14 t c1 c2 c3 c4 c5 c6 c7 c8 c9 c10 c11 c12 c13 c14.
(0 ≤ t) ∧ (prob_space p) ∧
(∀ x'. MEM x' (fail_event_list p
  ([x1; x2; x3; x4; x5;
    x6; x6; x7; x8; x9; x10; x11; x12; x13; x14]) t)) ⇒ x' ∈ events p) ∧
(mutual_indep p ((fail_event_list p
    ([x1; x2; x3; x4; x5; x6; x7; x8; x9; x10; x11; x12; x13; x14]) x))) ∧
list_exp p
  ([c1; c2; c3; c4; c5; c6; c7; c8; c9; c10; c11; c12; c13; c14])
    ([x1; x2; x3; x4; x5; x6; x7; x8; x9; x10; x11; x12; x13; x14]) ⇒
  (prob p (Solar_FT p
    x1 x2 x3 x4 x5 x6 x7 x8 x9 x10 x11 x12 x13 x14 t) =
  (1 - (exp -(t*(list_sum [c1;c2;c3;c4])))) +
    list_prod(one_minus_exp t [c5;c6;c7]) +
    (1 - (exp -(t*(list_sum
      [c7; c8; c9; c10; c11; c12; c13; c14])))) -
    (1 - list_prod(one_minus_exp_prod t
      [[c1;c5;c6];[c2;c5;c6];[c3;c5;c6];[c4;c5;c6]])) -
    (1 - (exp -(t*(list_sum [c1;c2;c3;c4])))) *
      (1 - (exp -(t*(list_sum
        [c7; c8; c9; c10; c11; c12; c13; c14])))) -
    (1 - list_prod(one_minus_exp_prod t
      [[c5;c6;c7];[c5;c6;c8];[c5;c6;c9];[c5;c6;c10];
        [c5;c6;c11];[c5;c6;c12];[c5;c6;c13];[c5;c6;c14]])) +
    (1 - list_prod(one_minus_exp_prod t
      [[c1;c5;c6];[c2;c5;c6];[c3;c5;c6];[c4;c5;c6]])) *

```
(1 - (exp -(t*
    (list_sum [c7; c8; c9; c10; c11; c12; c13; c14]))))))
```

The first assumption ensures the variable t that models time can acquire positive values only. The second assumption ensure that p is a valid probability space based on the probability theory in HOL4 [29]. The next two assumptions ensure that the events corresponding to the failures modeled by the random variables x1 to x14 are valid events from the probability space p and they are mutually exclusive. Finally, the last assumption characterizes the random variables x1 to x14 as exponential random variables with failure rates c1 to c14, respectively. The conclusion of the Theorem 4 represents the failure probability of the given solar array in terms of the failure rates of its components as follows:

$$(1 - e^{-(c1+c2+c3+c4)t}) + \prod_{i=5}^{6}(1 - e^{-(c_i t)}) +$$

$$(1 - e^{-(c7+c8+c9+c10+c11+c12+c13+c14)t}) - (1 - \prod_{i=1}^{4}(1 - \prod_{j=5}^{6}[(1 - e^{-c_i t})(1 - e^{-c_j t})])) -$$

$$(1 - e^{-(c1+c2+c3+c4)t}) * (1 - e^{-(c7+c8+c9+c10+c11+c12+c13+c14)t}) -$$

$$(1 - \prod_{i=7}^{14}(1 - \prod_{j=5}^{6}[(1 - e^{-c_i t})(1 - e^{-c_j t})])) +$$

$$(1 - \prod_{i=1}^{4}(1 - \prod_{j=5}^{6}[(1 - e^{-c_i t})(1 - e^{-c_j t})])) * (1 - e^{-(c7+c8+c9+c10+c11+c12+c13+c14)t})$$

$$(5)$$

where the function exp represents a exponential function, the function list_sum is used to sum all the element of the given list of failure rates, the function one_minus_exp accepts a list of failure rates and returns a one minus list of exponentials and the function one_minus_exp_prod accepts a two dimensional list of failure rates and returns a list with one minus product of one minus exponentials of every sub-list. For example, one_minus_exp_prod$[[c1; c2; c3]; [c4; c5]; [c6; c7; c8]]$ $x = [1 - ((1 - e^{-(c1)x}) * (1 - e^{-(c2)x}) * (1 - e^{-(c3)x})); (1 - (1 - e^{-(c4)x}) * (1 - e^{-(c5)x})); (1 - (1 - e^{-(c6)x}) * (1 - e^{-(c7)x}) * (1 - e^{-(c8)x}))].$

The proof of the above theorem utilizes the failure probabilities of AND and OR FT gates, given in Table 2, along with Lemma 2 and Theorem 3 and some fundamental facts and axioms of probability theory. Due to the universally quantified variables in Theorem 3, the proof of Theorem 4 is quite straightforward (about 800 lines of HOL code) as compared to that of Theorem 3. The distinguishing features of the formally verified Theorem 4 includes its generic nature, i.e., all the variables are universally quantified and thus can be specialized to obtain the failure probability for any given failure rates, and its guaranteed correctness due to the involvement of a sound theorem prover in its verification, which ensures that all the required assumptions for the validity of the result are accompanying the theorem.

A fuzzy reasoning Petri Net (FRPN), which is a combination of fuzzy logic [30] and Petri Nets [31], based failure analysis for the above-mentioned solar

array is presented in [5]. In this work, the FT of Fig. 1 is first represented as a
Petri Net such that the gates are represented by transitions and the failure events
are modeled as places. The possibility of fault occurrence is then evaluated by
using fuzzy degree of truth on the basis of petri nets transitions. However, the
truth degree values evaluated using these FRPN models cannot be regarded as
precise and sound as the formally verified expression using the HOL theorem
prover due to the involvement of numerical techniques and pseudo randomness.
On the other hand, our analysis result, i.e., Theorem 4, is based on a probability theoretic formal reasoning, verified in a sound theorem prover and is valid
for all possible values of the failure rates. These features constitute the main
motivations of the work presented in this paper.

6   Conclusion

The accuracy of failure analysis is a dire need for safety and mission-critical
applications, where an incorrect failure analysis may lead to disastrous situations
including the loss of human lives or heavy financial setbacks. In this paper,
we presented an accurate FTA approach, based on higher-order-logic theorem
proving, to tackle the analysis of such critical systems. In particular the paper
presents a formalization of commonly used FT gates and the PIE principle, which
are the foremost foundations for formal reasoning about FTA within a sound core
of theorem prover. As a case-study, the paper also presents the formal failure
analysis of a satellite's solar array.

Building upon the results, presented in this paper, other FT gates, such as
priority AND and voting OR gate, can also be formally modeled and thus the
scope of FTA-based formal reliability analysis [32] can be further enhanced.
Some interesting real-world applications that can benefit from our work include
transportation systems [3], healthcare systems [4] and avionics [33]. Moreover,
we also plan to further facilitate the formal FT-based failure analysis by incorporating the automatic simplification capabilities of CAS, such as Mathmatica,
for MCS calculation. This obtained MCS can then be validated within the sound
environment of the HOL theorem prover.

References

1. IEC: International Electrotechnical Commission, 61025 Fault Tree Analysis (2006)
2. Roberts, N.H., Vesely, W.E.: Fault Tree Handbook. Government Printing,
   Washington (1987)
3. Huang, H.Z., Yuan, X., Yao, X.S.: Fuzzy fault tree analysis of railway traffic safety.
   In: Conference on Traffic and Transportation Studies, American Society of Civil
   Engineers, pp. 107–112 (2000)
4. Hyman, W.A., Johnson, E.: Fault tree analysis of clinical alarms. J. Clin. Eng.
   33(2), 85–94 (2008)
5. Wu, J., Yan, S., Xie, L.: Reliability analysis method of a solar array by using
   fault tree analysis and fuzzy reasoning Petri net. Acta Astronaut. **69**(11), 960–968
   (2011)

6. Trivedi, K.S.: Probability and Statistics with Reliability, Queuing and Computer Science Applications, 2nd edn. Wiley, New York (2002)
7. Epstein, S., Rauzy, A.: Can we trust PRA? Reliab. Eng. Syst. Saf. **88**(3), 195–205 (2005)
8. ReliaSoft (2015). http://www.reliasoft.com/
9. ASENT (2015). https://www.raytheoneagle.com/asent/rbd.htm
10. Long, W., Sato, Y., Horigome, M.: Quantification of sequential failure logic for fault tree analysis. Reliab. Eng. Syst. Saf. **67**(3), 269–274 (2000)
11. Ortmeier, F., Schellhorn, G.: Formal Fault Tree Analysis-Practical Experiences, vol. 185, pp. 139–151. Elsevier, Amsterdam (2007)
12. Xiang, J., Futatsugi, K., He, Y.: Fault tree and formal methods in system safety analysis. In: Computer and Information Technology, pp. 1108–1115. IEEE (2004)
13. Futatsugi, K., Nakagawa, A.T., Tamai, T.: CAFE: An Industrial-Strength Algebraic Formal Method. Elsevier, Amsterdam (2000)
14. Bozzano, M., Cimatti, A., Katoen, J.P., Nguyen, V.Y., Noll, T., Roveri, M.: The COMPASS approach: correctness, modelling and performability of aerospace systems. In: Buth, B., Rabe, G., Seyfarth, T. (eds.) SAFECOMP 2009. LNCS, vol. 5775, pp. 173–186. Springer, Heidelberg (2009)
15. Harrison, J.: Handbook of Practical Logic and Automated Reasoning. Cambridge University Press, Cambridge (2009)
16. Hurd, J.: Formal verification of probabilistic algorithms. Ph.D. thesis, University of Cambridge, UK (2002)
17. Hasan, O., Tahar, S.: Formalization of continuous probability distributions. In: Pfenning, F. (ed.) CADE 2007. LNCS (LNAI), vol. 4603, pp. 3–18. Springer, Heidelberg (2007)
18. Abbasi, N., Hasan, O., Tahar, S.: An approach for lifetime reliability analysis using theorem proving. J. Comput. Syst. Sci. **80**(2), 323–345 (2014)
19. Mhamdi, T., Hasan, O., Tahar, S.: On the formalization of the lebesgue integration theory in HOL. In: Kaufmann, M., Paulson, L.C. (eds.) ITP 2010. LNCS, vol. 6172, pp. 387–402. Springer, Heidelberg (2010)
20. Hölzl, J., Heller, A.: Three chapters of measure theory in Isabelle/HOL. In: Eekelen, M., Geuvers, H., Schmaltz, J., Wiedijk, F. (eds.) ITP 2011. LNCS, vol. 6898, pp. 135–151. Springer, Heidelberg (2011)
21. Ahmed, W., Hasan, O., Tahar, S., Hamdi, M.S.: Towards the formal reliability analysis of oil and gas pipelines. In: Watt, S.M., Davenport, J.H., Sexton, A.P., Sojka, P., Urban, J. (eds.) CICM 2014. LNCS, vol. 8543, pp. 30–44. Springer, Heidelberg (2014)
22. Brandhorst Jr., H.W., Rodiek, J.A.: Space Solar Array Reliability: A Study and Recommendations, vol. 63, pp. 1233–1238. Elsevier, Amsterdam (2008)
23. Airclaims Ascend Spacetrak Database (2015). www.ascendspacetrak.com/Home
24. Ahmad, W.: Formal fault tree analysis of Satellite's Solar Array (2015). http://save.seecs.nust.edu.pk/projects/fta.html
25. Halmos, P.R.: Naive Set Theory. Springer, Heidelberg (1960)
26. Todor, A., Gabr, H., Dobra, A., Kahveci, T.: Large scale analysis of signal reachability. Bioinformatics **30**(12), 96–104 (2014)
27. Gao, F., Liu, X., Liu, H.: A rapid algorithm for computing ST reliability of radio-communication networks. In: Apolloni, B., Howlett, R.J., Jain, L. (eds.) KES 2007, Part II. LNCS (LNAI), vol. 4693, pp. 167–174. Springer, Heidelberg (2007)
28. Jianing, W., Shaoze, Y.: Reliability analysis of the solar array based on fault tree analysis. J. Phys. **305**, 012006 (2011). IOP Publishing

29. Mhamdi, T., Hasan, O., Tahar, S.: On the formalization of the Lebesgue integration theory in HOL. In: Kaufmann, M., Paulson, L.C. (eds.) ITP 2010. LNCS, vol. 6172, pp. 387–402. Springer, Heidelberg (2010)
30. Zadeh, L.A.: Toward a Theory of Fuzzy Information Granulation and its Centrality in Human Reasoning and Fuzzy Logic, vol. 90. Elsevier, Amsterdam (1997)
31. Peterson, J.L.: Petri Net Theory and the Modeling of Systems. Prentice Hall, Englewood Cliffs (1981)
32. Volkanovski, A., Čepin, M., Mavko, B.: Application of the fault tree analysis for assessment of power system reliability. Reliab. Eng. Syst. Saf. Elsevier **94**(6), 1116–1127 (2009)
33. Lefebvre, A., Simeu-Abazi, Z., Derain, J.P., Glade, M., et al.: Diagnostic of the avionic equipment based on dynamic fault tree. In: IFAC-CEA Conference (2007)

Optimizing a Certified Proof Checker
for a Large-Scale Computer-Generated Proof

Luís Cruz-Filipe[(✉)] and Peter Schneider-Kamp

Department of Mathematics and Computer Science,
University of Southern Denmark, Campusvej 55, 5230 Odense M, Denmark
{lcf,petersk}@imada.sdu.dk

Abstract. In recent work, we formalized the theory of optimal-size sorting networks with the goal of extracting a verified checker for the large-scale computer-generated proof that 25 comparisons are optimal when sorting 9 inputs, which required more than a decade of CPU time and produced 27 GB of proof witnesses. The checker uses an untrusted oracle based on these witnesses and is able to verify the smaller case of 8 inputs within a couple of days, but it did not scale to the full proof for 9 inputs. In this paper, we describe several non-trivial optimizations of the algorithm in the checker, obtained by appropriately changing the formalization and capitalizing on the symbiosis with an adequate implementation of the oracle. We provide experimental evidence of orders of magnitude improvements to both runtime and memory footprint for 8 inputs, and actually manage to check the full proof for 9 inputs.

1 Introduction

Sorting networks are hardware-oriented algorithms to sort a fixed number of inputs using a predetermined sequence of comparisons between them. They are built from a primitive operator – the *comparator* –, which reads the values on two channels, and interchanges them if necessary to guarantee that the smallest one is always on a predetermined channel. Comparisons between independent pairs of values can be performed in parallel, and the two main optimization problems one wants to addressed are: how many comparators do we need to sort n inputs (the *optimal size* problem); and how many computation steps do we need to sort n inputs (the *optimal depth* problem).

In previous work [2], we proposed a generate-and-prune algorithm to show size optimality of sorting networks, and used it to show that 25-comparator sorting networks have optimal size for 9 inputs. The proof was performed on a massively parallel cluster and consumed more than 10 years of computational time. During execution we recorded the results of *successful* search routines that allowed for reduction of the search space, resulting in approx. 27 GB of witnesses.

Subsequently [5], we formalized the relevant theory of sorting networks in Coq, therefrom extracting a certified checker able to confirm the validity of our informal computer-generated proof. The checker bypasses the original search steps by means of an untrusted oracle, implemented by reading the log file

© Springer International Publishing Switzerland 2015
M. Kerber et al. (Eds.): CICM 2015, LNAI 9150, pp. 55–70, 2015.
DOI: 10.1007/978-3-319-20615-8_4

produced by the original program, and could verify the proof for the smaller case of 8 inputs, thereby constituting the first computer-validated proof of the results in [6]. However, due to the much larger dimension of the oracle, verifying the full proof for 9 inputs was estimated to require approx. 20 years of (non-parallelizable) computation.

In this paper, we show how careful optimizations of the formalization result in runtime improvement of several orders of magnitude, as well as drastic reductions of the memory footprint for the checker. Throughout the paper, we benchmark the impact of the individual improvements using the feasible case of 8 inputs, until we are able to check the full proof for 9 inputs using around one week of computation on a single thread on an Intel Xeon E5 clocked at 2.4 GHz with 64 GB of RAM.

Section 2 shortly introduces the basic of sorting networks, the generate-and-prune algorithm from [2], and our formalization from [5] to the degree necessary to understand the improvements. In Sect. 3 we change the checker algorithm in the formalization in order to bring runtime down by at least an order of magnitude, while we reduce memory footprint by a factor of 3 in Sect. 4. Further substantial improvements to runtime and memory footprint are described in Sects. 5 and 6, respectively. We conclude in Sect. 7 with a summary of the results and an outlook to possible future work.

1.1 Related Work

The Curry–Howard correspondence states that every constructive proof of an existential statement embodies an algorithm to produce a witness of the required property. This correspondence has been made more precise by the development of program extraction mechanisms for the most popular theorem provers. In this paper, we focus on extracting a program from a Coq formalization, using the mechanism described in [11].

Early experiments of program extraction from a large-scale formalization that was built from a purely mathematical perspective showed however that it is unreasonable to expect *efficient* program extraction as a side result of formalizing textbook proofs [4]. In spite of that, one can actually develop mathematically-minded formalizations that yield efficient extracted programs with only minor attention to definitions [9,12]. This is in contrast with formalizations built with extraction as a primary goal, such as those in the CompCert project [10], or with strategies that potentially compromise the validity of the extracted program (e.g. using imperative data structures as in [13]).

In this work we go one step further, and show that if the extracted program does not perform well enough, we can optimize it by tweaking the formalization without significantly changing it. The latter means less work reproving lemmas and theorems and ensures that the formalization remains understandable, in turn giving us confidence that we actually prove what we wish to prove.

Our contributions rely on the idea of an *untrusted oracle* [7,10], where the extracted program checks the result of computations obtained through the oracle. More specifically, we use an *offline untrusted oracle*, where computation and checking are separated by logging the results of computations to a file. This separation allows the use of massively parallel clusters for computation and the cheap reuse of the results during the development of the formalization and the checker. In particular, we capitalize on the ability to pre-process the computational results offline to optimize the checker.

This offline approach to untrusted oracles is found in work on termination proofs [3,15], where the separation is necessary as informal proof tools and checkers are modular programs developed by different research units. The difference to our work is the scale of the proofs: typical termination proofs have 10–100 proof witnesses and total at most a few MB of data. Recent work mentions that problems were encountered when considering proofs of "several hundred megabytes" [14]. In contrast, verifying the proof of size-optimality of sorting networks with 9 inputs uses nearly 70 million proof witnesses, totalling 27 GB of oracle data.

2   Background

We briefly summarize the key notions relevant to this work. The interested reader is referred to [8] for a more extensive introduction to sorting networks, and to [2] for a detailed description of the proof we verify.

A *comparator network* C with n channels and size k is a sequence of *comparators* $C = (i_1, j_1); \ldots; (i_k, j_k)$, where each comparator (i_ℓ, j_ℓ) is a pair of channels $1 \leq i_\ell < j_\ell \leq n$. If C_1 and C_2 are comparator networks with n channels, then $C_1; C_2$ denotes the comparator network obtained by concatenating C_1 and C_2. An input $x = x_1 \ldots x_n \in \{0,1\}^n$ propagates through C as follows: $x^0 = x$, and for $0 < \ell \leq k$, x^ℓ is the permutation of $x^{\ell-1}$ obtained by interchanging $x_{i_\ell}^{\ell-1}$ and $x_{j_\ell}^{\ell-1}$ whenever $x_{i_\ell}^{\ell-1} > x_{j_\ell}^{\ell-1}$. The output of the network for input x is $C(x) = x^k$, and $\mathsf{outputs}(C) = \{C(x) \mid x \in \{0,1\}^n\}$. The comparator network C is a *sorting network* if all elements of $\mathsf{outputs}(C)$ are sorted (in ascending order). The zero-one principle [8] implies that a sorting network also sorts sequences over any other totally ordered set, e.g. integers. The image on the right depicts a sorting network on 4 channels, consisting of 6 comparators. The channels are indicated as horizontal lines (with channel 4 at the bottom), comparators are indicated as vertical lines connecting a pair of channels, and input values propagate from left to right. The sequence of comparators associated with a picture representation is obtained by a left-to-right, top-down traversal. For example, the network depicted above is $(1,2); (3,4); (1,4); (1,3); (2,4); (2,3)$.

The optimal-size sorting network problem is about finding the smallest size, $S(n)$, of a sorting network on n channels. In 1964, Floyd and Knuth presented sorting networks of optimal size for $n \leq 8$ and proved their optimality [6]. For nearly fifty years there was no further progress on this problem, until we established that $S(9) = 25$ [2] and, consequently, using a theoretical result on lower bounds [16], that $S(10) = 29$. Currently, the best known bounds for $S(n)$ are:

n	1	2	3	4	5	6	7	8	9	10	11	12	13	14	15	16
Upper bound for $S(n)$	0	1	3	5	9	12	16	19	25	29	35	39	45	51	56	60
Lower bound for $S(n)$	0	1	3	5	9	12	16	19	25	29	33	37	41	45	49	53

Our proof relies on a program that checks that there is no sorting network on 9 channels with only 24 comparators. The algorithm exploits symmetries in comparator networks, in particular the notion of *subsumption*. Given two comparator networks on n channels C_a and C_b and a permutation π on $\{1, \ldots, n\}$, we say that C_a subsumes C_b by π, and write $C_a \leq_\pi C_b$, if there exists a permutation π such that $\pi(\mathsf{outputs}(C_a)) \subseteq \mathsf{outputs}(C_b)$. We will write simply $C_a \preceq C_b$ to denote that $C_a \leq_\pi C_b$ for some π.

Subsumption is a powerful mechanism for reducing candidate sequences of comparators when looking for sorting networks: if C_a and C_b have the same size, $C_a \preceq C_b$ and there is a sorting network $C_b; C$ of size k, then there also is a sorting network $C_a; C'$ of size k. This motivated the *generate-and-prune* approach to the optimal-size sorting network problem: starting with the empty network, alternately add one comparator in all possible ways and reduce the result by eliminating subsumptions. More precisely, the algorithm iteratively builds two sets R_k^n and N_k^n of n channel networks of size k. First, it initializes R_0^n to contain only the empty comparator network. Then, it repeatedly applies two types of steps, `Generate` and `Prune`.

1. `Generate`: Given R_k^n, construct N_{k+1}^n by adding one comparator to each element of R_k^n in all possible ways.
2. `Prune`: Given N_{k+1}^n, construct R_{k+1}^n such that every element of N_{k+1}^n is subsumed by an element of R_{k+1}^n.

The algorithm stops when a sorting network is found, in which case $|R_k^n| = 1$.

Soundness of the algorithm relies on the fact that N_k^n (and R_k^n) are *complete* for the optimal size sorting network problem on n channels: if there exists an optimal size sorting network on n channels, then there exists one of the form $C; C'$ for some $C \in N_k^n$ (or $C \in R_k^n$), for every k.

Computationally, the big bottleneck is the pruning step, where to find subsumptions we test all pairs of networks by looking at $9! \approx 3.6 \times 10^5$ permutations – and at the peak the set N_k^9 contains around 1.8×10^7 networks, so there are potentially 3.2×10^{14} tests. By extending generate-and-prune with the optimizations and extensive parallelization described in [2], we were able to show that $S(9) = 25$ in around three weeks of computation on 288 threads.

However, the same optimizations that made the program work made it less trustworthy. Therefore, we formalized the soundness of generate-and-prune in the theorem prover Coq with the goal of extracting a provenly correct checker of the same result [5] to Haskell.[1] In order to eliminate the search step in `Prune`, this

---

[1] The choice of Haskell as target language is pragmatic: preliminary experiments suggested that it was the fastest one for this project.

formalization is parameterized on an oracle, which produces triples $\langle C_a, C_b, \pi \rangle$ such that $C_a \leq_\pi C_b$. This oracle is untrusted, so the checker will validate this subsumption and discard it if it cannot do so; but using it allows us to remove all search, while simultaneously making the number of tests linear in N_n^k, rather than quadratic. It is implemented by reading the logs produced by the original execution of generate-and-prune, in which all successful subsumptions were recorded. They amount to a total of 27 GB, making this one of the largest computer-generated proofs ever.

The formalization defines `comparator` to be a pair of natural numbers and the type CN of comparator networks to be `list comparator`. We then specify what it means for a comparator network to be a sorting network on n channels, and show that this is a decidable predicate. The details of the formalization of the theory of sorting networks can be found in [5]. The formalization itself is available online at http://imada.sdu.dk/~petersk/sn/.

This part of the formalization was developed closely following the mathematical sources. Yet several results yield meaningful extracted functions that will be used in the development of the checker; for example, the result stating decidability of whether a comparator network is a sorting network on n channels yields a function sN_dec such that sN_dec n C evaluates to True iff C represents a sorting network on n channels. In this sense, we are taking serious advantage of the Curry–Howard isomorphism, unlike in the next stage, where we will implement the checker in Coq and prove its soundness.

The implementation of generate-and-prune proceeds in several steps. We translate `Generate` directly into Coq code, which we omit since it is straightforward and we will not discuss it further. As for `Prune`, we closely follow the original pseudo-code in [2].[2]

```
Definition Oracle := list (CN * CN * (list nat)).

Function Prune (O:Oracle) (R:list CN) (n:nat)
                {measure length R} : list CN := match O with
  | nil => R
  | cons (C,C',pi) O' => match (CN_eq_dec C C') with
    | left _ => R
    | right _ => match (In_dec CN_eq_dec C R) with
      | right _ => R
      | left _ => match (pre_permutation_dec n pi) with
        | right _ => R
        | left A => match (subsumption_dec n C C' pi' Hpi) with
          | right _ => R
          | left _ => Prune O' (remove CN_eq_dec C' R) n
end end end end end.
```

Prune processes each subsumption $\langle C, C', \pi \rangle$ given by the oracle sequentially and makes all the relevant checks: that $C \neq C'$ (`left` extracts as True, `right` as

---

[2] Throughout this presentation we will always show transcribed Coq code, which is almost completely computational and preserved by extraction.

False), that $C \in R$, that π represents a valid permutation, and that $C \leq_\pi C'$. If all checks succeed, C' is removed from R, otherwise the subsumption is discarded. For legibility, we write pi' for the translation of π into our representation of permutations, and Hpi for the proof term needed for the subsumption test.

Both Generate and Prune are proven to take complete sets of filters into complete sets of filters, as well as to satisfy some aditional properties necessary for the soundness of the algorithm. These functions are then incorporated in a larger loop that applies them alternately. The code uses OGenerate, an optimized version of Generate that removes some networks using known results about redundant comparators that were implemented in the original algorithm and that are easily shown to be sound [2,8]. This loop receives as inputs the number of channels m and the number of iterations n, and returns an answer: (yes m k) if a sorting network of size k was found; (no m k R) if a set R of comparator networks of size k is constructed that is complete and contains no sorting network; or maybe if an error occurs. The answer no contains some extra proof terms necessary for the correctness proof. These are removed in the extracted checker, and since they make the code quite complex to read, we replace them by _ below.

```
Fixpoint Generate_and_Prune (m n:nat) (O:list Oracle) : Answer :=
  match n with
  | 0 => match m with
         | 0 => yes 0 0
         | 1 => yes 1 0
         | _ => no m 0 (nil :: nil) _ _ _
         end
  | S k => match O with
           | nil => maybe
           | X::O' => let GP := (Generate_and_Prune m k O') in match GP with
                      | maybe => maybe
                      | yes p q => yes p q
                      | no p q R _ _ _ => let GP' := Prune X (OGenerate R p) p in
                                          match (exists_SN_dec p GP' _) with
                                          | left _ => yes p (S q)
                                          | right _ => no p (S q) GP' _ _ _
end end end end.
```

Here Answer is the suitably defined inductive type of answers. The elimination over exists_SN_dec uses the fact that we can decide whether a set contains a sorting network. Correctness of the result is shown in the two theorems below. In these, the oracle O is universally quantified, reflecting that they hold regardless of whether the oracle is giving right or wrong information.

```
Theorem GP_yes : forall m n O k, Generate_and_Prune m n O = yes m k ->
                 (forall C, sorting_network m C -> length C >= k) /\
                 exists C, sorting_network m C /\ length C = k.

Theorem GP_no : forall m n O R HRO HR1 HR2,
                 Generate_and_Prune m n O = no m n R HRO HR1 HR2 ->
                 forall C, sorting_network m C -> length C > n.
```

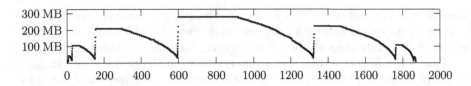

Fig. 1. Memory usage (MB/min) during the verification of the proof for 8 channels.

The extracted code for `Generate_and_Prune` is a function that takes two natural numbers m and n and a list of oracles, applies generate-and-prune on m channels for n iterations using the oracles, and returns `yes m k` or `no m k R`. The soundness theorems guarantee that these answers have a mathematical meaning.

3   Reducing Runtime of the Pruning Step

Figure 1 displays the memory usage during the validation of the proof for 8 channels. The exact values are immaterial, but we can very easily trace the execution of the algorithm by noting that every upwards jump corresponds to `Generate`, whereas the descending curve corresponds to `Prune`. The picture shows that the three most costly iterations account for almost 90 % of the execution time. For 9 channels, there are four costly iterations, and the imbalance will be even greater, as the differences in size between the sets N_k^9 are much more significant.

The biggest cost in the execution of the checker is in the pruning step, as was already the case with the original, uncertified program. The use of the oracle allows us to bypass the original search, but the algorithm is still very inefficient: for every subsumption, it iterates through the set being pruned to verify that the subsuming network is there and to remove it. Due to lazy evaluation in Haskell, these verifications are made in a single pass; but execution time is still quadratic on the number of generated networks.

In this section we take advantage of the offline nature of our oracle, and show that we can greatly improve the algorithm using the fact that we already know all the subsumptions we will make. Indeed, we need to do three things.

1. Check that all subsumptions are valid.
2. Remove all subsumed networks.
3. Check that all networks used in subsumptions are kept.

Each subsumption in step 1 is checked individually, so this step scales linearly in the number of networks. The other two steps can be significantly improved.

3.1   Optimizing the Removal Step

In theory, step 2 could be substantially optimized by delaying the removals until all subsumptions have been read: if we obtained the networks to be removed from the oracle in the same order as we generate them in the checker, then we could remove all subsumed networks with one single pass over the whole set, instead of having to iterate through the set of networks for each subsumed network.

This is the first time that the symbiosis between the prune algorithm and the implementation of the untrusted oracle becomes a key ingredient for optimization. As we use an *offline* oracle [5], we can actually reorder the oracle information to suit the needs of the checker with an efficient (untrusted) preprocessor. An inspection of the definition of Generate shows that comparators are added in lexicographic order, and we can pre-process the oracle information such that the subsumptions are provided in the same order.

Then we can define a function remove_all to complete step 2 in linear time by simultaneously traversing the list of subsumed networks and the list of all networks and removing all elements of the former from the latter.

3.2   Optimizing the Presence Check

Unfortunately, one cannot do a similar optimization to step 3 immediately, since sorting the oracle information by the subsumed networks will yield an unsorted sequence of subsuming networks. However, we can proceed in a different way: rather than checking that the subsuming networks are kept at each step, only check that they are present in the final (reduced) set. This will still be a quadratic algorithm, but relative to the size of the final set – which, in the most time-consuming steps, is only around 5 % of the size of the original one.

This idea again requires an important change to the oracle implementation, this time in the subsumptions presented by the oracle. As it happens, there are often chains of subsumptions $C_1 \preceq C_2 \preceq \ldots \preceq C_n$, which pose no problem for the original algorithm, but would result in a false negative result of the checker, if we were to check the presence of the subsuming networks in the final set. Consider e.g. C_2, which is used to remove C_3, but which is itself removed by C_1.

However, we can benefit from the offline character of the oracle and use the transitivity of subsumption to transform such chains of subsumptions into "reduced" subsumptions $C_1 \preceq C_2$, $C_1 \preceq C_3$, ..., $C_1 \preceq C_n$. This again requires pre-processing the oracle information, identifying such chains and computing adequate permutations for the new resulting subsumptions.

In order to achieve this, we implemented a data structure in the pre-processor that we term a subsumption graph: a labeled directed graph whose nodes are comparator networks, and where there is a edge from C' to C labeled by π if $C \leq_\pi C'$. Once we have built the full graph for one pruning step, we can obtain the reduced oracle information as follows: (i) find all non-empty paths in the graph ending in a node without outgoing edges; (ii) starting with the identity permutation, traverse each such path while composing the permutations on the edges; (iii) the start- and end-node of each path, together with the resulting permutation, describe one reduced subsumption. The oracle then provides the reduced subsumptions instead of the original ones.

The formalized definitions for the improved pruning step now look as follows. Functions `oracle_ok_1` and `oracle_ok_2` perform steps 1 and 3 above, and `Prune` uses `remove_all` to perform step 2.

```
Fixpoint oracle_ok_1 (n:nat) (O:Oracle) : bool := match O with
  | nil => true
  | (C,C',pi) :: O' => match (pre_permutation_dec n pi) with
            | right _ => false
            | left A => match (subsumption_dec n C C' pi' Hpi) with
                 | right _ => false
                 | left _ => oracle_ok_1 n O'
end end end.

Fixpoint oracle_ok_2 (O:Oracle) (R:list CN) : bool := match O with
  | nil => true
  | (C,_,_)::O' => match (In_dec CN_eq_dec C R) with
                 | left _ => oracle_ok_2 O' R
                 | right _ => false
end end.

Definition Prune (O:Oracle) (R:list CN) (n:nat) : list CN :=
  match (oracle_ok_1 n O) with
  | false => R
  | true => let R' := remove_all CN_eq_dec (map snd (map fst O)) R in
            match (oracle_ok_2 O R') with
            | false => R
            | true => R'
end end.
```

This approach is completely modular: after we reprove the lemmas regarding the correctness of `Prune`, the proofs for the whole algorithm mostly go through unchanged, and where tweaking of the proofs is necessary, the changes are trivial and require no deep insights into the proofs.

3.3   Practical Impact on Runtime

With these optimizations, we essentially eliminate the dominating quadratic step of the original algorithm. The presence check (step 3) is still quadratic, but on such a smaller set of networks that the overall behaviour of the algorithm becomes linear for practical purposes.

In the following table, we compare the runtime of the original implementaton of the proof checker with the improved one presented in this section. We focus on the case of 8 inputs, the largest case that we can systematically handle.

Configuration	Original algorithm	Improved algorithm
Runtime	1863 m	167 m

Clearly, we see an order of magnitude improvement for 8 inputs. We also ran the first 10 pruning steps of 9 and infer an even larger improvement for 9 inputs, bringing down the expected runtime from two decades to several months. The much lower weight of `Prune` is patent in the new memory trace (Fig. 2).

4   Reducing Memory Footprint by Tuning the Extraction

The contributions of the previous section left us with a checker that was nearly fast enough, but that had too large memory requirements due to reading all subsumptions at once, rather than processing them one by one. Our attempts to run the checker for 9 inputs quickly drained the available computing resources, and we estimated that more than 200 GB of RAM would be needed. Profiling showed that most of the memory was being taken up by lists and natural numbers – not surprising, since the checker is producing millions of comparator networks. But when, at the peak, we potentially need to store 18 million networks ×15 comparators ×2 channels, the Peano representation of natural numbers in Coq is extremely expensive, even with all numbers ranging from 0 to 8.

The most natural idea was to extract natural numbers to Haskell native types. In general, this loses the guaranteed correctness of the extracted program; but in this particular example it not pose significant risks, as natural numbers are identifiers for channels and not objects with which to do computations. This means that only five Haskell functions are needed: `succ`, `(=)`, `(<)`, `(-)` and `max`, besides the recursor

```
(\ f0 fS n -> if n==0 then (f0 __) else fS (n-1))
```

(Function `max` is used only in the definition of predecessor, while `{-}` is used only in the recursor.) Furthermore, they only operate on the numbers 0 to 8 (except for `succ`, which goes up to 25), so it is easy to verify exhaustively that they are correct. As a side-effect, we also need to extract booleans to the native `Bool` type (which has exactly the same definition as extracting from the Coq type), and since we do not use any functions on `Bool` this is also not a problem.

4.1   Practical Impact on Memory Usage

In the following table, we compare the memory usage of the extracted Peano numerals (an algebraic data structure with constructors O and S) against several native representations. Once again, we consider the case of 8 inputs.

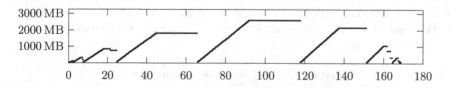

Fig. 2. Memory usage (MB/min) verifying the proof for 8 channels, after optimizations.

Representation	Peano naturals	64-bit integers	8-bit integers	Enum
Memory usage (MB)	2536	844	1669	999

We clearly see that native 64-bit integers (int) take significantly less memory than the algebraic data structure for Peano numerals. Interestingly, other datatypes perform worse: although enum or int8 in general use less memory than int, the Haskell compiler keeps a small store of "reusable" integers in the heap. This means that all comparators are pointers to the same location, whereas by using int8 or enum they become different copies of the same number. For this reason, int actually performs better, memory-wise, than either int8 or enum.

We also experimented using Haskell lists instead of extracted Coq lists, but this does not help: these datatypes are isomorphic, and we still use a recursor instead of pattern-matching.

5   Optimizing Data Structures

With all these optimizations in place, the task of checking the full proof for 9 inputs became just beyond reachable. Experiments that the memory consumption for each iteration of generate-and-prune was linear on the number of comparators in the subsumptions in the oracle; this allowed us to estimate the total memory required at $80 - 90$ GB. Likewise, the execution times for the first 12 steps showed a linear dependency on the total number of generated nets (which seemed reasonable and hard to improve, since we need to generate them explicitly and then prune them) and a quadratic dependency on the size of the pruned set (due to the check that all networks used in subsumptions are kept). A rough estimate based on a least squares fit of the data yielded around four months for the whole execution.

We therefore focused on more localized aspects of the formalization in order to bring these requirements down and actually verify the complete proof. Our decision on what constitutes "reasonable" is directly related to the available resources: 500 h of computation on a computer with 64 GB of RAM memory. In this section we focus on runtime.

5.1   Using Binary Search Trees to Decide Membership

The step that we felt was most inefficient was the verification in oracle_ok_2, where we iterate over all networks used in subsumptions and check that they occur in the pruned set.

There are two reasons why the implementation of this step is not satisfactory. First, since we are iterating over all subsumptions, we repeatedly test the same network many times: at peak, there are about 20 times as many subsumptions as networks in the pruned set. Secondly, these subsumptions are unordered, but the list of pruned networks is ordered; however, since Coq lists do not have direct access, we are still forced to look for them in linear time. This means that this step takes time proportional to both the number of subsumptions and the number of pruned networks, and is thus roughly quadractic on the latter.

Ideally, we would like to do something similar to the optimization of the pruning step itself, where by ensuring the list of all networks and the list of networks to be removed are ordered in the same way we can solve the problem in linear time. However, the trick we used before is no longer applicable, since we cannot change the order of the oracle.

Instead, we pursued the idea of sorting the networks used in subsumptions (and removing duplicates as we do so). In order to do this efficiently, we changed the data structure storing these networks, from a list to a search tree. This required enriching our formalization with a type of binary trees and operations for adding and retrieving the minimum element of such a tree.

In keep with the remainder of the formalization [5], we defined binary trees without any restrictions, together with a predicate stating that a binary tree is a search tree. This is similar to the formalization of binary trees in Chap. 11 of [1]; however, that formalization only considered trees over Coq integers, whereas we formalize binary trees over an arbitrary type T over which we have a comparison function. This straightforward generalization poses no difficulties.

```
Inductive BinaryTree (T:Type) : Type :=
  nought : BinaryTree
| node : Tree -> BinaryTree -> BinaryTree -> BinaryTree.
```

We then define predicates BT_in to test that an element occurs in a binary tree, BT_wf to check that a binary tree is a search tree, and the usual function BT_add to add an element to a tree. For efficiency, we also define a function BT_split that simultaneously computes the minimum element of a search tree *and* the tree obtained by removing it.

```
Fixpoint BT_split (T:Type) (BT:BinaryTree T) (val:T) : T * BinaryTree :=
  match BT with
  | nought => (val,nought)
  | node t nought R => (t,R)
  | node t L R => let (t',L') := BT_split L val in (t',node t L' R)
  end.
```

We show that the functions defined work correctly on search trees; in particular, any object of type BinaryTree built from nought by repeated application of BT_add satisfies BT_wf. Then, we changed the implementation of oracle_ok_1 to return also a binary tree, proved that this is a search tree containing all networks used in the subsumptions given by the oracle, and rewrote oracle_ok_2 to run in only slightly superlinear time.

```
Fixpoint oracle_ok_2 (BT:BinaryTree CN) (R:list CN) := match BT,R with
  | nought, _ => true
  | _, nil => false
  | _, C' :: R' => let (C,BT') := (BT_split BT nil) in
                   match (OCN_eq_dec C C') with
                   | left _ => oracle_test BT' R'
```

```
                    | right _ => oracle_test BT R'
end end.
```

Some of the proofs in the pruning step required a bit of adaptation, since they now rely on lemmas over `BinaryTrees` instead of `lists`, but the changes were localized to this part of the formalization.

The recursive call in `oracle_ok_2` is on the remainder of the list, so the total execution time depends on the length of this list and the depth of the search tree `BT`. Before experimenting with the newly extracted program, we exhaustively ran the oracle sources through a small Java program to check how balanced the constructed search trees would be. The maximum depth is only 94 (corresponding to a very unbalanced tree, but much better than the previous list), and for the two biggest sets of subsumptions we actually obtain trees of depth 69, storing 848,914 networks, in one case, or 568,287, in the other.

5.2 Using Binary Search Trees for Subsumption Checking

The availability of binary trees unexpectedly opened the door to yet another improvement in the program: the subsumption test itself. As we mentioned, Lemma `subsumption_dec` states that $C \leq_\pi C'$ is decidable, and the proof simply proceeds by computing outputs(C) and outputs(C') and directly checking that $\pi(\text{outputs}(C)) \subseteq \text{outputs}(C')$. Since the number of outputs is fixed, this check takes almost constant time (computing the outputs becomes slightly more time-consuming as the networks grow bigger, but this is not noticeable), but on 9 channels the lists of outputs contain 512 elements, and again they have many repetitions and are reasonably unordered.

Therefore, we experimented with reproving `subsumption_dec` by storing the computed outputs in a search tree rather than in a list. The impact on performance was stunning: since the execution time was now dominated by the validation of all the subsumptions, we were able to check the proof for 8 inputs in less than half the time.

5.3 Practical Impact on Runtime

The following table summarizes the impact of the contributions in this section on the verification of the proof for 8 inputs.

Configuration	Original	Tree-based presence check	Everything tree-based
Runtime	126 m	111 m	48 m

Using trees for checking for the presence of subsuming networks has a moderate impact on 8 inputs. However, this impact becomes greater as the number of inputs grows: experiments with the initial pruning steps for 9 inputs gave an estimated runtime reduction of 30 %. Experiments suggest that using both optimizations yields approx. 70 % reduction of runtime on 9 inputs.

One might wonder whether we could not use search trees in the original formalization and gain a similar speedup. The answer is negative: the improvement stems both from the numerous repetitions among the subsuming networks and from their failure to be ordered. The generation step produces networks that are both ordered and without repetitions, whence the result of storing them in a search tree would be isomorphic to a list.

6   Gödelizing Comparators to Reduce Memory Footprint

At this point, the remaining bottleneck was memory, and we again shifted focus from runtime to reducing the memory footprint. We decided to take advantage of Haskell's caching of small integers by using a Gödelization of comparators: represent each comparator (a pair of natural numbers) by a single natural number, using the bijection $\varphi(i,j) = \frac{1}{2}j \times (j-1) + i$. This happens to map very nicely to the function all_st_comps described earlier, since the comparator (i,j) is exactly the $\varphi(i,j)$-th element of all_st_comps n (as long as $i,j < n$).

We then defined a type OCN := list nat of optimized comparator networks and a mapping to CN. Using this mapping, it was possible to reimplement Generate and Prune to run on lists of OCN, while reusing all the old theory about comparator networks. From a formalization point of view, it was also the most reasonable option, as it keeps a consistent theory of comparator networks formalized according to intuition, and uses a more efficient representation only for implementation purposes.

The following table compares memory usage of representing comparators by a pair of int or by one Gödelized int, for the case of 8 inputs.

Comparators	Explicit	Gödelized
Memory usage (MB)	844	541

Assymptotically, this change reduces memory consumption to just over one half: for each comparator we are now just storing one number instead of a pair of numbers. Again, experiments suggest that the improvement for 9 inputs is greater than for the case detailed. There is some overhead of mapping from CN to OCN to test subsumptions, but it is offset by an improvement in pruning times due to testing for equality directly on OCN.

With all these optimizations in place, our checker was able to verify the original proof of optimality of 25 comparators for sorting 9 inputs, using the available proof witnesses. The verification took 163.8 h, or just under one week, required a maximum of 50.05 GB of RAM, and returned the answer yes 9 25. The formalization, extracted code and oracle source files (before and after preprocessing) can be found at http://imada.sdu.dk/~petersk/sn/.

7   Conclusion

The contributions of Sects. 3–6 allowed us to run a formal validation of the proof from [2] that 25 comparators suffice for sorting 9 inputs, using the formalization of the theory of sorting networks described in [5].

We also showed that it is feasible to optimize extracted code without significantly changing the underlying formalized theory, and therefore the latter can be developed without excessive concerns about the extracted code. Indeed, the original formalization closely follows Knuth [8], with the new theoretical results from [2] and a straightforward implementation of the algorithm therein proposed. While this theory took three months to formalize, each of the changes described in this paper required only around one day, as they amounted to changing localized parts of the checker and reproving their properties. In other words, the optimizations were obtained by concentrating on the computational aspects of the checker without needing to worry about the underlying theory.

These results support our choice of an *offline* untrusted oracle for the original formalization [5] as it allows for a nice separation between the development of the theory and the optimization of the checker, as well as giving us the possibily of exploring the interplay between the checker and oracle.

The cornerstone of our work is that having an offline oracle makes it possible to validate such proofs much more efficiently by bypassing the search steps – using the well-known principle that *checking* the correctness of a witness is usually much easier than *finding* it. We plan to test this principle and use our approach to validate other search-intensive, large-scale computer-generated proofs.

Acknowledgements. We would like to thank Pierre Letouzey for suggesting and helping with extracting to Haskell native types, Søren Haagerup for helping with profiling, and Michael Codish for his support and his enthusiasm about sorting networks.

The authors were supported by the Danish Council for Independent Research, Natural Sciences. Computational resources were generously provided by the Danish Center for Scientific Computing.

References

1. Bertot, Y., Castéran, P.: Interactive Theorem Proving and Program Development. Texts in Theoretical Computer Science. Springer, Heidelberg (2004)
2. Codish, M., Cruz-Filipe, L., Frank, M., Schneider-Kamp, P.: Twenty-five comparators is optimal when sorting nine inputs (and twenty-nine for ten). In: ICTAI 2014, pp. 186–193. IEEE (2014)
3. Contejean, E., Courtieu, P., Forest, J., Pons, O.: Automated certified proofs with CiME3. In: Schmidt-Schauß, M. (ed.) RTA 2011. LIPIcs, vol. 10, pp. 21–30. Schloss Dagstuhl, Germany (2011)
4. Cruz-Filipe, L., Letouzey, P.: A large-scale experiment in executing extracted programs. Electron. Notes Comput. Sci. **151**(1), 75–91 (2006)
5. Cruz-Filipe, L., Schneider-Kamp, P.: Formalizing size-optimal sorting networks: extracting a certified proof checker. In: Proceedings of ITP 2015, LNCS, Springer (2015, Submitted for Publication). CoRR. abs/1502.05209

6. Floyd, R.W., Knuth, D.E.: The Bose-Nelson sorting problem. In: Srivastava, J.N. (ed.) A Survey of Combinatorial Theory, pp. 163–172. North-Holland, Amsterdam (1973)
7. Fouilhe, A., Monniaux, D., Périn, M.: Efficient generation of correctness certificates for the abstract domain of polyhedra. In: Logozzo, F., Fähndrich, M. (eds.) Static Analysis. LNCS, vol. 7935, pp. 345–365. Springer, Heidelberg (2013)
8. Knuth, D.E.: The Art of Computer Programming, Volume III: Sorting and Searching. Addison-Wesley, Redwood City (1973)
9. Krebbers, R., Spitters, B.: Computer certified efficient exact reals in Coq. In: Davenport, J.H., Farmer, W.M., Urban, J., Rabe, F. (eds.) MKM 2011 and Calculemus 2011. LNCS, vol. 6824, pp. 90–106. Springer, Heidelberg (2011)
10. Leroy, X.: Formal verification of a realistic compiler. Commun. ACM **52**(7), 107–115 (2009)
11. Letouzey, P.: Extraction in Coq: an overview. In: Beckmann, A., Dimitracopoulos, C., Löwe, B. (eds.) CiE 2008. LNCS, vol. 5028, pp. 359–369. Springer, Heidelberg (2008)
12. O'Connor, R.: Certified exact transcendental real number computation in Coq. In: Mohamed, O.A., Muñoz, C., Tahar, S. (eds.) TPHOLs 2008. LNCS, vol. 5170, pp. 246–261. Springer, Heidelberg (2008)
13. Oury, N.: Observational equivalence and program extraction in the Coq proof assistant. In: Hofmann, M.O. (ed.) TLCA 2003. LNCS, vol. 2701, pp. 271–285. Springer, Heidelberg (2003)
14. Sternagel, C., Thiemann, R.: The certification problem format. In: Benzmüller, C., Paleo, B.W. (eds.) UITP 2014. EPTCS, vol. 167, pp. 61–72. ACM Press, New York (2014)
15. Thiemann, R.: Formalizing bounded increase. In: Blazy, S., Paulin-Mohring, C., Pichardie, D. (eds.) ITP 2013. LNCS, vol. 7998, pp. 245–260. Springer, Heidelberg (2013)
16. van Voorhis, D.C.: Toward a lower bound for sorting networks. In: Miller, R.E., Thatcher, J.W. (eds.) Complexity of Computer Computations. The IBM Research Symposia Series, pp. 119–129. Plenum Press, New York (1972)

A First Class Boolean Sort in First-Order Theorem Proving and TPTP

Evgenii Kotelnikov[1]([⊠]), Laura Kovács[1], and Andrei Voronkov[2]

[1] Chalmers University of Technology, Gothenburg, Sweden
{evgenyk,laura.kovacs}@chalmers.se
[2] The University of Manchester, Manchester, UK
andrei@voronkov.com

Abstract. To support reasoning about properties of programs operating with boolean values one needs theorem provers to be able to natively deal with the boolean sort. This way, program pro perties can be translated to first-order logic and theorem provers can be used to prove program properties efficiently. However, in the TPTP language, the input language of automated first-order theorem provers, the use of the boolean sort is limited compared to other sorts, thus hindering the use of first-order theorem provers in program analysis and verification. In this paper, we present an extension FOOL of many-sorted first-order logic, in which the boolean sort is treated as a first-class sort. Boolean terms are indistinguishable from formulas and can appear as arguments to functions. In addition, FOOL contains `if-then-else` and `let-in` constructs. We define the syntax and semantics of FOOL and its model-preserving translation to first-order logic. We also introduce a new technique of dealing with boolean sorts in superposition-based theorem provers. Finally, we discuss how the TPTP language can be changed to support FOOL.

1   Introduction

Automated program analysis and verification requires discovering and proving program properties. Typical examples of such properties are loop invariants or Craig interpolants. These properties usually are expressed in combined theories of various data structures, such as integers and arrays, and hence require reasoning with both theories and quantifiers. Recent approaches in interpolation and loop invariant generation [10,12,14] present initial results of using first-order theorem provers for generating quantified program properties. First-order theorem provers can also be used to generate program properties with quantifier alternations [12]; such properties could not be generated fully automatically by any previously known method. Using first-order theorem prover to generate, and not

L. Kovács—The first two authors were partially supported by the Wallenberg Academy Fellowship 2014, the Swedish VR grant D0497701, and the Austrian research project FWF S11409-N23.

A. Voronkov—Partially supported by the EPSRC grant "Reasoning in Verification and Security".

M. Kerber et al. (Eds.): CICM 2015, LNAI 9150, pp. 71–86, 2015.
DOI: 10.1007/978-3-319-20615-8_5

only prove program properties, opens new directions in analysis and verification of real-life programs.

First-order theorem provers, such as iProver [11], E [18], and Vampire [13], lack however various features that are crucial for program analysis. For example, first-order theorem provers do not yet efficiently handle (combinations of) theories; nevertheless, sound but incomplete theory axiomatisations can be used in a first-order prover even for theories having no finite axiomatisation. Another difficulty in modelling properties arising in program analysis using theorem provers is the gap between the semantics of expressions used in programming languages and expressiveness of the logic used by the theorem prover. A similar gap exists between the language used in presenting mathematics. For example, a standard way to capture assignment in program analysis is to use a `let-in` expression, which introduces a local binding of a variable, or a function for array assignments, to a value. There is no local binding expression in first-order logic, which means that any modelling of imperative programs using first-order theorem provers at the backend, should implement a translation of `let-in` expressions. Similarly, mathematicians commonly use local definitions within definitions and proofs. Some functional programming languages also contain expressions introducing local bindings. In all three cases, to facilitate the use of first-order provers, one needs a theorem prover implementing `let-in` constructs natively.

Efficiency of reasoning-based program analysis largely depends on how programs are translated into a collection of logical formulas capturing the program semantics. The boolean structure of a program property that can be efficiently treated by a theorem prover is however very sensitive to the architecture of the reasoning engine of the prover. Deriving and expressing program properties in the "right" format therefore requires solid knowledge about how theorem provers work and are implemented — something that a user of a verification tool might not have. Moreover, it can be hard to efficiently reason about certain classes of program properties, unless special inference rules and heuristics are added to the theorem prover, see e.g. [8] when it comes to prove properties of data collections with extensionality axioms.

In order to increase the expressiveness of program properties generated by reasoning-based program analysis, the language of logical formulas accepted by a theorem prover needs to be extended with constructs of programming languages. This way, a straightforward translation of programs into first-order logic can be achieved, thus relieving users from designing translations which can be efficiently treated by the theorem prover. One example of such an extension is recently added to the TPTP language [19] of first-order theorem provers, resembling `if-then-else` and `let-in` expressions that are common in programming languages. Namely, special functions `$ite_t` and `$ite_f` can respectively be used to express a conditional statement on the level of logical terms and formulas, and `$let_tt`, `$let_tf`, `$let_ff` and `$let_ft` can be used to express local variable bindings for all four possible combinations of logical terms (`t`) and formulas (`f`). While satisfiability modulo theory (SMT) solvers, such as Z3 [6] and CVC4 [2], integrate `if-then-else` and `let-in` expressions, in the first-order theorem proving community so far only Vampire supports such expressions. To illustrate the advantage of using `if-then-else` and `let-in` expressions in automated provers, let us consider the following simple example. We are interested in verifying the partial correctness of the code fragment below:

```
if (r(a)) {
  a := a + 1
} else {
  a := a + q(a)
}
```

using the pre-condition $((\forall x)P(x) \Rightarrow x \geq 0) \wedge ((\forall x)q(x) > 0) \wedge P(a)$ and the post-condition $a > 0$. Let $a1$ denote the value of the program variable a after the execution of the if-statement. Using `if-then-else` and `let-in` expressions, the next state function for a can naturally be expressed by the following formula:

```
a1 = if r(a) then let a = a + 1 in a
              else let a = a + q(a) in a
```

This formula can further be encoded in TPTP, and hence used by a theorem prover as a hypothesis in proving partial correctness of the above code snippet. We illustrate below the TPTP encoding of the first-order problem corresponding to the partial program correctness problem we consider. Note that the pre-condition becomes a hypothesis in TPTP, whereas the proof obligation given by the post-condition is a TPTP conjecture. All formulas below are typed first-order formulas (`tff`) in TPTP that use the built-in integer sort (`$int`).

```
tff(1, type, p : $int > $o).
tff(2, type, q : $int > $int).
tff(3, type, r : $int > $o).
tff(4, type, a : $int).
tff(5, hypothesis, ! [X : $int] : (p(X) => $greatereq(X, 0))).
tff(6, hypothesis, ! [X : $int] : ($greatereq(q(X), 0))).
tff(7, hypothesis, p(a)).
tff(8, hypothesis,
    a1 = $ite_t(r(a), $let_tt(a, $sum(a, 1), a),
                      $let_tt(a, $sum(a, q(a)), a))).
tff(9, conjecture, $greater(a1, 0)).
```

Running a theorem prover that supports `$ite_t` and `$let_tt` on this TPTP problem would prove the partial correctness of the program we considered. Note that without the use of `if-then-else` and `let-in` expressions, a more tedious translation is needed for expressing the next state function of the program variable a as a first-order formula. When considering more complex programs containing multiple conditional expressions assignments and composition, computing the next state function of a program variable results in a formula of size exponential in the number of conditional expressions. This problem of computing the next state function of variables is well-known in the program analysis community, by computing so-called static single assignment (SSA) forms. Using the `if-then-else` and `let-in` expressions recently introduced in TPTP and already implemented in Vampire [7], one can have a linear-size translation instead.

Let us however note that the usage of conditional expressions in TPTP is somewhat limited. The first argument of $\texttt{\$ite\_t}$ and $\texttt{\$ite\_f}$ is a logical formula, which means that a boolean condition from the program definition should be translated as such. At the same time, the same condition can be treated as a value in the program, for example, in a form of a boolean flag, passed as an argument to a function. Yet we cannot mix terms and formulas in the same way in a logical statement. A possible solution would be to map the boolean type of programs to a user-defined boolean sort, postulate axioms about its semantics, and manually convert boolean terms into formulas where needed. This approach, however, suffers the disadvantages mentioned earlier, namely the need to design a special translation and its possible inefficiency.

Handling boolean terms as formulas is needed not only in applications of reasoning-based program analysis, but also in various problems of formalisation of mathematics. For example, if one looks at two largest kinds of attempts to formalise mathematics and proofs: those performed by interactive proof assistants, such as Isabelle [16], and the Mizar project [21], one can see that first-order theorem provers are the main workhorses behind computer proofs in both cases – see e.g. [5,22]. Interactive theorem provers, such as Isabelle routinely use quantifiers over booleans. Let us illustrate this by the following examples, chosen among 490 properties about (co)algebraic datatypes, featuring quantifiers over booleans, generated by Isabelle and kindly found for us by Jasmin Blanchette. Consider the distributivity of a conditional expression (denoted by the *ite* function) over logical connectives, a pattern that is widely used in reasoning about properties of data structures. For lists and the $\texttt{contains}$ function that checks that its second argument contains the first one, we have the following example:

$$(\forall p : bool)(\forall l : list_A)(\forall x : A)(\forall y : A)$$
$$\texttt{contains}(l, \text{ite}(p, x, y)) \doteq \tag{1}$$
$$(p \Rightarrow \texttt{contains}(l, x)) \wedge (\neg p \Rightarrow \texttt{contains}(l, y))$$

A more complex example with a heavy use of booleans is the unsatisfiability of the definition of $\texttt{subset\_sorted}$. The $\texttt{subset\_sorted}$ function takes two sorted lists and checks that its second argument is a sublist of the first one.

$$(\forall l_1 : list_A)(\forall l_2 : list_A)(\forall p : bool)$$
$$\neg(\texttt{subset\_sorted}(l_1, l_2) \doteq p \wedge$$
$$(\forall l_2' : list_A)\neg(l_1 \doteq \texttt{nil} \wedge l_2 \doteq l_2' \wedge p) \wedge$$
$$(\forall x_1 : A)(\forall l_1' : list_A)\neg(l_1 \doteq \texttt{cons}(x_1, l_1') \wedge l_2 \doteq \texttt{nil} \wedge \neg p) \wedge$$
$$(\forall x_1 : A)(\forall l_1' : list_A)(\forall x_2 : A)(\forall l_2' : list_A) \tag{2}$$
$$\neg(l_1 \doteq \texttt{cons}(x_1, l_1') \wedge l_2 \doteq \texttt{cons}(x_2, l_2') \wedge$$
$$p \doteq \text{ite}(x_1 < x_2, \textit{false},$$
$$\text{ite}(x_1 \doteq x_2, \texttt{subset\_sorted}(l_1', l_2'),$$
$$\texttt{subset\_sorted}(\texttt{cons}(x_1, l_1'), l_2')))))$$

Formulas with boolean terms are also common in the SMT-LIB project [3], the collection of benchmarks for SMT-solvers. Its core logic is a variant of first-order logic that treats boolean terms as formulas, in which logical connectives and conditional expressions are defined in the core theory.

In this paper we propose a modification FOOL of first-order logic, which includes a first-class boolean sort and if-then-else and let-in expressions, aimed for being used in automated first-order theorem proving. It is the smallest logic that contains both the SMT-LIB core theory and the monomorphic first-order subset of TPTP. The syntax and semantics of the logic are given in Sect. 2. We further describe how FOOL can be translated to the ordinary many-sorted first-order logic in Sect. 3. Section 4 discusses superposition-based theorem proving and proposes a new way of dealing with the boolean sort in it. In Sect. 5 we discuss the support of the boolean sort in TPTP and propose changes to it required to support a first-class boolean sort. We point out that such changes can also partially simplify the syntax of TPTP. Section 6 discusses related work and Sect. 7 contains concluding remarks.

The main contributions of this paper are the following:

1. the definition of FOOL and its semantics;
2. a translation from FOOL to first-order logic, which can be used to support FOOL in existing first-order theorem provers;
3. a new technique of dealing with the boolean sort in superposition theorem provers, allowing one to replace boolean sort axioms by special rules;
4. a proposal of a change to the TPTP language, intended to support FOOL and also simplify if-then-else and let-in expressions.

2   First-Order Logic with Boolean Sort

First-order logic with the boolean sort (FOOL) extends many-sorted first-order logic (FOL) in two ways:

1. formulas can be treated as terms of the built-in boolean sort; and
2. one can use if-then-else and let-in expressions defined below.

FOOL is the smallest logic containing both the SMT-LIB core theory and the monomorphic first-order part of the TPTP language. It extends the SMT-LIB core theory by adding let-in expressions defining functions and TPTP by the first-class boolean sort.

2.1   Syntax

We assume a countable infinite set of *variables*.

Definition 1. A *signature* of first-order logic with the boolean sort is a triple $\Sigma = (S, F, \eta)$, where:

1. S is a set of *sorts*, which contains a special sort *bool*. A *type* is either a sort or a non-empty sequence $\sigma_1, \ldots, \sigma_n, \sigma$ of sorts, written as $\sigma_1 \times \ldots \times \sigma_n \to \sigma$. When $n = 0$, we will simply write σ instead of $\to \sigma$. We call a *type assignment* a mapping from a set of variables and function symbols to types, which maps variables to sorts.

2. F is a set of *function symbols*. We require F to contain binary function symbols $\vee, \wedge, \Rightarrow$ and $\Leftrightarrow$, used in infix form, a unary function symbol $\neg$, used in prefix form, and nullary function symbols $true, false$.

3. η is a *type assignment* which maps each function symbol f into a type τ. When the signature is clear from the context, we will write $f : \tau$ instead of $\eta(f) = \tau$ and say that f is of the type τ.

   We require the symbols $\vee, \wedge, \Rightarrow, \Leftrightarrow$ to be of the type $bool \times bool \to bool$, $\neg$ to be of the type $bool \to bool$ and $true, false$ to be of the type $bool$.   □

In the sequel we assume that $\Sigma = (S, F, \eta)$ is an arbitrary but fixed signature.

To define the semantics FOOL, we will have to extend the signature and also assign sorts to variables. Given a type assignment η, we define $\eta, x : \sigma$ to be the type assignment that maps a variable x to σ and coincides otherwise with η. Likewise, we define $\eta, f{:}\tau$ to be the type assignment that maps a function symbol f to τ and coincides otherwise with η.

Our next aim to define the set of terms and their sorts with respect to a type assignment η. This will be done using a relation $\eta \vdash t{:}\sigma$, where $\sigma \in S$, terms can then be defined as all such expressions t.

Definition 2. The relation $\eta \vdash t : \sigma$, where t is an expression and $\sigma \in S$ is defined inductively as follows. If $\eta \vdash t : \sigma$, then we will say that t is a *term of the sort* σ w.r.t. η.

1. If $\eta(x) = \sigma$, then $\eta \vdash x : \sigma$.
2. If $\eta(f) = \sigma_1 \times \ldots \times \sigma_n \to \sigma$, $\eta \vdash t_1 : \sigma_1, \ldots, \eta \vdash t_n : \sigma_n$, then $\eta \vdash f(t_1, \ldots, t_n) : \sigma$.
3. If $\eta \vdash \phi : bool$, $\eta \vdash t_1 : \sigma$ and $\eta \vdash t_2 : \sigma$, then $\eta \vdash (\texttt{if } \phi \texttt{ then } t_1 \texttt{ else } t_2) : \sigma$.
4. Let f be a function symbol and $x_1, \ldots, x_n$ pairwise distinct variables. If $\eta, x_1 : \sigma_1, \ldots, x_n : \sigma_n \vdash s : \sigma$ and $\eta, f : (\sigma_1 \times \ldots \times \sigma_n \to \sigma) \vdash t : \tau$, then $\eta \vdash (\texttt{let } f(x_1 : \sigma_1, \ldots, x_n : \sigma_n) = s \texttt{ in } t) {:} \tau$.
5. If $\eta \vdash s : \sigma$ and $\eta \vdash t : \sigma$, then $\eta \vdash (s \doteq t) : bool$.
6. If $\eta, x : \sigma \vdash \phi : bool$, then $\eta \vdash (\forall x : \sigma)\phi : bool$ and $\eta \vdash (\exists x : \sigma)\phi : bool$.   □

We only defined a `let-in` expression for a single function symbol. It is not hard to extend it to a `let-in` expression that binds multiple pairwise distinct function symbols in parallel, the details of such an extension are straightforward.

When η is the type assignment function of Σ and $\eta \vdash t : \sigma$, we will say that t is a Σ-*term of the sort* σ, or simply that t is *a term of the sort* σ. It is not hard to argue that every Σ-term has a unique sort.

According to our definition, not every term-like expression has a sort. For example, if x is a variable and η is not defined on x, then x is a not a *term* w.r.t. η. To make the relation between term-like expressions and terms clear, we introduce a notion of free and bound occurrences of variables and function symbols. We call the following occurrences of variables and function symbols *bound*:

1. any occurrence of x in $(\forall x : \sigma)\phi$ or in $(\exists x : \sigma)\phi$;
2. in the term let $f(x_1 : \sigma_1, \ldots, x_n : \sigma_n) = s$ in t any occurrence of a variable x_i in $f(x_1 : \sigma_1, \ldots, x_n : \sigma_n)$ or in s, where $i = 1, \ldots, n$.
3. in the term let $f(x_1 : \sigma_1, \ldots, x_n : \sigma_n) = s$ in t any occurrence of the function symbol f in $f(x_1 : \sigma_1, \ldots, x_n : \sigma_n)$ or in t.

All other occurrences are called *free*. We say that a variable or a function symbol is *free* in a term t if it has at least one free occurrence in t. A term is called *closed* if it has no occurrences of free variables.

Theorem 1. Suppose $\eta \vdash t : \sigma$. Then

1. for every free variable x of t, η is defined on x;
2. for every free function symbol f of t, η is defined on f;
3. if x is a variable not free in t, and σ' is an arbitrary sort, then $\eta, x : \sigma' \vdash t : \sigma$;
4. if f is a function symbol not free in t, and τ is an arbitrary type, then $\eta, f : \tau \vdash t : \sigma$.   □

Definition 3. A *predicate symbol* is any function symbol of the type $\sigma_1 \times \ldots \times \sigma_n \to bool$. A *$\Sigma$-formula* is a Σ-term of the sort *bool*. All Σ-terms that are not Σ-formulas are called *non-boolean terms*.   □

Note that, in addition to the use of `let-in` and `if-then-else`, FOOL is a proper extension of first-order logic. For example, in FOOL formulas can be used as arguments to terms and one can quantify over booleans. As a consequence, every quantified boolean formula is a formula in FOOL.

2.2 Semantics

As usual, the semantics of FOOL is defined by introducing a notion of *interpretation* and defining how a term is evaluated in an interpretation.

Definition 4. Let η be a type assignment. A *η-interpretation I* is a map, defined as follows. Instead of $I(e)$ we will write $[e]_I$, for every element e in the domain of I.

1. Each sort $\sigma \in S$ is mapped to a nonempty domain $[\sigma]_I$. We require $[bool]_I = \{0, 1\}$.
2. If $\eta \vdash x : \sigma$, then $[x]_I \in [\sigma]_I$.
3. If $\eta(f) = \sigma_1 \times \ldots \times \sigma_n \to \sigma$, then $[f]_I$ is a function from $[\sigma_1]_I \times \ldots \times [\sigma_n]_I$ to $[\sigma]_I$.
4. We require $[true]_I = 1$ and $[false]_I = 0$. We require $[\wedge]_I$, $[\vee]_I$, $[\Rightarrow]_I$, $[\Leftrightarrow]_I$ and $[\neg]_I$ respectively to be the logical conjunction, disjunction, implication, equivalence and negation, defined over $\{0, 1\}$ in the standard way.

Given a η-interpretation I and a function symbol f, we define I_f^g to be the mapping that maps f to g and coincides otherwise with I. Likewise, for a variable x and value a we define I_x^a to be the mapping that maps x to a and coincides otherwise with I.

Definition 5. Let I be a η-interpretation, and $\eta \vdash t : \sigma$. The *value of t in I*, denoted as $\mathrm{eval}_I(t)$, is a value in $[\sigma]_I$ inductively defined as follows:

$$\mathrm{eval}_I(x) = [x]_I.$$
$$\mathrm{eval}_I(f(t_1, \ldots, t_n)) = [f]_I(\mathrm{eval}_I(t_1), \ldots, \mathrm{eval}_I(t_n)).$$
$$\mathrm{eval}_I(\texttt{if } \phi \texttt{ then } s \texttt{ else } t) = \begin{cases} \mathrm{eval}_I(s), \text{ if } \mathrm{eval}_I(\phi) = 1; \\ \mathrm{eval}_I(t), \text{ otherwise.} \end{cases}$$
$$\mathrm{eval}_I(\texttt{let } f(x_1 : \sigma_1, \ldots, x_n : \sigma_n) = s \texttt{ in } t) = \mathrm{eval}_{I_f^g}(t),$$

where g is such that for all $i = 1, \ldots, n$ and $a_i \in [\sigma_i]_I$, we have $g(a_1, \ldots, a_n) = \mathrm{eval}_{I_{x_1 \ldots x_n}^{a_1 \ldots a_n}}(s)$.

$$\mathrm{eval}_I(s \doteq t) = \begin{cases} 1, \text{ if } \mathrm{eval}_I(s) = \mathrm{eval}_I(t); \\ 0, \text{ otherwise.} \end{cases}$$
$$\mathrm{eval}_I((\forall x : \sigma)\phi) = \begin{cases} 1, \text{ if } \mathrm{eval}_{I_x^a}(\phi) = 1 \\ \qquad \text{for all } a \in [\![\sigma]\!]_I; \\ 0, \text{ otherwise.} \end{cases}$$
$$\mathrm{eval}_I((\exists x : \sigma)\phi) = \begin{cases} 1, \text{ if } \mathrm{eval}_{I_x^a}(\phi) = 1 \\ \qquad \text{for some } a \in [\![\sigma]\!]_I; \\ 0, \text{ otherwise.} \end{cases}$$

Theorem 2. Let $\eta \vdash \phi : bool$ and I be a η-interpretation. Then

1. for every free variable x of ϕ, I is defined on x;
2. for every free function symbol f of ϕ, I is defined on f;
3. if x is a variable not free in ϕ, σ is an arbitrary sort, and $a \in [\sigma]_I$ then $\mathrm{eval}_I(\phi) = \mathrm{eval}_{I_x^a}(\phi)$;
4. if f is a function symbol not free in ϕ, $\sigma_1, \ldots, \sigma_n, \sigma$ are arbitrary sorts and $g \in [\sigma_1]_I \times \ldots \times [\sigma_n]_I \to [\sigma]_I$, then $\mathrm{eval}_I(\phi) = \mathrm{eval}_{I_f^g}(\phi)$. □

Let $\eta \vdash \phi : bool$. A η-interpretation I is called a *model* of ϕ, denoted by $I \models \phi$, if $\mathrm{eval}_I(\phi) = 1$. If $I \models \phi$, we also say that I *satisfies* ϕ. We say that ϕ is *valid*, if $I \models \phi$ for all η-interpretations I, and *satisfiable*, if $I \models \phi$ for at least one η-interpretation I. Note that Theorem 2 implies that any interpretation, which coincides with I on free variables and free function symbols of ϕ is also a model of ϕ.

3   Translation of FOOL to FOL

FOOL is a modification of FOL. Every FOL formula is syntactically a FOOL formula and has the same models, but not the other way around. In this section we present a translation from FOOL to FOL, which preserves models of ϕ. This translation can be used for proving theorems of FOOL using a first-order theorem prover. We do not claim that this translation is efficient – more research is required on designing translations friendly for first-order theorem provers.

We do not formally define many-sorted FOL with equality here, since FOL is essentially a subset of FOOL, which we will discuss now.

We say that an occurrence of a subterm s of the sort *bool* in a term t is in a *formula context* if it is an argument of a logical connective or the occurrence in either $(\forall x : \sigma)s$ or $(\exists x : \sigma)s$. We say that an occurrence of s in t is in a *term context* if this occurrence is an argument of a function symbol, different from a logical connective, or an equality. We say that a formula of FOOL is *syntactically first order* if it contains no if-then-else and let-in expressions, no variables occurring in a formula context and no formulas occurring in a term context. By restricting the definition of terms to the subset of syntactically first-order formulas, we obtain the standard definition of many-sorted first-order logic, with the only exception of having a distinguished boolean sort and constants *true* and *false* occurring in a formula context.

Let ϕ be a closed Σ-formula of FOOL. We will perform the following steps to translate ϕ into a first-order formula. During the translation we will maintain a set of formulas D, which initially is empty. The purpose of D is to collect a set of formulas (definitions of new symbols), which guarantee that the transformation preserves models.

1. Make a sequence of translation steps obtaining a syntactically first order formula ϕ'. During this translation we will introduce new function symbols and add their types to the type assignment η. We will also add formulas describing properties of these symbols to D. The translation will guarantee that the formulas ϕ and $\bigwedge_{\psi \in D} \psi \wedge \phi'$ are equivalent, that is, have the same models restricted to Σ.
2. Replace the constants *true* and *false*, standing in a formula context, by nullary predicates $\top$ and $\bot$ respectively, obtaining a first-order formula.
3. Add special boolean sort axioms.

During the translation, we will say that a function symbol or a variable is *fresh* if it neither appears in ϕ nor in any of the definitions, nor in the domain of η.

We also need the following definition. Let $\eta \vdash t : \sigma$, and x be a variable occurrence in t. The *sort of this occurrence of* x is defined as follows:

1. any free occurrence of x in a subterm s in the scope of $(\forall x : \sigma')s$ or $(\exists x : \sigma')s$ has the sort σ'.
2. any free occurrence of x_i in a subterm s_1 in the scope of let $f(x_1 : \sigma_1, \ldots, x_n : \sigma_n) = s_1$ in s_2 has the sort σ_i, where $i = 1, \ldots, n$.
3. a free occurrence of x in t has the sort $\eta(x)$.

If $\eta \vdash t : \sigma$, s is a subterm of t and x a free variable in s, we say that x has a sort σ' in s if its free occurrences in s have this sort.

The translation steps are defined below. We start with an empty set D and an initial FOOL formula ϕ, which we would like to change into a syntactically first-order formula. At every translation step we will select a formula χ, which is either ϕ or a formula in D, which is not syntactically first-order, replace a subterm in χ it by another subterm, and maybe add a formula to D. The translation steps can be applied in any order.

1. Replace a boolean variable x occurring in a formula context, by $x \doteq true$.
2. Suppose that ψ is a formula occurring in a term context such that (i) ψ is different from $true$ and $false$, (ii) ψ is not a variable, and (iii) ψ contains no free occurrences of function symbols bound in χ. Let $x_1, \ldots, x_n$ be all free variables of ψ and $\sigma_1, \ldots, \sigma_n$ be their sorts. Take a fresh function symbol g, add the formula $(\forall x_1 : \sigma_1) \ldots (\forall x_n : \sigma_n)(\psi \Leftrightarrow g(x_1, \ldots, x_n) \doteq true)$ to D and replace ψ by $g(x_1, \ldots, x_n)$. Finally, change η to $\eta, g : \sigma_1 \times \ldots \times \sigma_n \to bool$.
3. Suppose that $if\ \psi\ then\ s\ else\ t$ is a term containing no free occurrences of function symbols bound in χ. Let $x_1, \ldots, x_n$ be all free variables of this term and $\sigma_1, \ldots, \sigma_n$ be their sorts. Take a fresh function symbol g, add the formulas $(\forall x_1 : \sigma_1) \ldots (\forall x_n : \sigma_n)(\psi \Rightarrow g(x_1, \ldots, x_n) \doteq s)$ and $(\forall x_1 : \sigma_1) \ldots (\forall x_n : \sigma_n)(\neg\psi \Rightarrow g(x_1, \ldots, x_n) \doteq t)$ to D and replace this term by $g(x_1, \ldots, x_n)$. Finally, change η to $\eta, g : \sigma_1 \times \ldots \times \sigma_n \to \sigma_0$, where σ_0 is such that $\eta, x_1 : \sigma_1, \ldots, x_n : \sigma_n \vdash s : \sigma_0$.
4. Suppose that $let\ f(x_1 : \sigma_1, \ldots, x_n : \sigma_n) = s\ in\ t$ is a term containing no free occurrences of function symbols bound in χ. Let $y_1, \ldots, y_m$ be all free variables of this term and $\tau_1, \ldots, \tau_m$ be their sorts. Note that the variables in $x_1, \ldots, x_n$ are not necessarily disjoint from the variables in $y_1, \ldots, y_m$.

   Take a fresh function symbol g and fresh sequence of variables $z_1, \ldots, z_n$. Let the term s' be obtained from s by replacing all free occurrences of $x_1, \ldots, x_n$ by $z_1, \ldots, z_n$, respectively. Add the formula $(\forall z_1 : \sigma_1) \ldots (\forall z_n : \sigma_n)(\forall y_1 : \tau_1) \ldots (\forall y_m : \tau_m)(g(z_1, \ldots, z_n, y_1, \ldots, y_m) \doteq s')$ to D. Let the term t' be obtained from t by replacing all bound occurrences of $y_1, \ldots, y_m$ by fresh variables and each application $f(t_1, \ldots, t_n)$ of a free occurrence of f in t by $g(t_1, \ldots, t_n, y_1, \ldots, y_m)$. Then replace $let\ f(x_1 : \sigma_1, \ldots, x_n : \sigma_n) = s\ in\ t$ by t'. Finally, change η to $\eta, g : \sigma_1 \times \ldots \times \sigma_n \times \tau_1 \times \ldots \times \tau_m \to \sigma_0$, where σ_0 is such that $\eta, x_1 : \sigma_1, \ldots, x_n : \sigma_n, y_1 : \tau_1, \ldots, y_m : \tau_m \vdash s : \sigma_0$.

The translation terminates when none of the above rules apply.

We will now formulate several of properties of this translation, which will imply that, in a way, it preserves models. These properties are not hard to prove, we do not include proofs in this paper.

Lemma 1. Suppose that a single step of the translation changes a formula ϕ_1 into ϕ_2, δ is the formula added at this step (for step 1 we can assume $true = true$ is added), η is the type assignment before this step and η' is the type assignment after. Then for every η'-interpretation I we have $I \models \delta \Rightarrow (\phi_1 \Leftrightarrow \phi_2)$. $\square$

By repeated applications of this lemma we obtain the following result.

Lemma 2. Suppose that the translation above changes a formula ϕ into ϕ', D is the set of definitions obtained during the translation, η is the initial type assignment and η' is the final type assignment of the translation. Let I' be any interpretation of η'. Then $I' \models \bigwedge_{\psi \in D} \psi \Rightarrow (\phi \Leftrightarrow \phi')$.   □

We also need the following result.

Lemma 3. Any sequence of applications of the translation rules terminates. □

The lemmas proved so far imply that the translation terminates and the final formula is equivalent to the initial formula in every interpretation satisfying all definitions in D. To prove model preservation, we also need to prove some properties of the introduced definitions.

Lemma 4. Suppose that one of the steps 2–4 of the translation translates a formula ϕ_1 into ϕ_2, δ is the formula added at this step, η is the type assignment before this step, η' is the type assignment after, and g is the fresh function symbol introduced at this step. Let also I be η-interpretation. Then there exists a function h such that $I_g^h \models \delta$.   □

These properties imply the following result on model preservation.

Theorem 3. Suppose that the translation above translates a formula ϕ into ϕ', D is the set of definitions obtained during the translation, η is the initial type assignment and η' is the final type assignment of the translation.

1. Let I be any η-interpretation. Then there is a η'-interpretation I' such that I' is an extension of I and $I' \models \bigwedge_{\psi \in D} \psi \wedge \phi'$.
2. Let I' be a η'-interpretation and $I' \models \bigwedge_{\psi \in D} \psi \wedge \phi'$. Then $I' \models \phi$.   □

This theorem implies that ϕ and $\bigwedge_{\psi \in D} \psi \wedge \phi'$ have the same models, as far as the original type assignment (the type assignment of Σ) is concerned. The formula $\bigwedge_{\psi \in D} \psi \wedge \phi'$ in this theorem is syntactically first-order. Denote this formula by γ. Our next step is to define a model-preserving translation from syntactically first-order formulas to first-order formulas.

To make γ into a first-order formula, we should get rid of *true* and *false* occurring in a formula context. To preserve the semantics, we should also add axioms for the boolean sort, since in first-order logic all sorts are uninterpreted, while in FOOL the interpretations of the boolean sort and constants *true* and *false* are fixed.

To fix the problem, we will add axioms expressing that the boolean sort has two elements and that *true* and *false* represent the two distinct elements of this sort.

$$\forall (x : bool)(x \doteq true \vee x \doteq false) \wedge true \neq false. \tag{3}$$

Note that this formula is a tautology in FOOL, but not in FOL.

Given a syntactically first-order formula γ, we denote by $fol(\gamma)$ the formula obtained from γ by replacing all occurrences of *true* and *false* in a formula context by logical constants $\top$ and $\bot$ (interpreted as always true and always false), respectively and adding formula (3).

Theorem 4. Let η is a type assignment and γ be a syntactically first-order formula such that $\eta \vdash \gamma : bool$.

1. Suppose that I is a η-interpretation and $I \models \gamma$ in FOOL. Then $I \models fol(\gamma)$ in first-order logic.
2. Suppose that I is a $\eta\eta$-interpretation and $I \models fol(\gamma)$ in first-order logic. Consider the FOOL-interpretation I' that is obtained from I by changing the interpretation of the boolean sort $bool$ by $\{0, 1\}$ and the interpretations of $true$ and $false$ by the elements 1 and 0, respectively, of this sort. Then $I' \models \gamma$ in FOOL. ☐

Theorems 3 and 4 show that our translation preserves models. Every model of the original formula can be extended to a model of the translated formulas by adding values of the function symbols introduced during the translation. Likewise, any first-order model of the translated formula becomes a model of the original formula after changing the interpretation of the boolean sort to coincide with its interpretation in FOOL.

4   Superposition for FOOL

In Sect. 3 we presented a model-preserving syntactic translation of FOOL to FOL. Based on this translation, automated reasoning about FOOL formulas can be done by translating a FOOL formula into a FOL formula, and using an automated first-order theorem prover on the resulting FOL formula. State-of-the-art first-order theorem provers, such as Vampire [13], E [18] and Spass [23], implement superposition calculus for proving first-order formulas. Naturally, we would like to have a translation exploiting such provers in an efficient manner.

Note however that our translation adds the two-element domain axiom $\forall (x : bool)(x \doteq true \lor x \doteq false)$ for the boolean sort. This axioms will be converted to the clause

$$x \doteq true \lor x \doteq false, \tag{4}$$

where x is a boolean variable. In this section we explain why this axiom requires a special treatment and propose a solution to overcome problems caused by its presence.

We assume some basic understanding of first-order theorem proving and superposition calculus, see, e.g. [1,15]. We fix a superposition inference system for first-order logic with equality, parametrised by a simplification ordering $\succ$ on literals and a well-behaved literal selection function [13], that is a function that guarantees completeness of the calculus. We denote selected literals by underlining them. We assume that equality literals are treated by a dedicated inference rule, namely, the ordered paramodulation rule [17]:

$$\frac{\underline{l \doteq r} \lor C \quad \underline{L[s]} \lor D}{(L[r] \lor C \lor D)\theta} \text{ if } \theta = \text{ mgu}(l, s),$$

where C, D are clauses, L is a literal, l, r, s are terms, $\text{mgu}(l, s)$ is a most general unifier of l and s, and $r\theta \not\succeq l\theta$. The notation $L[s]$ denotes that s is a subterm of L, then $L[r]$ denotes the result of replacement of s by r.

Suppose now that we use an off-the-shelf superposition theorem prover to reason about FOL formulas obtained by our translation. W.l.o.g, we assume that $true \succ false$ in the term ordering used by the prover. Then self-paramodulation (from $true$ to $true$) can be applied to clause (4) as follows:

$$\frac{x \doteq true \vee x \doteq false \qquad y \doteq true \vee y \doteq false}{x \doteq y \vee x \doteq false \vee y \doteq false}$$

The derived clause $x \doteq y \vee x \doteq false \vee y \doteq false$ is a recipe for disaster, since the literal $x \doteq y$ must be selected and can be used for paramodulation into every non-variable term of a boolean sort. Very soon the search space will contain many clauses obtained as logical consequences of clause (4) and results of paramodulation from variables applied to them. This will cause a rapid degradation of performance of superposition-based provers.

To get around this problem, we propose the following solution. First, we will choose term orderings $\succ$ having the following properties: $true \succ false$ and $true$ and $false$ are the smallest ground terms w.r.t. $\succ$. Consider now all ground instances of (4). They have the form $s \doteq true \vee s \doteq false$, where s is a ground term. When s is either $true$ or $false$, this instance is a tautology, and hence redundant. Therefore, we should only consider instances for which $s \succ true$. This prevents self-paramodulation of (4).

Now the only possible inferences with (4) are inferences of the form

$$\frac{x \doteq true \vee x \doteq false \qquad C[s]}{C[true] \vee s \doteq false},$$

where s is a non-variable term of the sort $bool$. To implement this, we can remove clause (4) and add as an extra inference rule to the superposition calculus the following rule:

$$\frac{C[s]}{C[true] \vee s \doteq false},$$

where s is a non-variable term of the sort $bool$ other than $true$ and $false$.

5   TPTP Support for FOOL

The typed monomorphic first-order formulas subset, called TFF0, of the TPTP language [20], is a representation language for many-sorted first-order logic. It contains if-then-else and let-in constructs (see below), which is useful for applications, but is inconsistent in its treatment of the boolean sort. It has a predefined atomic sort symbol $o denoting the boolean sort. However, unlike all other sort symbols, $o can only be used to declare the return type of predicate symbols. This means that one cannot define a function having a boolean argument, use boolean variables or equality between booleans.

Such an inconsistent use of the boolean sort results in having two kinds of if-then-else expressions and four kinds of let-in expressions. For example, a FOOL-term let $f(x_1 : \sigma_1, \ldots, x_n : \sigma_n) = s$ in t can be represented using one of the four TPTP alternatives \$let_tt, \$let_tf, \$let_ft and \$let_ff, depending on whether s and t are terms or formulas.

Since the boolean type is second-class in TPTP, one cannot directly represent formulas coming from program analysis and interactive theorem provers, such as formulas (1) and (2) of Sect. 1.

We propose to modify the TFF0 language of TPTP to coincide with FOOL. It is not late to do so, since there is no general support for if-then-else and let-in. To the best of our knowledge, Vampire is currently the only theorem prover supporting full TFF0. Note that such a modification of TPTP would make multiple forms of if-then-else and let-in redundant. It will also make it possible to directly represent the SMT-LIB core theory.

We note that our changes and modifications on TFF0 can also be applied to the TFF1 language of TPTP [4]. TFF1 is a polymorphic extension of TFF0 and its formalisation does not treat the boolean sort. Extending our work to TFF1 should not be hard but has to be done in detail.

6   Related Work

Handling boolean terms as formulas is common in the SMT community. The SMT-LIB project [3] defines its core logic as first-order logic extended with the distinguished first-class boolean sort and the let-in expression used for local bindings of variables. The core theory of SMT-LIB defines logical connectives as boolean functions and the ad-hoc polymorphic if-then-else (*ite*) function, used for conditional expressions. The language FOOL defined here extends the SMT-LIB core language with local function definitions, using let-in expressions defining functions of arbitrary, and not just zero, arity. This, FOOL contains both this language and the TFF0 subset of TPTP. Further, we present a translation of FOOL to FOL and show how one can improve superposition theorem provers to reason with the boolean sort.

Efficient superposition theorem proving in finite domains, such as the boolean domain, is also discussed in [9]. The approach of [9] sometimes falls back to enumerating instances of a clause by instantiating finite domain variables with all elements of the corresponding domains. We point out here that for the boolean (i.e., two-element) domain there is a simpler solution. However, the approach of [9] also allows one to handle domains with more than two elements. One can also generalise our approach to arbitrary finite domains by using binary encodings of finite domains, however, this will necessarily result in loss of efficiency, since a single variable over a domain with 2^k elements will become k variables in our approach, and similarly for function arguments.

7   Conclusion

We defined first-order logic with the first class boolean sort (FOOL). It extends ordinary many-sorted first-order logic (FOL) with (i) the boolean sort such that terms of this sort are indistinguishable from formulas and (ii) if-then-else and let-in expressions. The semantics of let-in expressions in FOOL is essentially their semantics in functional programming languages, when they are not used for recursive definitions. In particular, non-recursive local functions can be defined and function symbols can be bound to a different sort in nested let-in expressions.

We argued that these extensions are useful in reasoning about problems coming from program analysis and interactive theorem proving. The extraction of properties from certain program definitions (especially in functional programming languages) into FOOL formulas is more straightforward than into ordinary FOL formulas and potentially more efficient. In a similar way, a more straightforward translation of certain higher-order formulas into FOOL can facilitate proof automation in interactive theorem provers.

FOOL is a modification of FOL and reasoning in it reduces to reasoning in FOL. We gave a translation of FOOL to FOL that can be used for proving theorems in FOOL in a first-order theorem prover. We further discussed a modification of superposition calculus that can reason efficiently in presence of the boolean sort. Finally, we pointed out that the TPTP language can be changed to support FOOL, which will also simplify some parts of the TPTP syntax.

Implementation of theorem proving support for FOOL, including its superposition-friendly translation to CNF, is an important task for future work. Further, we are also interested in extending FOOL with theories, such as the theory of integer linear arithmetic and arrays.

References

1. Bachmair, L., Ganzinger, H.: Resolution theorem proving. Handbook of Automated Reasoning, pp. 19–99. Elsevier and MIT Press, Cambridge (2001)
2. Barrett, C., Conway, C.L., Deters, M., Hadarean, L., Jovanović, D., King, T., Reynolds, A., Tinelli, C.: CVC4. In: Gopalakrishnan, G., Qadeer, S. (eds.) CAV 2011. LNCS, vol. 6806, pp. 171–177. Springer, Heidelberg (2011)
3. Barrett, C., Stump, A., Tinelli, C.: The SMT-LIB standard: version 2.0. Technical report, Department of Computer Science, The University of Iowa (2010). Available at www.SMT-LIB.org
4. Blanchette, J.C., Paskevich, A.: TFF1: The TPTP typed first-order form with Rank-1 polymorphism. In: Bonacina, M.P. (ed.) CADE 2013. LNCS, vol. 7898, pp. 414–420. Springer, Heidelberg (2013)
5. Böhme, S., Nipkow, T.: Sledgehammer: judgement day. In: Giesl, J., Hähnle, R. (eds.) IJCAR 2010. LNCS, vol. 6173, pp. 107–121. Springer, Heidelberg (2010)
6. de Moura, L., Bjørner, N.S.: Z3: An efficient SMT solver. In: Ramakrishnan, C.R., Rehof, J. (eds.) TACAS 2008. LNCS, vol. 4963, pp. 337–340. Springer, Heidelberg (2008)

7. Dragan, I., Kovács, L.: Lingva: generating and proving program properties using symbol elimination. In: Voronkov, A., Virbitskaite, I. (eds.) PSI 2014. LNCS, vol. 8974, pp. 67–75. Springer, Heidelberg (2015)
8. Gupta, A., Kovács, L., Kragl, B., Voronkov, A.: Extensional crisis and proving identity. In: Cassez, F., Raskin, J.-F. (eds.) ATVA 2014. LNCS, vol. 8837, pp. 185–200. Springer, Heidelberg (2014)
9. Hillenbrand, T., Weidenbach, C.: Superposition for bounded domains. In: Bonacina, M.P., Stickel, M.E. (eds.) Automated Reasoning and Mathematics. LNCS, vol. 7788, pp. 68–100. Springer, Heidelberg (2013)
10. Hoder, K., Kovács, L., Voronkov, A.: Playing in the grey area of proofs. In: Proceedings of POPL, pp. 259–272 (2012)
11. Korovin, K.: iProver – an instantiation-based theorem prover for first-order logic (System description). In: Armando, A., Baumgartner, P., Dowek, G. (eds.) IJCAR 2008. LNCS (LNAI), vol. 5195, pp. 292–298. Springer, Heidelberg (2008)
12. Kovács, L., Voronkov, A.: Finding loop invariants for programs over arrays using a theorem prover. In: Chechik, M., Wirsing, M. (eds.) FASE 2009. LNCS, vol. 5503, pp. 470–485. Springer, Heidelberg (2009)
13. Kovács, L., Voronkov, A.: First-order theorem proving and Vampire. In: Sharygina, N., Veith, H. (eds.) CAV 2013. LNCS, vol. 8044, pp. 1–35. Springer, Heidelberg (2013)
14. McMillan, K.L.: Quantified invariant generation using an interpolating saturation prover. In: Ramakrishnan, C.R., Rehof, J. (eds.) TACAS 2008. LNCS, vol. 4963, pp. 413–427. Springer, Heidelberg (2008)
15. Nieuwenhuis, R., Rubio, A.: Paramodulation-based theorem proving. In: Robinson, A., Voronkov, A. (eds.) Handbook of Automated Reasoning, vol. I, pp. 371–443. Elsevier Science, Cambridge (2001). Chap. 7
16. Nipkow, T., Paulson, L.C., Wenzel, M.: Isabelle/HOL - A Proof Assistant for Higher-Order Logic. Springer, Heidelberg (2002)
17. Robinson, G., Wos, L.: Paramodulation and theorem-proving in first-order theories with equality. Mach. Intell. 4, 135–150 (1969)
18. Schulz, S.: System description: E 1.8. In: McMillan, K., Middeldorp, A., Voronkov, A. (eds.) LPAR-19 2013. LNCS, vol. 8312, pp. 735–743. Springer, Heidelberg (2013)
19. Sutcliffe, G.: The TPTP problem library and associated infrastructure. J. Autom. Reason. 43(4), 337–362 (2009)
20. Sutcliffe, G., Schulz, S., Claessen, K., Baumgartner, P.: The TPTP typed first-order form with arithmetic. In: Bjørner, N., Voronkov, A. (eds.) LPAR-18 2012. LNCS, vol. 7180, pp. 406–419. Springer, Heidelberg (2012)
21. Trybulec, A.: Mizar. In: Wiedijk, F. (ed.) The Seventeen Provers of the World. LNCS (LNAI), vol. 3600, pp. 20–23. Springer, Heidelberg (2006)
22. Urban, J., Hoder, K., Voronkov, A.: Evaluation of automated theorem proving on the Mizar mathematical library. In: Fukuda, K., Hoeven, J., Joswig, M., Takayama, N. (eds.) ICMS 2010. LNCS, vol. 6327, pp. 155–166. Springer, Heidelberg (2010)
23. Weidenbach, C., Dimova, D., Fietzke, A., Kumar, R., Suda, M., Wischnewski, P.: SPASS version 3.5. In: Schmidt, R.A. (ed.) CADE-22. LNCS, vol. 5663, pp. 140–145. Springer, Heidelberg (2009)

Type Inference for ZFH

Steven Obua[✉], Jacques Fleuriot, Phil Scott, and David Aspinall

School of Informatics, Edinburgh University, 10 Crichton Street,
EH8 9AB Edinburgh, Scotland, UK
steven.obua@googlemail.com
http://www.proofpeer.net

Abstract. *ZFH* stands for Zermelo-Fraenkel set theory implemented in higher-order logic. It is a descendant of Agerholm's and Gordon's HOL-ST but does not allow the use of type variables nor the definition of new types. We first motivate why we are using ZFH for *ProofPeer*, the collaborative theorem proving system we are building. We then focus on the type inference algorithm we have developed for ZFH. In ZFH's syntax, function application, written as juxtaposition, is overloaded to be either set-theoretic or higher-order. Our algorithm extends Hindley-Milner type inference to cope with this particular overloading of function application. We describe the algorithm, prove its correctness, and discuss why prior general approaches to type inference in the presence of coercions or overloading do not cover our particular case.

1   Introduction

The *ProofPeer* project [1,2] is our attempt to combine interactive theorem proving (ITP) and the modern web, making ITP technology more accessible than it has been. We will first explain why we have chosen ZFH as the logic of ProofPeer, and then introduce the problem this paper solves.

1.1   Why ZFH?

Despite a few prominent counter examples [3,4] it is particularly astonishing how few mathematicians are aware of or even use ITP systems. We believe that one reason for this is that traditionally the development and application of ITP technology has been driven by computer scientists, not mathematicians. Major successful ITP systems like Isabelle and Coq are based on variants of type theory, while most mathematicians feel more familiar with set theory. Simple mathematical standards like point set topology cannot be formalized in either system without the result feeling alien to most mathematicians.

We have therefore decided that the logic used in the ProofPeer system should be based on Zermelo-Fraenkel set theory which is more or less familiar to all mathematicians. At the same time we want to build on the considerable technical advances that contemporary ITP systems have achieved. Therefore we embed set theory within simply-typed classical higher-order logic by introducing a special type $\mathscr{U}$ which forms the universe of Zermelo-Fraenkel sets, additional constants

© Springer International Publishing Switzerland 2015
M. Kerber et al. (Eds.): CICM 2015, LNAI 9150, pp. 87–101, 2015.
DOI: 10.1007/978-3-319-20615-8_6

like the element-of operator $\in$: $\mathscr{U} \to \mathscr{U} \to \mathbb{P}$, and additional axioms describing the properties of these new constants. The symbol $\mathbb{P}$ denotes the type of propositions/booleans, and for any two types α and β we can form the type of higher-order functions $\alpha \to \beta$. For a full list of all new constants and axioms see theory root [9] in the ProofPeer system.

Because technically, all we have done is add an additional type together with a few new constants and axioms, all of the machinery present in systems like HOL-4 or Isabelle/HOL can be ported to work in our system. For example, Isabelle/HOL's facilities for defining partial and nested recursive functions [8] could be translated to ProofPeer.

This approach was first advocated by Agerholm and Gordon [5]. They called the resulting logic HOL-ST. A related approach is pursued by Isabelle/ZF which embeds set theory within its intuitionistic higher-order meta logic [7]. The Isabelle/ZF approach seems more involved than our approach: it begins at base with *intuitionistic* higher-order logic, over which first-order classical logic is introduced, which in turn, is used to formalise set theory. We instead skip the middle step and base set theory directly on *classical* higher-order logic, obtaining a more powerful logic by simpler means. This is just how HOL-ST works as well, but that opens up a new dilemma: HOL-ST is so powerful that often it is not clear how concepts should best be formalised. Take for example the natural numbers: should they be formalised as a type, or should they be formalised as a set, i.e. as an element of $\mathscr{U}$? Or take lists: should they be formalised as a type α list together with polymorphic operations like cons : $\alpha \to \alpha$ list $\to \alpha$ list, or should they be formalised as a constant list : $\mathscr{U} \to \mathscr{U}$ such that list α denotes the set of lists over elements of α? Note how in the latter case we can extend our discussion to the *class* of all (heterogeneous) lists by defining

$$\text{isList } l = \exists \alpha. \, l \in \text{list } \alpha$$

The type of cons would now be $\mathscr{U} \to \mathscr{U} \to \mathscr{U}$ and for reasonable definitions of list we could prove theorems like

$$\forall l. \text{ isList } l \to \forall x. \text{ isList}(\text{cons } x \, l)$$
$$\forall l \, \alpha. \, l \in \text{list } \alpha \to \forall x \in \alpha. \text{ cons } x \, l \in \text{list } \alpha$$

We want people to perceive ProofPeer as a system based on set theory; the only reason we also employ simply-typed higher-order logic is because of its technical convenience and simplicity. Therefore for us there is an easy and coherent way out of the dilemma that HOL-ST has: we forbid the introduction of new types besides the ones we already described, and we furthermore do not use type variables as part of our internal term representation. The only polymorphic constants in our logic are equality ($=$), universal quantification ($\forall$) and existential quantification ($\exists$), and we do not provide any means for defining additional ones.

Abstaining from polymorphic terms in favour of monomorphic ones has a further advantage noticed already by Gordon [6, Sect. 3]: We can treat theories as simple (albeit large) theorems. The axioms of the theory become the antecedents of the theorem, and constants declared in the theory can be treated as universally quantified variables. This doesn't work in polymorphic simply-typed higher order logic because polymorphic constants can appear with different types in the theory but variables must appear always with the same type in the theorem.

We choose the name *ZFH* for the logical system we obtain by embedding set theory into classical higher-order logic in the way outlined above. ZFH represents the same logic as HOL-ST minus type variables and minus a mechanism for defining custom types. In particular this means that ZFH and HOL-ST are equiconsistent, and that both HOL and ZFC can be formalized and proven to be consistent within ZFH.

1.2 Set-Theoretic vs. Higher-Order Function Application

There are two kinds of function application in ZFH:

– application of a higher-order function $f : \alpha \to \beta$ to its argument $x : \alpha$, and
– application of a set-theoretic function $f : \mathscr{U}$ to its argument $x : \mathscr{U}$.

In ZFH, set-theoretic functions are governed by two properties:

$$\forall X\ f.\ \mathsf{fun}\ X\ f = \{(x,\ f\ x) \mid x \in X\}$$
$$\forall X\ f.\ \forall x \in X.\ \mathsf{apply}\ (\mathsf{fun}\ X\ f)\ x = f\ x$$

Here $\mathsf{fun} : \mathscr{U} \to (\mathscr{U} \to \mathscr{U}) \to \mathscr{U}$ takes a domain $X : \mathscr{U}$ and a higher-order function $f : \mathscr{U} \to \mathscr{U}$ as its arguments and produces the corresponding set-theoretic function on that domain. Set-theoretic functions created thus can then be applied via $\mathsf{apply} : \mathscr{U} \to \mathscr{U} \to \mathscr{U}$.

In the actual ProofPeer theory [9], the second property is written like this:

$$\forall X\ f.\ \forall x \in X.\ \mathsf{fun}\ X\ f\ x = f\ x$$

Instead of explicitly mentioning apply we write application of a set-theoretic function in exactly the same way as application of a higher-order function! This is possible because in the above $\mathsf{fun}\ X\ f$ is a set, which leads type inference to conclude that set-theoretic function application must be meant, not higher-order function application.

In general, the situation is not so clear-cut. Consider the following term:

$$\forall x.\ \exists f.\ f\ x = x$$

Informally the above says that every x is the fixpoint of some function f. But which types should we assign to f and x? There are infinitely many valid ones:

1. $f : \mathscr{U}$ and $x : \mathscr{U}$
2. $f : \mathbb{P} \to \mathbb{P}$ and $x : \mathbb{P}$
3. $f : \mathscr{U} \to \mathscr{U}$ and $x : \mathscr{U}$
4. $f : (\mathscr{U} \to \mathscr{U}) \to (\mathscr{U} \to \mathscr{U})$ and $x : \mathscr{U} \to \mathscr{U}$
5. $f : (\mathbb{P} \to \mathscr{U}) \to (\mathbb{P} \to \mathscr{U})$ and $x : \mathbb{P} \to \mathscr{U}$

...and so on

Even if we had type variables at our disposal to formulate the typing (which we don't) there would still be two equally valid typings to choose from:

1. $f : \mathscr{U}$ and $x : \mathscr{U}$
2. $f : \alpha \to \alpha$ and $x : \alpha$

Which one should we pick?

In the next section we will present a type inference algorithm for ZFH with the following properties:

– If there is a valid typing at all, the algorithm will find one, and will otherwise fail. In particular, all function applications will be resolved to be either set-theoretic or higher-order.
– Preference is given to the type $\mathscr{U}$ over all other types, and to set-theoretic function application over higher-order function application.

Note that the second property is a desirable one in our case, as this again emphasises the set theory focus of ProofPeer.

In our above example the algorithm yields then the typing $f : \mathscr{U}$ and $x : \mathscr{U}$.

2   The Type Inference Algorithm

We first introduce the types and terms our algorithm operates on. Then we introduce the type equations which guide the algorithm, and recall how to solve type equations. After highlighting the basic difficulties of the problem we state the algorithm. Finally we prove that the algorithm terminates, that it is sound, and in what sense it is complete.

2.1   Types and Terms

Although we do not allow type variables as part of proper ZFH terms, we do allow them for type inference purposes. In particular a *pretype* τ is either the universal type $\mathscr{U}$, the propositional/boolean type $\mathbb{P}$, a function type $\tau_1 \to \tau_2$, or a type variable α:

$$\tau ::= \mathscr{U} \mid \mathbb{P} \mid \tau_1 \to \tau_2 \mid \alpha.$$

A *type* is a pretype which does not contain any type variables. A *preterm* t is either a constant c, a polymorphic constant $p[\tau]$, an explicit typing $t : \tau$, a higher-order function $x : \tau_1 \mapsto t : \tau_2$, a variable x, a higher-order function application $t_1 \diamond_\mathsf{H} t_2 : \tau$, a set-theoretic function application $t_1 \diamond_\mathsf{ZF} t_2 : \tau$, or a function application $t_1 \diamond_? t_2 : \tau$ where it is unspecified if it is of higher-order or set-theoretic kind:

$$t ::= c \mid p[\tau] \mid t : \tau \mid x : \tau_1 \mapsto t : \tau_2 \mid x \mid t_1 \diamond_\mathsf{H} t_2 : \tau \mid t_1 \diamond_\mathsf{ZF} t_2 : \tau \mid t_1 \diamond_? t_2 : \tau.$$

A *term* is a preterm which does not contain any type variables, nor any function applications of unspecified kind.

Example 1. Our introductory example $\forall x.\ \exists f.\ fx = x$ corresponds to the preterm:

$$\forall[\alpha_1]\ \Diamond_{\mathsf{H}}\ (x : \alpha_2 \mapsto (\exists[\alpha_3]\ \Diamond_{\mathsf{H}}$$
$$(f : \alpha_4 \mapsto ((= [\alpha_5]\ \Diamond_{\mathsf{H}}\ (f \Diamond_? x : \alpha_6) : \alpha_7)\ \Diamond_{\mathsf{H}}\ x : \alpha_8) : \alpha_9) : \alpha_{10}) : \alpha_{11}) : \alpha_{12}.$$

Note that everywhere our preterm format requires a type, we simply used a fresh type variable.

2.2 Type Equations

A *substitution* σ associates every type variable α with a pretype σ_α. Applying a substitution to a pretype τ means replacing every type variable in τ by its associated pretype (Fig. 1), and applying a substitution to a preterm t means applying the substitution to every pretype in t (Fig. 2).

With each constant c a fixed type $\mathcal{C}(c)$ is associated, e.g. $\mathcal{C}(\mathsf{apply}) = \mathscr{U} \to \mathscr{U} \to \mathscr{U}$. Assuming also a partial map $\mathcal{V}$ from variables to pretypes we can associate with each preterm t its type $\Gamma_{\mathcal{C},\mathcal{V}}(t)$ and a set of equations between pretypes $\mathcal{E}_{\mathcal{C},\mathcal{V}}(t)$ as shown in Fig. 3. In the following we will assume an implicitly given $\mathcal{C}$ and define

$$\mathcal{E}(t) = \mathcal{E}_{\mathcal{C},\emptyset}(t)$$

where $\emptyset$ in this context denotes the empty map.

A substitution σ is a *unifier* of a set $\mathcal{E}$ of equations of pretypes iff for all equations $l \equiv r \in \mathcal{E}$ the left hand side and the right hand side of the equation become identical after substitution, i.e. $\sigma(l) = \sigma(r)$ holds. We call $\mathcal{E}$ *solvable* if it has a unifier. Defining $\sigma(\mathcal{E}) = \{\sigma(l) \equiv \sigma(r) \mid l \equiv r \in \mathcal{E}\}$ allows the following rephrasing: σ is a unifier of $\mathcal{E}$ iff $\sigma(\mathcal{E})$ is a set of identities.

$$\sigma(\mathscr{U}) = \mathscr{U}$$
$$\sigma(\mathbb{P}) = \mathbb{P}$$
$$\sigma(\tau_1 \to \tau_2) = \sigma(\tau_1) \to \sigma(\tau_2)$$
$$\sigma(\alpha) = \sigma_\alpha$$

Fig. 1. Applying a substitution σ to a pretype

$$\sigma(c) = c$$
$$\sigma(p[\tau]) = p[\sigma(\tau)]$$
$$\sigma(t : \tau) = \sigma(t) : \sigma(\tau)$$
$$\sigma(x : \tau_1 \mapsto t : \tau_2) = x : \sigma(\tau_1) \mapsto \sigma(t) : \sigma(\tau_2)$$
$$\sigma(x) = x$$
$$\sigma(t_1 \Diamond_{\mathsf{H}} t_2 : \tau) = \sigma(t_1) \Diamond_{\mathsf{H}} \sigma(t_2) : \sigma(\tau)$$
$$\sigma(t_1 \Diamond_{\mathsf{ZF}} t_2 : \tau) = \sigma(t_1) \Diamond_{\mathsf{ZF}} \sigma(t_2) : \sigma(\tau)$$
$$\sigma(t_1 \Diamond_? t_2 : \tau) = \sigma(t_1) \Diamond_? \sigma(t_2) : \sigma(\tau)$$

Fig. 2. Applying a substitution σ to a preterm

$$\Gamma_{C,\mathcal{V}}(c) = \mathcal{C}(c)$$
$$\mathcal{E}_{C,\mathcal{V}}(c) = \emptyset$$
$$\Gamma_{C,\mathcal{V}}(p[\tau]) = \begin{cases} (\tau \to \mathbb{P}) \to \mathbb{P} & \text{if } p \in \{\forall, \exists\} \\ \tau \to \tau \to \mathbb{P} & \text{if } p \in \{=\} \end{cases}$$
$$\mathcal{E}_{C,\mathcal{V}}(p[\tau]) = \emptyset$$
$$\Gamma_{C,\mathcal{V}}(t : \tau) = \tau$$
$$\mathcal{E}_{C,\mathcal{V}}(t : \tau) = \mathcal{E}_{C,\mathcal{V}}(t) \cup \{\Gamma_{C,\mathcal{V}}(t) \equiv \tau\}$$
$$\Gamma_{C,\mathcal{V}}(x : \tau_1 \mapsto t : \tau_2) = \tau_1 \to \tau_2$$
$$\mathcal{E}_{C,\mathcal{V}}(x : \tau_1 \mapsto t : \tau_2) = \mathcal{E}_{C,\mathcal{W}}(t) \cup \{\Gamma_{C,\mathcal{W}}(t) \equiv \tau_2\} \text{ where } \mathcal{W} = \mathcal{V}[x := \tau_1]$$
$$\Gamma_{C,\mathcal{V}}(x) = \begin{cases} \mathcal{V}(x) & \text{if } \mathcal{V} \text{ is defined at } x \\ \mathcal{U} & \text{otherwise} \end{cases}$$
$$\mathcal{E}_{C,\mathcal{V}}(x) = \begin{cases} \emptyset & \text{if } \mathcal{V} \text{ is defined at } x \\ \{\mathcal{U} \equiv \mathbb{P}\} & \text{otherwise} \end{cases}$$
$$\Gamma_{C,\mathcal{V}}(t_1 \diamond_{\mathsf{H}} t_2 : \tau) = \tau$$
$$\mathcal{E}_{C,\mathcal{V}}(t_1 \diamond_{\mathsf{H}} t_2 : \tau) = \mathcal{E}_{C,\mathcal{V}}(t_1) \cup \mathcal{E}_{C,\mathcal{V}}(t_2) \cup \{\Gamma_{C,\mathcal{V}}(t_1) \equiv \Gamma_{C,\mathcal{V}}(t_2) \to \tau\}$$
$$\Gamma_{C,\mathcal{V}}(t_1 \diamond_{\mathsf{ZF}} t_2 : \tau) = \tau$$
$$\mathcal{E}_{C,\mathcal{V}}(t_1 \diamond_{\mathsf{ZF}} t_2 : \tau) = \mathcal{E}_{C,\mathcal{V}}(t_1) \cup \mathcal{E}_{C,\mathcal{V}}(t_2) \cup \{\Gamma_{C,\mathcal{V}}(t_1) \equiv \mathcal{U}, \Gamma_{C,\mathcal{V}}(t_2) \equiv \mathcal{U}, \tau \equiv \mathcal{U}\}$$
$$\Gamma_{C,\mathcal{V}}(t_1 \diamond_? t_2 : \tau) = \tau$$
$$\mathcal{E}_{C,\mathcal{V}}(t_1 \diamond_? t_2 : \tau) = \mathcal{E}_{C,\mathcal{V}}(t_1) \cup \mathcal{E}_{C,\mathcal{V}}(t_2)$$

Fig. 3. Definition of $\Gamma_{C,\mathcal{V}}$ and $\mathcal{E}_{C,\mathcal{V}}$

Substitutions can be composed. The composition $\delta \circ \sigma$ of a substitution σ with a substitution δ is defined via

$$(\delta \circ \sigma)_\alpha = \delta(\sigma_\alpha).$$

A unifier σ_1 is called more general than a unifier σ_2, in symbols $\sigma_1 \geq \sigma_2$, iff there is a substitution δ such that $\sigma_2 = \delta \circ \sigma_1$.

Lemma 1. *If $\mathcal{E}$ is solvable then it has an idempotent most general unifier* $\mathrm{mgu}_\mathcal{E}$, *i.e. the following two properties hold for* $\mathrm{mgu}_\mathcal{E}$:

1. $\mathrm{mgu}_\mathcal{E} \geq \sigma$ *for any unifier* σ *of* $\mathcal{E}$, *and*
2. $\mathrm{mgu}_\mathcal{E} \circ \mathrm{mgu}_\mathcal{E} = \mathrm{mgu}_\mathcal{E}$.

Proof. See [10, Sect. 4.5]. □

If $\mathcal{E}(t)$ is solvable for a given preterm t, then we define

$$\mathcal{S}(t) = \mathrm{mgu}_{\mathcal{E}(t)}(t).$$

Note that $\mathcal{S}(t)$ is unique up to a renaming of type variables. Computation of $\mathcal{S}(t)$ is known as *Hindley-Milner type inference* [11].

Example 2. Given the preterm t from Example 1, the type equations $\mathcal{E}(t)$ are:

$$(\alpha_1 \to \mathbb{P}) \to \mathbb{P} \equiv (\alpha_2 \to \alpha_{11}) \to \alpha_{12}$$

$$\alpha_{10} \equiv \alpha_{11}$$

$$(\alpha_3 \to \mathbb{P}) \to \mathbb{P} \equiv (\alpha_4 \to \alpha_9) \to \alpha_{10}$$

$$\alpha_8 \equiv \alpha_9$$

$$\alpha_7 \equiv \alpha_2 \to \alpha_8$$

$$\alpha_5 \to \alpha_5 \to \mathbb{P} \equiv \alpha_6 \to \alpha_7$$

A most general unifier for these equations of pretypes is given by

$$\mathsf{mgu}_{\mathcal{E}(t)}(\alpha_i) = \begin{cases} \alpha & \text{if } i \in \{1,2,5,6\} \\ \beta & \text{if } i \in \{3,4\} \\ \alpha \to \mathbb{P} & \text{if } i = 7 \\ \mathbb{P} & \text{if } i \in \{8,9,10,11,12\} \end{cases}$$

and therefore

$$\mathcal{S}(t) = \forall[\alpha] \diamond_\mathsf{H} (x : \alpha \mapsto (\exists[\beta] \diamond_\mathsf{H}$$
$$(f : \beta \mapsto ((= [\alpha] \diamond_\mathsf{H} (f \diamond_? x : \alpha) : \alpha \to \mathbb{P}) \diamond_\mathsf{H} x : \mathbb{P}) : \mathbb{P}) : \mathbb{P}) : \mathbb{P}) : \mathbb{P}.$$

2.3 A First Attempt

An obvious first attempt to solve our type inference problem for a given preterm t would be to single out all n occurrences of $\diamond_?$ in t and to form all 2^n possibilities t_i by replacing $\diamond_?$ with either $\diamond_\mathsf{H}$ or $\diamond_\mathsf{ZF}$.

If none of the sets of type equations $\mathcal{E}(t_i)$ is solvable, type inference fails. Otherwise let t_j denote those t_i for which $\mathcal{E}(t_i)$ is solvable. This gives us up to 2^n almost-solutions s_j where

$$s_j = \mathcal{S}(t_j).$$

Because the s_j possibly contain type variables, but proper ZFH terms may not contain type variables, we need to somehow eliminate all type variables from the s_j. One rather arbitrary way of doing so would be to replace all type variables by $\mathscr{U}$, i.e. to form

$$r_j = \mathcal{U}(s_j)$$

where $\mathcal{U}$ is the substitution which replaces all type variables by $\mathscr{U}$:

$$\mathcal{U}_\alpha = \mathscr{U} \text{ for all type variables } \alpha.$$

This leaves us finally with up to 2^n possible solutions r_j to our type inference problem. Computing all of these solutions is not practical for obvious performance reasons; furthermore, even if we *did* compute all of them, it is not clear which one among them we should pick as the result of the type inference.

2.4   The Algorithm

In our above attempt at a type inference algorithm we computed $\mathcal{S}(t_i)$ only for
preterms t_i which did not contain any occurrences of $\diamond_?$. This was an arbitrary
choice we made and it did not pay off.

Instead, given a preterm t which may still contain occurrences of $\diamond_?$, let us
directly compute $t_0 = \mathcal{S}(t)$ if $\mathcal{E}(t)$ is solvable. If t contained any occurrences of
$\diamond_?$, then so will t_0, *but we now might have more type information available to
decide whether an occurrence of $\diamond_?$ should really be replaced by $\diamond_H$ or $\diamond_{ZF}$!*

To exploit type information present in a preterm t we define a function $\mathcal{D}(t)$
which is able to decide in certain situations whether an occurrence of $\diamond_?$ in t
should be converted into $\diamond_H$ or into $\diamond_{ZF}$ (Fig. 4). Analogously to the definition of
$\mathcal{E}(t)$ in terms of $\mathcal{E}_{C,V}(t)$ we define $\mathcal{D}(t)$ in terms of $\mathcal{D}_{C,V}(t)$. The main work in $\mathcal{D}$
is done by the function

$$\diamond(\tau_1, \tau_2, \tau_3) \in \{\diamond_H, \diamond_{ZF}, \diamond_?\}$$

which takes three pretypes τ_1, τ_2, τ_3 as arguments and tries to determine which
kind of function application fx must be when the type of f is known to be
τ_1, the type of x is known to be τ_2 and the type of fx is known to be τ_3. If
any of the τ_i cannot be the type $\mathcal{U}$, in symbols $\neg^{\mathcal{U}}(\tau_i)$, then we know that
fx cannot be set-theoretic function application and therefore can only be (if
any at all) higher-order function application. On the other hand, if the type of
τ_1 cannot be a function type, in symbols $\neg^{\rightarrow}(\tau_1)$, then fx cannot be higher-
order function application and can therefore be only (if any at all) set-theoretic
function application (Fig. 5).

We have now gathered all the pieces to formulate our type inference algorithm
as shown in Fig. 6.

Example 3. Continuing Example 2 we compute now $\mathsf{TypeInfer}(t)$. Having already
computed $s = \mathcal{S}(t)$ we now need to compute $\mathcal{D}(s)$. There is only one occurrence
of $\diamond_?$ in s and the corresponding invocation of $\diamond$ yields

$$\diamond(\beta, \alpha, \alpha) = \diamond_?$$

$$\mathcal{D}(t) = \mathcal{D}_{C,\emptyset}(t)$$
$$\mathcal{D}_{C,V}(c) = c$$
$$\mathcal{D}_{C,V}(p[\tau]) = p[\tau]$$
$$\mathcal{D}_{C,V}(t : \tau) = \mathcal{D}_{C,V}(t) : \tau$$
$$\mathcal{D}_{C,V}(x : \tau_1 \mapsto b : \tau_2) = x : \tau_1 \mapsto \mathcal{D}_{C,W}(b) : \tau_2 \text{ where } W = V[x := \tau_1]$$
$$\mathcal{D}_{C,V}(x) = x$$
$$\mathcal{D}_{C,V}(t_1 \diamond_H t_2 : \tau) = \mathcal{D}_{C,V}(t_1) \diamond_H \mathcal{D}_{C,V}(t_2) : \tau$$
$$\mathcal{D}_{C,V}(t_1 \diamond_{ZF} t_2 : \tau) = \mathcal{D}_{C,V}(t_1) \diamond_{ZF} \mathcal{D}_{C,V}(t_2) : \tau$$
$$\mathcal{D}_{C,V}(t_1 \diamond_? t_2 : \tau) = \mathcal{D}_{C,V}(t_1) \diamond (\Gamma_{C,V}(t_1), \Gamma_{C,V}(t_2), \tau) \mathcal{D}_{C,V}(t_2) : \tau$$

Fig. 4. Definition of $\mathcal{D}$

$$\neg^{\mathcal{U}}(\tau) = \begin{cases} \text{true} & \text{if } \tau = \mathbb{P} \\ \text{true} & \text{if } \tau = \omega_1 \to \omega_2 \text{ for some pretypes } \omega_1 \text{ and } \omega_2 \\ \text{false} & \text{otherwise} \end{cases}$$

$$\neg^{\to}(\tau) = \begin{cases} \text{true} & \text{if } \tau = \mathbb{P} \\ \text{true} & \text{if } \tau = \mathcal{U} \\ \text{false} & \text{otherwise} \end{cases}$$

$$\diamond(\tau_1, \tau_2, \tau_3) = \begin{cases} \diamond_\mathsf{H} & \text{if } \neg^{\mathcal{U}}(\tau_1) \text{ or } \neg^{\mathcal{U}}(\tau_2) \text{ or } \neg^{\mathcal{U}}(\tau_3) \\ \diamond_\mathsf{ZF} & \text{else if } \neg^{\to}(\tau_1) \\ \diamond_? & \text{otherwise} \end{cases}$$

Fig. 5. Definition of $\diamond(\tau_1, \tau_2, \tau_3)$

```
TypeInfer(t) =
    if E(t) is not solvable then
        fail
    else
        s = S(t)
        d = D(s)
        if s = d then
            D(U(d))
        else
            TypeInfer(d)
        end
    end
```

Fig. 6. The Type Inference Algorithm

and thus $\mathcal{D}(s) = s$. This means that no recursive call to TypeInfer is necessary and therefore

$$\text{TypeInfer}(t) = \mathcal{D}(\mathcal{U}(s)) = \forall[\mathcal{U}] \diamond_\mathsf{H} (x : \mathcal{U} \mapsto (\exists[\mathcal{U}] \diamond_\mathsf{H}$$
$$(f : \mathcal{U} \mapsto ((= [\mathcal{U}] \diamond_\mathsf{H} (f \diamond_\mathsf{ZF} x : \mathcal{U}) : \mathcal{U} \to \mathbb{P}) \diamond_\mathsf{H} x : \mathbb{P}) : \mathbb{P}) : \mathbb{P}) : \mathbb{P}) : \mathbb{P}.$$

2.5 Termination

Let us first show that our algorithm actually terminates. There are only finitely many occurrences of $\diamond_?$ in a preterm t, let us denote the number of such occurrences by $N(t)$. For two preterms s and t let us write $s \sqsubseteq t$ if s arises from t by replacing some (or none) of the occurrences of $\diamond_?$ in t by either $\diamond_\mathsf{H}$ or $\diamond_\mathsf{ZF}$. Obviously $s \sqsubseteq t$ together with $s \neq t$ implies $N(s) < N(t)$.

Lemma 2. TypeInfer(t) *terminates for every preterm* t.

Proof. Given some preterm s, $\mathcal{D}(s) \sqsubseteq s$ holds. Therefore $s \neq \mathcal{D}(s)$ implies

$$N(\mathcal{D}(s)) < N(s).$$

We also know that $N(t) = N(\mathcal{S}(t))$ because $\mathcal{S}$ only possibly instantiates type variables and leaves occurrences of $\diamond_?$ unchanged. Together this means that for each recursive call to TypeInfer its argument strictly decreases as measured by N and therefore the algorithm must terminate. □

2.6   Soundness and Completeness

Given two preterms t and t' we say that t' is an instance of t, in symbols

$$t' \leq t$$

iff there is a substitution σ such that $t' \sqsubseteq \sigma(t)$.

What does it mean for our type inference algorithm to be sound? Given a preterm t as input it should output a preterm t' such that

1. t' is a term,
2. $t' \leq t$, and
3. $\mathcal{E}(t')$ is solvable.

If there is no such t' the algorithm should fail. If there are several possible candidates for t' it would also be good to have a simple and sensible criterion for which of the candidates the algorithm will pick. Our algorithm fulfills such a criterion: it will pick the unique candidate t' which is minimal with respect to the relation $\preceq$ which is first defined on types (Fig. 7) and then lifted to terms (Fig. 8). The reflexive, transitive and antisymmetric relation $\preceq$ expresses formally what we referred to earlier as "$\mathcal{U}$ *is preferred over any other type, and set-theoretic function application is preferred over higher-order function application*".

$$\frac{}{\mathbb{P} \preceq \mathbb{P}} \qquad \frac{\tau \text{ is a type}}{\mathcal{U} \preceq \tau} \qquad \frac{\tau_1 \preceq \omega_1 \text{ and } \tau_2 \preceq \omega_2}{\tau_1 \to \tau_2 \preceq \omega_1 \to \omega_2}$$

Fig. 7. Definition of $\preceq$ for Types

Lemma 3. *Let σ be a substitution and t a preterm. Then $\mathcal{E}(\sigma(t)) = \sigma(\mathcal{E}(t))$.*

Proof. Immediate from the definitions. □

Lemma 4. *Let t be a preterm such that $\mathcal{E}(t)$ is solvable. Then $\mathcal{E}(\mathcal{S}(t))$ is a set of identities.*

Proof. $\mathcal{E}(\mathcal{S}(t)) = \mathcal{E}(\mathsf{mgu}_{\mathcal{E}(t)}(t)) = \mathsf{mgu}_{\mathcal{E}(t)}(\mathcal{E}(t))$. □

Lemma 5. *Let t be a preterm such that $\mathcal{E}(t)$ is solvable and $\mathcal{S}(t) = t$. Then $\mathsf{mgu}_{\mathcal{E}(t)} = \mathsf{id}$ and $\mathcal{E}(t)$ is a set of identities.*

Proof. $\mathsf{id}(\mathcal{E}(t)) = \mathcal{E}(t) = \mathcal{E}(\mathcal{S}(t))$. □

Lemma 6. *Let s and t be preterms such that $s \sqsubseteq t$. Then $\mathcal{E}(t) \subseteq \mathcal{E}(s)$. In particular, if $\mathcal{E}(s)$ is solvable then so is $\mathcal{E}(t)$.*

$$\frac{}{c \preceq c} \qquad \frac{}{x \preceq x} \qquad \frac{\tau_1 \preceq \tau_2}{p[\tau_1] \preceq p[\tau_2]} \qquad \frac{t_1 \preceq t_2 \text{ and } \tau_1 \preceq \tau_2}{t_1 : \tau_1 \preceq t_2 : \tau_2}$$

$$\frac{\tau_1 \preceq \tau_2 \text{ and } t_1 \preceq t_2 \text{ and } \omega_1 \preceq \omega_2}{x : \tau_1 \mapsto t_1 : \omega_1 \preceq x : \tau_2 \mapsto t_2 : \omega_2}$$

$$\frac{t_1 \preceq s_1 \text{ and } t_2 \preceq s_2 \text{ and } \tau_1 \preceq \tau_2}{t_1 \diamond_H t_2 : \tau_1 \preceq s_1 \diamond_H s_2 : \tau_2}$$

$$\frac{t_1 \preceq s_1 \text{ and } t_2 \preceq s_2 \text{ and } \tau_1 \preceq \tau_2 \text{ and } \diamond \in \{\diamond_H, \diamond_{ZF}\}}{t_1 \diamond_{ZF} t_2 : \tau_1 \preceq s_1 \diamond s_2 : \tau_2}$$

Fig. 8. Definition of $\preceq$ for Terms

Proof. Immediate from the definitions. □

Lemma 7. *If t is a preterm without any type variables then $\mathcal{D}(t)$ is a term.*

Proof. If τ_1 does not contain any type variables then either $\neg^{\mathcal{U}}(\tau_1)$ or $\neg^{\rightarrow}(\tau_1)$ is true, and therefore $\diamond(\tau_1, \tau_2, \tau_3) \in \{\diamond_H, \diamond_{ZF}\}$. □

Lemma 8. *If t is a preterm without any type variables, and t' is a term such that $\mathcal{E}(t')$ is solvable and $t' \sqsubseteq t$ then $t' = \mathcal{D}(t)$.*

Proof. The terms t' and $\mathcal{D}(t)$ could only possibly differ in places where t has an occurrence of $\diamond_?$. In those places, choosing differently from $\mathcal{D}$ would make the resulting equations unsolvable; however, $\mathcal{E}(t')$ is solvable. □

Lemma 9. *For any preterm t and any substitution σ*

$$\mathcal{D}(\sigma(t)) \sqsubseteq \sigma(\mathcal{D}(t)).$$

Proof. This follows from the fact that $\diamond(\tau_1, \tau_2, \tau_3) \in \{\diamond_H, \diamond_{ZF}\}$ implies

$$\diamond(\sigma(\tau_1), \sigma(\tau_2), \sigma(\tau_3)) = \diamond(\tau_1, \tau_2, \tau_3). \qquad \square$$

Lemma 10. *Let t be a preterm and t' a term such that $t' \leq t$ and $\mathcal{E}(t')$ is solvable. Then $\mathcal{E}(t)$ is solvable and both $t' \leq \mathcal{S}(t)$ and $t' \leq \mathcal{D}(t)$ hold.*

Proof. Because $t' \leq t$ there exist σ and t'' such that $t'' = \sigma(t)$ and $t' \sqsubseteq t''$. Because $\mathcal{E}(t')$ is solvable so is $\mathcal{E}(t'')$. Because t' is a term, neither t' nor t'' contain any type variables and thus $\mathcal{S}(t'') = t''$ which implies that $\mathcal{E}(t'') = \mathcal{E}(\sigma(t)) = \sigma(\mathcal{E}(t))$ are all sets of identities, and therefore σ is a unifier of $\mathcal{E}(t)$. This means there is a substitution δ such that $\sigma = \delta \circ \mathrm{mgu}_{\mathcal{E}(t)}$ which implies $t'' = \sigma(t) = \delta(\mathcal{S}(t))$. Thus $t' \leq \mathcal{S}(t)$. Furthermore,

$$t' = \mathcal{D}(t'') = \mathcal{D}(\sigma(t)) \sqsubseteq \sigma(\mathcal{D}(t)),$$

and thus $t' \leq \mathcal{D}(t)$. □

Lemma 11. TypeInfer *is sound. It is also complete in the sense that it will compute the unique $\preceq$-minimal solution if there is any solution at all.*

Proof. Given a preterm t, TypeInfer will check if $\mathcal{E}(t)$ is solvable.

If it is not, it will fail; this is correct, because then there can be no solution t' with $t' \leq t$ and $\mathcal{E}(t')$ solvable because otherwise $\mathcal{E}(t)$ would be solvable as well because of Lemma 10.

If on the other hand $\mathcal{E}(t)$ is solvable it will either recursively call itself with argument d where $d = \mathcal{D}(\mathcal{S}(t))$ or perform a final calculation and return the result. In the case of a recursive call, we know because of Lemma 10 that every solution t' of t is also a solution of d.

So let us look at the final calculation now. We know that $d = s$ holds where $s = \mathcal{S}(t)$. In other words, d is a fixpoint of $\mathcal{D}$ which means that

$$\diamond(\tau_1, \tau_2, \tau_3) = \diamond_?$$

holds for all invocations of $\diamond$ during the computation of $\mathcal{D}(d)$ which implies that all of τ_1, τ_2 and τ_3 are either equal to $\mathcal{U}$ or equal to a type variable. The substitution $\mathcal{U}$ will therefore make all τ_i in those invocations equal to $\mathcal{U}$ and thus the effect of applying $\mathcal{D}$ to $\mathcal{U}(d)$ is to switch all occurrences of $\diamond_?$ to $\diamond_{ZF}$. In particular, $\mathcal{E}(\mathcal{D}(\mathcal{U}(d)))$ is solvable because $\mathcal{E}(d)$ is a set of identities and

$$\mathcal{E}(\mathcal{D}(\mathcal{U}(d))) = \mathcal{U}(\mathcal{E}(d)) \cup \{\mathcal{U} \equiv \mathcal{U}\}.$$

That means that t_0 is a solution where $t_0 = \mathcal{D}(\mathcal{U}(d))$. Furthermore t_0 is minimal with respect to $\preceq$ because for any solution t' we know $t' \leq d$ and because for any term a and any preterm b such that $a \leq b$ it follows that $\mathcal{D}^{ZF}(\mathcal{U}(b)) \preceq a$ where $\mathcal{D}^{ZF}$ replaces all occurrences of $\diamond_?$ in its argument by $\diamond_{ZF}$. Because of the antisymmetry of $\preceq$, minimality implies uniqueness. □

2.7 Examples

We present three more examples of applying TypeInfer. We will use abbreviated notations for preterms in the following.

Example 4. Let t be the preterm $\forall x : \alpha.\ \exists f : \beta.\ f \diamond_? x : \gamma$. Then

$$\mathcal{S}(t) = \forall x : \alpha.\ \exists f : \beta.\ f \diamond_? x : \mathbb{P}.$$

Because of $\diamond(\beta, \alpha, \mathbb{P}) = \diamond_H$ we know

$$t' = \mathcal{D}(\mathcal{S}(t)) = \forall x : \alpha.\ \exists f : \beta.\ f \diamond_H x : \mathbb{P} \neq \mathcal{S}(t)$$

Computing TypeInfer(t') yields first $\mathcal{S}(t') = \forall x : \alpha.\ \exists f : \alpha \to \mathbb{P}.\ f \diamond_H x$ and then

$$\mathsf{TypeInfer}(t) = \mathsf{TypeInfer}(t') = \mathcal{U}(\mathcal{S}(t')) = \forall x : \mathcal{U}.\ \exists f : \mathcal{U} \to \mathbb{P}.\ f \diamond_H x.$$

Example 5. Let t be $a : \alpha \mapsto b : \beta \mapsto c : \gamma \mapsto d : \delta \mapsto a \diamond_? b \diamond_? c \diamond_? d$. Then

$$t' = \mathcal{D}(\mathcal{S}(t)) = a : \alpha \mapsto b : \beta \mapsto c : \gamma \mapsto d : \delta \mapsto a \diamond_? b \diamond_? c \diamond_? d$$
$$\mathsf{TypeInfer}(t) = \mathcal{D}(\mathcal{U}(t')) = a : \mathcal{U} \mapsto b : \mathcal{U} \mapsto c : \mathcal{U} \mapsto d : \mathcal{U} \mapsto a \diamond_{\mathsf{ZF}} b \diamond_{\mathsf{ZF}} c \diamond_{\mathsf{ZF}} d.$$

Example 6. Let us modify the previous example and infer the type of

$$a : \alpha \mapsto b : \beta \mapsto c : \gamma \mapsto d : \delta \mapsto a \diamond_? b \diamond_? c \diamond_? d \wedge d.$$

This time the algorithm needs three recursive calls and yields finally

$$a : \mathcal{U} \to \mathcal{U} \to \mathbb{P} \to \mathbb{P} \mapsto b : \mathcal{U} \mapsto c : \mathcal{U} \mapsto d : \mathbb{P} \mapsto a \diamond_{\mathsf{H}} b \diamond_{\mathsf{H}} c \diamond_{\mathsf{H}} d \wedge d$$

This example can be generalized to produce for any n an example with n occurrences of $\diamond_?$ such that TypeInfer needs n recursive calls.

3   Related Work

In HOL-ST [5], set-theoretic and higher-order function application have different syntax; in particular, higher-order function application is written fx and set-theoretic function application is denoted by $f \diamond x$. Because HOL-ST has type variables and capabilities for defining new types, the type $\mathcal{U}$ is just one type besides many others; our type inference algorithm does not yield a desirable result in such a setting. Of course, as HOL-ST is a strict superset of ZFH, one could work in it as one works in ZFH; our type inference algorithm can be directly translated to HOL-ST to support such a scenario.

Isabelle/ZF [7] also uses two different notations, fx for higher-order and $f'x$ for set-theoretic function application. Although Isabelle/ZF is embedded in polymorphic intuitionistic higher-order logic it is used in an essentially monomorphic way using an identical type system to ZFH. Isabelle has a flexible mechanism for syntax extension by adding context-free grammar rules so it should be possible to introduce syntax to write set-theoretic function application via juxtaposition as well. Type information is used in Isabelle to disambiguate between several possible parse trees. Using this built-in mechanism would lead to a situation similar to what we described in Sect. 2.3: whenever there are multiple possible typings parsing would fail. But in principle it should be possible to write a system-level Isabelle extension which implements our type inference algorithm for Isabelle/ZF.

Our operator for set-theoretic function application apply : $\mathcal{U} \to \mathcal{U} \to \mathcal{U}$ could be viewed as a coercion from $\mathcal{U}$ to $\mathcal{U} \to \mathcal{U}$. There has been previous work with regard to the general problem of extending Hindley-Milner type inference in the presence of coercions. In [12] coercions between types which only differ in their base types but not in their type constructors are considered; because $\mathcal{U}$ does not contain the type constructor $\to$ but $\mathcal{U} \to \mathcal{U}$ does, their work is not applicable to our case. In [13] more general coercions are considered but their algorithm has the property that no coercions are inserted if Hindley-Milner type inference alone already yields a valid typing; this is not what we would like in our setting as this property means that their algorithm would choose the typing $f : \alpha \rightarrow \alpha$ and $x : \alpha$ over the typing $f : \mathscr{U}$ and $x : \mathscr{U}$ in our introductory example. And then there would still be the question of how that polymorphic type should be converted into a monomorphic one.

Another way of looking at our scenario is from an overloading point of view where the generic operator $\diamond_?$ of type $\alpha \rightarrow \beta \rightarrow \gamma$ has two different instances $\diamond_{ZF} : \mathscr{U} \rightarrow \mathscr{U} \rightarrow \mathscr{U}$ and $\diamond_H : (\alpha \rightarrow \beta) \rightarrow \alpha \rightarrow \beta$. But typical algorithms which extend Hindley-Milner to take overloading into account like in [14] compute a principal type of which all other possible valid typings are instances. This is not what our algorithm does; instead we minimize a preference relation $\preceq$ which is different from the is-an-instance-of relation a principal type maximizes.

4   Conclusion

We have implemented TypeInfer as part of the implementation of ProofScript, the proof language of ProofPeer. Combining the strengths of set theory with the strengths of higher-order logic has always had a certain appeal to ITP researchers. We believe that the answer has been staring into our faces for quite some time now in the form of ZFH; all we had to do to arrive at ZFH was to take HOL-ST and take away powers which HOL practitioners take for granted but which are of little use in the context of set theory. The existence of TypeInfer which allows us to fuse the notations for higher-order function application and set-theory function application into a single one *because of the absence of those powers* supports our belief.

References

1. ProofPeer. http://www.proofpeer.net
2. Obua, S., Fleuriot, J., Scott, P., Aspinall, D.: ProofPeer: Collaborative Theorem Proving. http://arxiv.org/abs/1404.6186
3. Hales, T., et al.: A formal proof of the Kepler conjecture. http://arxiv.org/abs/1501.02155
4. Homotopy Type Theory. http://homotopytypetheory.org/
5. Agerholm, S., Gordon, M.: Experiments with ZF set theory in HOL and Isabelle. In: Schubert, E.T., Alves-Foss, J., Windley, P. (eds.) HUG 1995. LNCS, vol. 971. Springer, Heidelberg (1995)
6. Gordon, M.: Set theory, higher order logic or both? In: von Wright, J., Harrison, J., Grundy, J. (eds.) TPHOLs 1996. LNCS, vol. 1125. Springer, Heidelberg (1996)
7. Paulson, L.C.: Set theory for verification: I. from foundations to functions. J. Autom. Reasoning **11**(3), 353–389 (1993). Springer
8. Krauss, A.: Partial and nested recursive function definitions in higher-order logic. J. Autom. Reasoning **44**(4), 303–336 (2010). Springer
9. ProofPeer Root Theory. http://proofpeer.net/repository?root.thy
10. Baader, F., Nipkow, T.: Term Rewriting and All That. Cambridge University Press, Cambridge (1999)

11. Milner, R.: A theory of type polymorphism in programming. J. Comput. Syst. Sci. **17**, 348–375 (1978)

12. Traytel, D., Berghofer, S., Nipkow, T.: Extending hindley-milner type inference with coercive structural subtyping. In: Yang, H. (ed.) APLAS 2011. LNCS, vol. 7078, pp. 89–104. Springer, Heidelberg (2011)

13. Luo, Z.: Coercions in a polymorphic type system. Math. Struct. Comput. Sci. **18**(04), 729–751 (2008). Cambridge Journals

14. Odersky, M., Wadler, P., Wehr, M.: A second look at overloading. In: Proceedings of the Seventh International Conference on Functional Programming Languages and Computer Architecture. ACM (1995)

Generic Literals

Florian Rabe$^{(\boxtimes)}$

Jacobs University, Bremen, Germany
f.rabe@jacobs-university.de

Abstract. MMT is a formal framework that combines the flexibility of knowledge representation languages like OPENMATH with the formal rigor of logical frameworks like LF. It systematically abstracts from theoretical and practical aspects of individual formal languages and tries to develop as many solutions as possible generically.

In this work, we allow MMT theories to declare user-defined literals, which makes literals as user-extensible as operators, axioms, and notations. This is particularly important for framework languages, which must be able to represent any choice of literals. Theoretically, our literals are introduced by importing a model that defines the denotations of some types and function symbols. Practically, MMT is coupled with a programming language, in which these models are defined.

Our results are implemented in the MMT system. In particular, literals and computation on them are integrated with the parser and type checker.

1 Introduction and Related Work

Even though literals (e.g., for booleans, integers, or strings) are a common feature of formal systems, there appears to be no general definition of what they are. Most languages simply use a fixed set of primitive types with built-in literals, which appear explicitly in the grammar, the semantics, and the implementation. To our knowledge, there is no system that is systematically parametric in the choice of literals – while users can declare new constants, functions, axioms, and notations, etc., the set of literals is usually fixed.

A fixed set of literals is often a reasonable choice. But it has some weird effects. For example, OPENMATH [BCC+04] is meant to subsume the languages of all other formal systems. But it fixes a set of literals and thus cannot represent any system that uses different literals. Another example, the proof assistant HOL Light [Har96] relegates as much as possible to its library, including, e.g., the definition of the type of natural numbers. But support for natural number literals must be built into the base system, anticipating the library's type definition.

The same observations apply to interpreted constants, which usually go along with literals. For example, languages with natural number literals "0", "1", . . . : nat usually also provide built-in constants such as $+ : \mathsf{nat} \to \mathsf{nat} \to \mathsf{nat}$. When applied to literals, their values can be directly computed, e.g., "1" + "1" $\leadsto$ "2". (To clarify the presentation, we will enclose all literals in quotes throughout this

© Springer International Publishing Switzerland 2015
M. Kerber et al. (Eds.): CICM 2015, LNAI 9150, pp. 102–117, 2015.
DOI: 10.1007/978-3-319-20615-8_7

paper.) In higher-order languages, we can think of these as literals of function type, but the general case requires a distinction between literals and interpreted functions.

State of the Art. We speak of **primitive literals** if a language fixes sets L_A for types A and adds the rule schema

$$\frac{}{\vdash \text{``}l\text{''} : A} \text{ for } l \in L_A$$

Note that L_A may be infinite, e.g., we could put $L_{\text{nat}} = 0, 1, \ldots$ Sets of (usually finitely many) interpreted constants are supplied accordingly.

This is typical for programming languages and computer algebra systems. Often the types A include booleans, bounded integers, floating point numbers, characters, and strings. Computer algebra systems prefer unlimited precision integers and rational numbers. Mathematica [Wol12] offers many primitive literals, e.g., for images, which users input as files.

Logics, on the other hand, use primitive literals very sparingly: firstly, because the interplay between reasoning and computation (when interpreted functions are applied to non-literals) is very difficult; secondly, because unverified computation endangers logical consistency. Martin-Löf type theory [ML74] uses a primitive type for natural numbers. Nuprl [CAB+86] uses a primitive type of integers. Recently, the TPTP family of interchange logics added support for primitive literals for integer, rational, and real numbers [SSCB12].

We speak of **literals as constants** if every literal is declared as a constant. This makes literals extensible but is only feasible for very small sets of literals. The most common example are the boolean constants `true` and `false`.

We speak of **enumeration literals** if languages allow defining new finite types by listing fresh identifiers for their elements. This is supported by C or any language with inductive data types. It is less impractical than literals as constants but still impractical in general.

We speak of **derived literals** if literals appear only in the concrete syntax and are converted into other representations in the abstract syntax. For example, we can use constants 0 and *succ* for the abstract representation of derived natural number literals. Then interpreted constants are not needed at all because they can be defined on the abstract representations, e.g., via induction or rewriting.

Examples of derived literals are strings in C (converted to character arrays) or XML literals in Scala (converted to objects). They are also used in various proof assistants. HOL Light [Har96] uses derived natural number literals, which are represented as (essentially) lists of bits. Somewhat similarly, Mizar [TB85] uses 16-bit integer literals as natural numbers.

Derived literals allow reducing a lot of the technical difficulties to logical reasoning (of course, at the expense of efficiency). However, they are still not extensible in practice because users are usually not able to modify the parser and printer that perform the conversion. That makes it difficult for users to add new derived literals themselves.

Literals present a particular **challenge for knowledge representation languages** like OPENMATH because there is no good canonical choice which literals to support. For example, OPENMATH [BCC+04] somewhat arbitrarily uses primitive literals for unbounded decimal integers (OMI), decimal or hexadecimal IEEE floating point numbers (OMF), strings (OMSTR), and bytearrays (OMB), and literals as constants for booleans (CD logic1). It lacks, e.g., literals for bounded or arbitrary-base integers as well as characters (not to mention more exotic types like URIs, dates, or colors).

They are also a **challenge for logical frameworks** like LF [HHP93], which represent other languages as theories. Because these other languages may use different literals, a logical framework should not fix any literals but allow theories to declare flexibly which literals are used. The Twelf system [PS99] supports constraint domains for this purpose, which subsume primitive literals. However, only developers but not users can add new constraint domains.

Contribution. We give a **general definition** of literal and interpreted constants. We base our definition on the MMT language [RK13], which already gives extremely general definitions of theory, declaration, and expression. Our MMT literals are **extensible** and **context-specific**: MMT theories can declare new primitive literals and new interpreted constants, and every theory sees only the ones it declares (or imports). We implement our literals as a part of the MMT system [Rab13].

We hold that this is the **only feasible design for generic languages** such as OPENMATH-style representation languages or logical frameworks.

Our key idea is to extend the MMT syntax with a single constructor for literals: "l" where l is an arbitrary element of some semantic domain. Clearly, in a declarative formal language, we cannot easily use a set theoretical domain. Therefore, in practice, we use the Scala programming language, which underlies the MMT system, as the semantic domain.

Building on the correspondence between the MMT module system and Scala inheritance [KMR13], we let MMT **theories and Scala classes import each other**. Then a typical use of our system proceeds in three steps:

1. We axiomatize an MMT theory T, e.g., with a type `nat` and a constant `plus`.
2. T behaves like an abstract Scala class. We implement it in a concrete Scala class C, e.g., using the positive integers for `nat` and addition for `plus`.
3. C behaves like an MMT theory that includes T and adds a definiens to every constant in T. The respective definiens is the literal representing the Scala value assigned by C. MMT theories including C may use the literals and the interpretations of the constants into scope. For example, that would make natural number literals available, and add addition to the MMT simplifier.

Our primary motivation is to use our extensible literals sparingly: The choice of literals will usually be made at the beginning of a development, and most users will never add new literals.

However, if used more aggressively, our approach also offers a framework in which we can **combine deduction** (which guarantees correctness) **and**

computation (which is much more efficient) and investigate their interplay. Our new MMT theories can mediate between the two paradigms by combining logical theories T and computational theories C. Even without further insights into how to verify the correctness of computations, this can help make them more trustworthy: Just putting the two side by side in a common formal framework is a step up compared to some current systems.

Therefore, we do not simply design our system as a foreign function interface to MMT (even though it can be seen as such). Instead, we first develop the semantics of combined theories for arbitrary models C, and then specialize to the case where C is an effectively given model, i.e., a set of computations. This can help ensure the correctness of the computations because it ties them closer to their expected logical properties. Moreover, this situation can serve as a good starting point for a rigorous treatment that would formally verify the implementations.

Finally, because MMT already includes a generic type-checker, our approach allows integrating computation not only with deduction but also with type-checking. Indeed, the combination of dependent types with literals allows defining languages in MMT where type-checking is carried out up to computational equality.

Overview. We recap the existing MMT language in Sect. 2. Then we extend it with a general notion of literals in Sect. 3 and specialize to Scala-based computational literals in Sect. 4. We present the practical aspects of our implementation in Sect. 5. In Sect. 6, we discuss further related work and conclude.

2 Preliminaries: The MMT Language

The MMT language and module system were originally introduced in [RK13]. The recap here follows the formulation of [Rab14] and focuses on a very small, non-modular fragment.

Grammar. The grammar for basic MMT theories is given in Fig. 1. A **theory** Σ is a list of **constant** declarations. These are of the form $c[: A][= t][\#N]$ where c is an identifier, A is its **type**, t its **definiens**, and N its notation, all of which are optional. A **context** Γ is very similar to a theory and declares typed variables.

Theory	Σ	::=	$(c[: t][= t][\#N])^*$
Context	Γ	::=	$(x : t)^*$
Terms	t	::=	$c \mid x \mid c(\Gamma; t^*)$
Notation	N	::=	omitted

Fig. 1. MMT grammar

Constant declarations subsume the most common declarations of formal systems including built-in symbols, universes, function/predicate/type symbols, and (by using propositions as types) axioms, theorems, and inference rules.

Type and definiens are terms, and **terms** are formed from constants c, variables x, and complex terms $c(\Gamma; t_1, \ldots, t_n)$. Complex terms bind the variables of Γ in the arguments t_i. This includes OPENMATH-style binding and application:

$\Gamma = \cdot$ yields application (OMA) of c to the t_i, and $\Gamma \neq \cdot$ and $n = 1$ yields binding (OMBIND) with binder c. Like OPENMATH objects, MMT terms subsume most expressions of formal systems such as terms, types, formulas, and proofs.

Like OPENMATH, MMT's syntax is **generic** in the sense that there are no predefined constants in the grammar. Therefore, to represent any language L in MMT, we first have to declare a special theory (which we usually also call L for convenience) with one untyped constant for each built-in symbol of L. Then we define L-**theories** as the MMT theories of the form L, Σ such that all Σ-constants have a type.

We usually think of L as a logical framework that is used to define other languages as L-theories. As running examples, we will use the dependently typed λ-calculus LF [HHP93] as a logical framework and the natural numbers as an LF-theory:

Example 1 (An MMT Theory for LF). LF is the theory shown on the left of Fig. 2. For example, the abstract syntax for a λ-abstraction is $\mathtt{lambda}(x : A; t)$ where $x : A$ the single bound variable, and t the single argument. The notation declares both the arity and binding-arity and the concrete syntax. So the concrete syntax of $\mathtt{lambda}(x : A; t)$ is $[x : A]t$. Similarly, in an application $\mathtt{apply}(\cdot; f, t)$, no variables are bound, and f and t are the arguments. The concrete syntax is $f\ t$.

The theory Nat on the right of Fig. 2 extends LF with some example declarations for the natural numbers. We can also represent axioms in the usual LF-style, but do not do that here because we do not need them later on.

theory LF		theory Nat	
type		prop	: type
Pi	$\#\ \{\, x_1 : t_1 \,\}\, t_2$	nat	: type
lambda	$\#\ [\, x_1 : t_1 \,]\, t_2$	succ	: nat $\to$ nat
apply	$\#\ t_1\ t_2$	plus	: nat $\to$ nat $\to$ nat $\#\ t_1 + t_2$
arrow	$\#\ t_1 \to t_2$	equal	: nat $\to$ nat $\to$ prop $\#\ t_1 \doteq t_2$

Fig. 2. LF in MMT (left) and natural numbers in LF (right)

Inference System. The main **judgments** of MMT are given in Fig. 3. There is no need to introduce the type system in detail here, and we refer to [Rab14] instead. For our purposes, it is sufficient to know that the type system is generic as well: MMT itself only provides the structural rules for declarations, congruence, α-equality, and constants.

$\Gamma \vdash_\Sigma \_ : A$	A may be a type
$\Gamma \vdash_\Sigma t : A$	t has type A
$\Gamma \vdash_\Sigma t \equiv t'$	t and t' are equal

Fig. 3. Judgments of MMT

All other rules are provided separately when representing a logical framework:

Example 2 (The Logical Framework LF). For LF, we add the usual rules for λ-abstraction including

$$\frac{\Gamma \vdash_\Sigma A : \mathsf{type} \quad \Gamma, x : A \vdash_\Sigma B : \mathsf{type}}{\Gamma \vdash_\Sigma \{x : A\}B : \mathsf{type}}$$

$$\frac{\Gamma, x : A \vdash_\Sigma t : B}{\Gamma \vdash_\Sigma [x : A]t : \{x : A\}B} \qquad \frac{\Gamma \vdash_\Sigma f : \{x : A\}B \quad \Gamma \vdash_\Sigma t : A}{\Gamma \vdash_\Sigma f\,t : B[x/t]}$$

where $B[x/t]$ denotes substitution of x with t in B. We do not need any rules for the LF-theory Nat because it inherits all rules from LF.

Once a logical framework L is represented in this way, we never have to add rules again – we represent further languages declaratively as L-theories.

Finally, consider an L-theory $\Sigma = \ldots, c_i : A_i[= t_i], \ldots$ and let $\Sigma^i = \ldots, c_{i-1} : A_{i-1}[= t_{i-1}]$. Σ is **valid** if $\vdash_{L,\Sigma^i} \_ : A_i$ [and $\vdash_{L,\Sigma^i} t_i : A_i$] for all i.

Implementation. The MMT system [Rab13] implements the above concepts. This includes generic implementations of parsing (according to the notations) and type reconstruction (according to the supplied rules). It adds a module system for theories, of which we will make modest use in some examples.

3   Literals as Semantic Values

We first introduce some general semantic notions for MMT before we define literals. The main intuition is that models are collections of values and functions, which we want to reflect into the syntax.

3.1   Models of MMT Theories

A **semantic domain** is a triple $(U, \hat{:}, \hat{=})$ where U is any collection of objects, and $\hat{:}$ and $\hat{=}$ are binary relations on U such that $\hat{=}$ is an equivalence and congruent with respect to $\hat{:}$.

Example 3 (Set Theory). To obtain a semantic domain for set theory, we put U to be the collection of all classes (in the sense of set theory). Then $\hat{:}$ is the $\in$ relation, and $\hat{=}$ is extensional equality of classes.

Let us now fix a logical framework L represented in MMT and a semantic domain $(U, \hat{:}, \hat{=})$. Given an L-theory Σ, a Σ-**model** M is a function that maps every Σ-constant c to an element c^M of U. **Assignments** α interpret the free variables of a context accordingly.

Example 4 (Standard Natural Numbers). Given the LF-theory for the natural numbers from Example 1, we define the standard model StdNat in set theory in the obvious way: $\text{nat}^{\text{StdNat}} = \mathbb{N}$, $\text{prop}^{\text{StdNat}} = \{0, 1\}$, and $\text{succ}^{\text{StdNat}}$, $\text{plus}^{\text{StdNat}}$, $\text{equal}^{\text{StdNat}}$ are defined as usual.

A **semantics** for an MMT theory L consists of a semantic domain and a set of interpretation rules. The interpretation rules must extend every Σ-model M to a (usually partial) interpretation function $[\![-]\!]^M$ that maps closed Σ-terms to elements of U.

We do not spell out the general shape of these interpretation rules, and instead only give a concrete example for the standard set theoretical semantics of LF. In fact, if we use LF as a logical framework in which other languages are specified, we only need to define the semantics of LF anyway:

Example 5 (Set Theoretical Semantics of LF). We use the domain from Example 3. Given a model M of an LF-theory Σ and an assignment α for the context Γ, we use the following interpretation rules to define $[\![t]\!]_\alpha^M \in U$ for terms t in context Γ:

$$[\![type]\!]_\alpha^M = \mathcal{SET} \qquad [\![A \to B]\!]_\alpha^M = ([\![B]\!]_\alpha^M)^{[\![A]\!]_\alpha^M}$$

$$[\![[x:A]t]\!]_\alpha^M = [\![A]\!]_\alpha^M \ni u \mapsto [\![t]\!]_{\alpha,x\mapsto u}^M \qquad [\![f\ t]\!]_\alpha^M = [\![f]\!]_\alpha^M([\![t]\!]_\alpha^M)$$

We omit the analogous rule for terms constructed using Pi.

Finally, a model M is **sound** if it preserves typing and equality, i.e.,

$$\text{if } \Gamma \vdash_\Sigma t : A \quad \text{then} \quad [\![t]\!]_\alpha^M \mathbin{\hat{:}} [\![A]\!]_\alpha^M$$

$$\text{if } \Gamma \vdash_\Sigma t \equiv t' \quad \text{then} \quad [\![t]\!]_\alpha^M \mathbin{\hat{=}} [\![t']\!]_\alpha^M$$

In particular, soundness requires models to interpret Σ-constants $c : A[= t]$ as values $c^M \mathbin{\hat{:}} [\![A]\!]^M$ [such that $c^M \mathbin{\hat{=}} [\![t]\!]^M$] (†). We call the semantics as a whole **sound** if the inverse holds too, i.e., if every model satisfying (†) is sound. This is the case for the semantics of Example 5.

3.2   Internalizing Models

We fix a logical framework L and a sound semantics for L that uses a semantic domain $(U, \mathbin{\hat{:}}, \mathbin{\hat{=}})$. We will now internalize the semantic domain, the semantics, and individual models into MMT. The main idea is to treat every element of $l \in U$ as a literal that may occur as a term "l". Of course, this may yield an uncountable set of terms, or even a proper class if we use the semantic domain from Example 3. However, this level of abstraction is well-suited to define the general concepts even if practical systems must be much more restricted.

To **internalize the semantic domain**, we extend the MMT grammar from Fig. 1 with **one new production** "l":

$$t ::= c \mid x \mid c(\Gamma; t^*) \quad \mid \text{``}l\text{'' for } l \in U$$

Moreover, we add **two new inference rules** to MMT:

$$\frac{}{\vdash_\Sigma \text{``}l\text{''} : \text{``}l'\text{''}} \text{ for } l \mathbin{\hat{:}} l' \qquad \frac{}{\vdash_\Sigma \text{``}l\text{''} \equiv \text{``}l'\text{''}} \text{ for } l \mathbin{\hat{=}} l' \quad (*)$$

Because we extended the grammar, we also need to extend the semantics with a **new interpretation rule** – models interpret every literal as itself:

$$[\![\text{``}l\text{''}]\!]^M = l \qquad (**)$$

We call t a **value** if $\vdash_\Sigma t \equiv \text{``}l\text{''}$ for some l. Then we have the following inverse of $(**)$:

Theorem 1. *If t is a value and M is a sound Σ-model, then $\vdash_\Sigma \text{``}[\![t]\!]^M\text{''} \equiv t$.*

Proof (Outline). Assume $\vdash_\Sigma t \equiv$ "l". The key step is to use soundness, $(**)$, and $(*)$ to obtain $\vdash_\Sigma$ "l" $\equiv$ "$[\![t]\!]$".

At this point, there are no typing rules that relate literals to non-literal terms. Thus, our literals are logically inconsequential. That changes with the next definition: We **internalize the semantics** by adding its interpretation rules as MMT typing rules. Again, it is sufficient for our purposes to give only an example for LF:

Example 6. We use the following typing rules to internalize the interpretation rules from Example 5:

$$\frac{}{\vdash_\Sigma \text{type} \equiv \text{"}\mathcal{SET}\text{"}} \qquad \frac{\vdash_\Sigma a \equiv \text{"}A\text{"} \qquad \vdash_\Sigma b \equiv \text{"}B\text{"}}{\vdash_\Sigma a \to b \equiv \text{"}B^A\text{"}}$$

$$\frac{\vdash_\Sigma f \equiv \text{"}F\text{"} \qquad \vdash_\Sigma t \equiv \text{"}T\text{"}}{\vdash_\Sigma f\, t \equiv \text{"}F(T)\text{"}} \qquad \frac{\vdash_\Sigma a \equiv \text{"}A\text{"} \qquad \vdash_\Sigma t[\text{"}u\text{"}] \equiv \text{"}T_u\text{"} \text{ for all } u \in A}{\vdash_\Sigma [x : A]t \equiv \text{"}A \ni u \mapsto T_u\text{"}}$$

Finally, we can **internalize models**. For an L-theory Σ and a Σ-model M, the MMT theory "M" contains for every Σ-constant $c : A[= t]$ the constant $c : A = $ "c^M". Intuitively, in "M", every constant is equal to its interpretation, and via the internalized semantics, so is every interpretable term. The following theorem shows that we can indeed treat models as special theories:

Theorem 2. *Consider a valid L-theory $\Sigma = c_1 : A_1, \ldots, c_n : A_n$. If all A_i are values, then a Σ-model M is sound iff "M" is a valid L-theory.*

Proof. Because we assume that Σ is valid, the validity of "M" is equivalent to $\vdash_{\text{"}M\text{"}}$ "c^M" : A for every Σ-constant $c : A$.

Assume M is sound. Then $c^M \hat{:} [\![A]\!]^M$ and thus $\vdash_{\text{"}M\text{"}}$ "c^M" : "$[\![A]\!]^M$". Then validity follows using Theorem 1.

Conversely, assume "M" is valid. We show that M is sound by induction on the declarations in Σ. For $c : A = $ "c^M" in "M", the induction hypothesis implies $[\![\text{"}c^M\text{"}]\!]^M \hat{:} [\![A]\!]^M$. Then $(**)$ yields $c^M \hat{:} [\![A]\!]^M$.

3.3   Literals Through Internalized Models

We can now show how MMT theories can declare specific sets of literals. First we give an L-theory Σ, which declares all types for which we want to supply literals and all function symbols for which we want to fix an interpretation. Then we give a Σ-model M, which supplies the literals and interpretations. Finally any theory of the form "M", Θ includes these literals and interpretations. In particular, every theory may use different literals.

Example 7. We work with the internalized semantics of LF from Example 6. Let Vec be the LF-theory of Fig. 4. It uses the MMT module system to include the theory "StdNat" from Example 4 (and thus also includes Nat from Example 1). The left part introduces vec n as the type of vectors of length n over a fixed type a (with a dummy value $c : a$). The right part introduces an operation for

```
theory Vec
include "StdNat"
c    : a
a    : type
vec  : nat → type
nil  : vec "0"
cons : {n : nat} a → vec n → vec (succ n)
```

$$\text{concat} : \{m : \text{nat}\}\{n : \text{nat}\}$$
$$\text{vec}\, m \to \text{vec}\, n \to \text{vec}\, (m + n)$$
$$\textit{test0} \quad : \text{vec}\, "2" = \text{cons}\, "1"\, c\, (\text{cons}\, "0"\, c\, \text{nil})$$
$$\textit{test1} \quad : \text{vec}\, "4" = \text{concat}\, "2"\, "2"\, \textit{test0}\, \textit{test0}$$

Fig. 4. Vectors in LF with natural number literals

concatenation (whose axioms we omit) and declares some example constants that use natural number literals.

Note that the example declarations are well-typed in LF only because the internalization of the semantics implies that $\vdash_{\text{Vec}} \text{vec}\, "2" \equiv \text{vec}\, (\text{succ}(\text{succ}("0")))$ and $\vdash_{\text{Vec}} \text{vec}\, "4" \equiv \text{vec}\, ("2" + "2")$. Such an integration of interpreted constants into a dependent type theory is non-trivial even for a fixed set of literals.

It remains to define theories like "StdNat" in practice. Obviously, this is only possible if we can internalize the semantics in a computationally effective way. We look at that in the next section.

4   Literals as Computational Values

4.1   Models as Implementations

While set theory is interesting theoretically, in practice we need to use a programming language or computer algebra system to define models. To do that, we can reuse all concepts from Sect. 3 – we only have to use a different semantic domain. As a concrete example, we will use the simply-typed functional programming language Scala [OSV07], but we could use any other computational language analogously:

Example 8 (Scala as a Semantic Domain). Scala permits top level declarations of new types (classes) and values (objects). Therefore, the set of Scala expressions depends on the top level declarations that are in scope. So let us fix a set G of top level declarations. We further assume all classes are immutable (which is the only case we need anyway).

Then we obtain a semantic domain $(U, \hat{:}, \hat{=})$ as follows. U consists of the symbol **type**, all Scala types over G, and all typed Scala expressions over G. $\hat{:}$ relates all types to **type** and all expressions to their type. $\hat{=}$ relates **type** to itself, two types if they expand to the same normalized type, and two expressions if they evaluate to the same value.

Scala is not dependently typed. Therefore, we use a straightforward type erasure translation to interpret LF in Scala – it interprets LF functions as Scala functions but removes all arguments from dependent types:

Example 9 (Semantics of LF in Scala). Using the semantic domain from Example 8, we define the following interpretation rules:

$$[\![type]\!]^M = \textbf{type} \quad [\![A \to B]\!]^M = [\![A]\!]^M \Rightarrow [\![B]\!]^M \quad [\![\{x : A\}B]\!]^M = [\![A]\!]^M \Rightarrow [\![B]\!]^M$$

$$\text{for terms :} \quad [\![[x : A]t]\!]^M = (x : [\![A]\!]^M) \Rightarrow [\![t]\!]^{M,x \mapsto x} \quad [\![f\ t]\!]^M = [\![f]\!]^M([\![t]\!]^M)$$

$$\text{for types :} \quad [\![[x : A]t]\!]^M = [\![t]\!]^M \quad [\![f\ t]\!]^M = [\![f]\!]^M$$

where $\Rightarrow$ is Scala's syntax for both function types and λ-abstraction.

Of course, the type erasure translation loses some precision if an LF-theory makes use of dependent types. More precisely, this semantics is not sound. This is harmless, however, because most interpreted functions that we want to implement in practice are simply-typed, usually even first-order. And the semantics is sound if we restrict attention to simply-typed models.

Giving models relative to this semantics means to implement MMT theories in Scala. Moreover, a model of a theory T has the same structure as a Scala object that implement an abstract class T. Therefore, we can directly use Scala syntax to write the models:

Example 10 (Integers in Scala). We give the abstract Scala class obtained by applying the interpretation rules from Example 9 to the theory `Nat` from Example 1, and a model of it:

```scala
abstract class Nat {
  type nat
  type prop
  def succ(x:nat): nat
  def plus(x:nat,y:nat): nat
  def equal(x:nat,y:nat): Boolean
}
object StdInt extends Nat {
  type nat = BigInt
  type prop = Boolean
  def succ(x:nat) = x+1
  def plus(x:nat,y:nat) = x+y
  def equal(x:nat,y:nat) = x == y
}
```

Here, we modeled the type `nat` as Scala's unbounded integers `BigInt`. We get back to that in Example 12.

Example 11 (OpenMath Literals). It is now straightforward to recover the 4 types of literals used by OPENMATH as special cases. `StdInt` already implements OMI, along with some interpreted constants. Floating point numbers, strings, and byte arrays are equally simple.

4.2   Types as Partial Equivalence Relations

We can generalize the semantic domain from Example 8 substantially if we use partial equivalence relations (PERs) instead of types. A PER consists of a Scala type A and a symmetric and transitive binary relation r on A.

It is well-known that a PER r on A defines a quotient of a subtype of A. To see that, let A^s be the subtype of A containing all elements that are in relation to any other element. Then the restriction of r to A^s is reflexive and thus an equivalence. The postulated quotient arises as the quotient of A^s by r.

This results in a more expressive semantic domain in which we can take subtypes and quotients to build the exact type we need for our literals:

Example 12 (Scala PER Domain). We define PERs using the Scala class:

```
abstract class PER {
  type univ
  def valid(u: univ): Boolean
  def normal(u: univ): univ
}
```

`valid` defines the subtype, and $\texttt{normal}(x) = \texttt{normal}(y)$ defines the relation r.

Then we obtain a semantic domain $(U, \hat{:}, \hat{=})$ as follows. U contains PER, all expressions p : PER, and all pairs (p, u) for p : PER and $u : p.\texttt{univ}$.

Then we put $p \mathbin{\hat{:}} \texttt{PER}$ if p : PER; and $p \mathbin{\hat{=}} p'$ if p and p' evaluate to the same value.

And we put $(p, u) \mathbin{\hat{:}} p$ if $p.\texttt{valid}(u)$; and $(p, u) \mathbin{\hat{=}} (p, v)$ if $p.\texttt{normal}(u) = p.\texttt{normal}(v)$.

We can now refine Example 10 by interpreting the type **nat** as a subtype of BigInt:

Example 13 (Natural and Rational Numbers). We use the semantic domain from Example 12 to define literals for natural numbers:

```
object StdNat extends PER {
  type univ = BigInt
  def valid(u: univ) = u >= 0
  def normal(u: univ) = u
}
```

Similarly, we can define rational numbers e/d using pairs (e, d) of integers:

```
object StdRat extends PER {
  type univ = (BigInt, BigInt)
  def valid(u: univ) = u._2 != 0
  def normal(u: univ) = {
    val (e, d) = u
    val g = (e gcd d) * d.signum
    return (e/g, d/g)
  }
}
```

Here validity ensures that the denominator is non-zero, and normalization cancels by the greatest common divider.

It is straightforward to adapt the semantics from Example 9 to this new semantic domain. The only subtlety are function types: Given two PERs p on A and q on B, it is easy to define the needed PER $p \Rightarrow q$ on $A \Rightarrow B$ in theory. However, validity and normalization in $p \Rightarrow q$ are not in general computable anymore, and the soundness of models that use functions on PERs is undecidable. This is acceptable because users anyway have to verify the correctness of their implementations manually.

5   Implementing Literals in the MMT System

5.1   Internalizing a Computational Semantics

Now we internalize the semantics of LF in Scala using the semantic domain of Example 12. Our reason for using Scala is that it underlies our implementation of MMT. Therefore, we can extend it with Scala-based literals seamlessly. We make the following changes to the MMT implementation:

(1) We add a feature to the MMT build tool: It exports all LF-theories T as abstract Scala classes T^*. Now the Scala-models of T are the Scala objects M that implement T^* (as in Example 10).

(2) Conversely, we allow include declarations that include Scala objects M (as in Example 7). If an MMT theory includes "M", MMT locates the Scala object M and dynamically adds its definitions to the type system.

(3) We implement the term constructor "l" for the special case where l is an element of the Scala PER domain from Example 12. In particular, l can be of the form (p, u).

At this point, we do not add concrete syntax for constructing Scala expressions u to the MMT grammar (but see Sect. 5.2). Therefore, users can reference these new terms only indirectly: If c is a constant in T and we have included a model "M" of T, then we can use c to refer to the Scala value "c^M".)

This restriction has two desirable effects: (i) Scala expressions can appear in MMT terms only if they have been explicitly imported. (ii) It remains transparent how a model M implements the theory T.

(4) We internalize the Scala semantics by adding the following rules to the MMT type checker:

$$\frac{p = c^M \text{ for some "}M\text{" imported into } T \qquad p.\mathtt{valid}(u) = true}{\vdash_T \text{ "}(p, u)\text{" : } c}$$

$$\frac{p.\mathtt{normal}(u) = p.\mathtt{normal}(u')}{\vdash_T \text{ "}(p, u)\text{" } \equiv \text{ "}(p, u')\text{"}} \qquad \frac{l = c^M(l_1, \ldots, l_n) \qquad \text{"}M\text{" imported into } \Sigma}{\vdash_\Sigma c \text{ "}l_1\text{" } \ldots \text{ "}l_n\text{" } \equiv \text{ "}l\text{"}}$$

This fully integrates literals and computation with MMT's type reconstruction. (If computations in M do not terminate, then type reconstruction times out.)

Notably, the models M can be written and included by users at run time just like normal MMT theories. In particular, the generated classes T^* hide details specific to the MMT code base, and users do not have to rebuild MMT after implementing M.

5.2 Lexing Rules

The changes of Sect. 5.1 only allow supplying interpreted function constants. They do not modify the parser, which is a major hurdle towards extensibility. However, MMT systematically uses rule-based lexing and parsing algorithms, whose rules are provided by the context. That makes it possible (and easy) to couple every type of literals with an appropriate lexing rule.

In our implementation, the Scala class PER actually has one additional field, which provides an optional lexing rule. Such a rule is a function that takes an input stream and returns either nothing or a literal that occurs at the beginning of the stream. We also provide some parametric lexing rules that can be instantiated to quickly create lexing rules for the most important cases. In particular, these include quoted and digit-based literals.

```
theory Vec′
include "StdNat"
c     : a
a     : type
vec   : nat → type
nil   : vec 0
cons  : {n : nat} a → vec n → vec (succ n)
```

$$\text{concat} : \{m : \text{nat}\}\{n : \text{nat}\}$$
$$\text{vec } m \to \text{vec } n \to \text{vec } (m + n)$$
$$\textit{test0} \quad : \text{vec } 2 = \text{cons } c \, (\text{cons } c \, \text{nil})$$
$$\textit{test1} \quad : \text{vec } 4 = \text{concat } \textit{test0} \, \textit{test0}$$

Fig. 5. Vectors in LF using Scala-based natural number literals

Example 14. Figure 5 shows a variant of Example 7, which used set theory to define the natural number literals. Now we use the model from Example 13 extended with an appropriate lexing rule for digit-based literals.

Figure 5 shows the concrete syntax that can be processed by MMT. In particular the interpreted functions of "StdNat" are integrated with the dependent type system. Note that we can also omit some of the arguments to cons and concat because they can be inferred by MMT.

5.3 Inversion Rules

A central difficulty of combining deduction and computation lies in terms that use both the usual function symbols and variables on the one hand and interpreted function symbols and literals on the other hand.

Example 15 (Unification). We extend Example 14 with the following declarations for the head of a non-empty vector:

$$\text{head} : \{n : \text{nat}\} \text{vec} \, (\text{succ } n) \to a \qquad \textit{test2} : a = \textit{head} \, \textit{test0}$$

Type-checking the declaration *test2* leads to the unification problem vec $(\text{succ}\,X) = \text{vec}\,2$ and thus $\text{succ}\,X = 2$, where X is a meta-variable representing the omitted first argument of head.

Without further help, the type checker is unable to solve this problem. In fact, because MMT knows nothing about how $\text{succ}^{\text{StdNat}}$ is implemented, it cannot even tell if the problem is solvable at all.

This is a general problem, for which we do not claim a complete solution. It is also related to the more general unification problems addressed by canonical structures in Coq and unification hints in Matita [ARCT09].

However, the rule-based and highly extensible type reconstruction algorithm of MMT provides a good setting for investigating possible solutions. As a first step, we allow Scala models to couple an interpretation c^M with a (possibly partial) implementation of the inverse function $c^{M^{-1}}$. If provided, MMT adds the following rule to the type checker:

$$\frac{(l_1,\ldots,l_n) = c^{M^{-1}}(l) \qquad \vdash_\Sigma t_i \equiv \text{``}l_i\text{''}}{\vdash_\Sigma c\,t_1\,\ldots\,t_n \equiv \text{``}l\text{''}}$$

This is already sufficient to solve many special cases in practice. For example, if we extend Example 15 with an implementation of $\text{succ}^{\text{StdNat}^{-1}}$ as the predecessor function, we can solve the above meta-variable as $X = \text{succ}^{\text{StdNat}^{-1}}(2) = 1$.

6   Conclusion and Further Related Work

Based on MMT, we have developed the syntax, semantics, and implementation of a formal language for mathematical content that offers extensible literals. Thus, no literals have to be built-in, and individual languages defined in MMT can fine-tune the set of literals freely. Our implementation uses partial equivalence relations on Scala types, which is expressive enough to cover all types of literals we are aware of.

Moreover, users can add interpreted functions that provide computation on literals. This computation is integrated into the equational theory of the MMT type system, including the use of computation in dependent types. Literals, interpreted functions, and the associated lexing and computation rules are subject to the MMT module system in the same way as constants, axioms, and notations. In particular, each MMT theory sees only the literals it declares or imports.

The applications of our work may go beyond our present focus on literals. Our design is a candidate for combining logical reasoning with efficient computation in the style of computer algebra systems. This is particularly interesting if we can generalize the concepts to allow structured literals (literals that may contain other terms as subterms). This would allow supplying literals for complex types such as polynomials over an arbitrary ring. Structured literals for functions would correspond to normalization by evaluation [BS91].

Related Work. The theoretical aspect of our work shares a basic idea with biform theories [FvM03]. In a theory "M", Θ, we can think of Θ as the axiomatic/intensional and of "M" as the algorithmic/extensional form of a theory. This corresponds to the distinction made in biform theories. More specifically, our Σ-models M mapping constants c to interpretations c^M can be seen as a set of transformers (c, c^M) in the sense of [Far07], in which case the axioms of Σ correspond to the meaning formulas of [Far07]. Thus, our work can be seen as generalizing biform theories to the MMT-level and providing an implementation and a module system for them.

In the context of rewrite systems, a similar idea was realized in [KN13]. There, sorted first-order rewrite theories consist of an interpreted and a free part, and computation on the interpreted part is relegated to an arbitrary model.

Our literals are also intriguingly similar to quotation in the sense of [Far13]. If we use the MMT language itself as the semantic domain, we obtain literals "t" for terms t, which can be seen as quoted terms. Structured literals would correspond to quasi-quotation.

References

[ARCT09] Asperti, A., Ricciotti, W., Sacerdoti Coen, C., Tassi, E.: Hints in unification. In: Berghofer, S., Nipkow, T., Urban, C., Wenzel, M. (eds.) TPHOLs 2009. LNCS, vol. 5674, pp. 84–98. Springer, Heidelberg (2009)

[BCC+04] Buswell, S., Caprotti, O., Carlisle, D., Dewar, M., Gaetano, M., Kohlhase, M.: The Open Math Standard, Version 2.0. Technical report, The Open Math Society (2004). http://www.openmath.org/standard/om20

[BS91] Berger, U., Schwichtenberg, H.: An inverse of the evaluation functional for typed λ-Calculus. In: Kahn, G. (ed.) Logic in Computer Science, pp. 203–211. IEEE Computer Society Press, Los Alamitos (1991)

[CAB+86] Constable, R., Allen, S., Bromley, H., Cleaveland, W., Cremer, J., Harper, R., Howe, D., Knoblock, T., Mendler, N., Panangaden, P., Sasaki, J., Smith, S.: Implementing Mathematics with the Nuprl Development System. Prentice-Hall, Upper Saddle River (1986)

[Far07] Farmer, W.M.: Biform theories in chiron. In: Kauers, M., Kerber, M., Miner, R., Windsteiger, W. (eds.) MKM/CALCULEMUS 2007. LNCS (LNAI), vol. 4573, pp. 66–79. Springer, Heidelberg (2007)

[Far13] Farmer, W.M.: The formalization of syntax-based mathematical algorithms using quotation and evaluation. In: Carette, J., Aspinall, D., Lange, C., Sojka, P., Windsteiger, W. (eds.) CICM 2013. LNCS, vol. 7961, pp. 35–50. Springer, Heidelberg (2013)

[FvM03] Farmer, W., von Mohrenschildt, M.: An overview of a formal framework for managing mathematics. Ann. Math. Artif. Intell. **38**(1–3), 165–191 (2003)

[Har96] Harrison, J.: HOL light: a tutorial introduction. In: Srivas, M., Camilleri, A. (eds.) FMCAD 1996. LNCS, vol. 1166, pp. 265–269. Springer, Heidelberg (1996)

[HHP93] Harper, R., Honsell, F., Plotkin, G.: A framework for defining logics. J. Assoc. Comput. Mach. **40**(1), 143–184 (1993)

[KMR13] Kohlhase, M., Mance, F., Rabe, F.: A universal machine for biform theory graphs. In: Carette, J., Aspinall, D., Lange, C., Sojka, P., Windsteiger, W. (eds.) CICM 2013. LNCS, vol. 7961, pp. 82–97. Springer, Heidelberg (2013)

[KN13] Kop, C., Nishida, N.: Term rewriting with logical constraints. In: Fontaine, P., Ringeissen, C., Schmidt, R.A. (eds.) FroCoS 2013. LNCS, vol. 8152, pp. 343–358. Springer, Heidelberg (2013)

[ML74] Martin-Löf, P.: An intuitionistic theory of types: predicative part. In: Proceedings of the 1973 Logic Colloquium, pp. 73–118. North-Holland (1974)

[OSV07] Odersky, M., Spoon, L., Venners, B.: Programming in Scala. Artima (2007)

[PS99] Pfenning, F., Schürmann, C.: System description: twelf - a meta-logical framework for deductive systems. In: Ganzinger, H. (ed.) CADE 1999. LNCS (LNAI), vol. 1632, pp. 202–206. Springer, Heidelberg (1999)

[Rab13] Rabe, F.: The MMT API: a generic MKM system. In: Carette, J., Aspinall, D., Lange, C., Sojka, P., Windsteiger, W. (eds.) CICM 2013. LNCS, vol. 7961, pp. 339–343. Springer, Heidelberg (2013)

[Rab14] Rabe, F.: How to identify, translate, and combine logics? J. Logic Comput. (2014). doi:10.1093/logcom/exu079

[RK13] Rabe, F., Kohlhase, M.: A scalable module system. Inf. Comput. **230**(1), 1–54 (2013)

[SSCB12] Sutcliffe, G., Schulz, S., Claessen, K., Baumgartner, P.: The TPTP typed first-order form with arithmetic. In: Bjørner, N., Voronkov, A. (eds.) LPAR-18 2012. LNCS, vol. 7180, pp. 406–419. Springer, Heidelberg (2012)

[TB85] Trybulec, A., Blair, H.: Computer assisted reasoning with MIZAR. In: Joshi, A. (ed.) Proceedings of the 9th International Joint Conference on Artificial Intelligence, pp. 26–28. Morgan Kaufmann, Los Angeles (1985)

[Wol12] Wolfram. Mathematica (2012)

Ranking/Unranking of Lambda Terms
with Compressed de Bruijn Indices

Paul Tarau[(✉)]

Department of Computer Science and Engineering,
University of North Texas, Denton, TX, USA
tarau@cse.unt.edu

Abstract. We introduce a compressed de Bruijn representation of
lambda terms and define its bijections to standard representations. Our
compressed terms facilitate derivation of size-proportionate ranking and
unranking algorithms of lambda terms and their inferred simple types.
We specify our algorithms as a literate Prolog program.

Keywords: Lambda calculus · de Bruijn indices · Lambda term com-
pression · Combinatorics of lambda terms · Ranking and unranking of
lambda terms · Bijective Gödel numberings of lambda terms

1 Introduction

Lambda terms [1] provide a foundation to modern functional languages, type
theory and proof assistants and have been lately incorporated into mainstream
programming languages including Java 8, C# and Apple's newly designed programming language Swift. Ranking and unranking of lambda terms (i.e., their
bijective mapping to unique natural number codes) has practical applications to
testing compilers that rely on lambda calculus as an intermediate language, as
well as in generation of random tests for user-level programs and data types. At
the same time, several instances of lambda calculus are of significant theoretical
interest given their correspondence with logic and proofs.

Of particular interest are lambda term representations that are canonical
up to alpha-conversion (variable renamings) among which are de Bruijn's indices
[2], representing each bound variable as the number of binders to traverse to the
lambda abstraction binding this variable.

A joke about the de Bruijn indices representation of lambda terms is that it can
be used to tell apart Cylons from humans [3]. Arguably, the compressed de Bruijn
representation that we introduce in this paper is taking their fictional use one
step further. To alleviate the legitimate fears of our (most likely, for now, human)
reader, these representations will be bijectively mapped to conventional ones.

A merit of our compressed representation is to simplify the underlying combinatorial structure of lambda terms, by exploiting their connection to the Catalan family of combinatorial objects [4]. This leads to algorithms that focus
on their (bijective) natural number encodings - known to combinatorialists as

© Springer International Publishing Switzerland 2015
M. Kerber et al. (Eds.): CICM 2015, LNAI 9150, pp. 118–133, 2015.
DOI: 10.1007/978-3-319-20615-8_8

ranking/unranking functions [5] and to logicians as *Gödel numberings* [6]. Among the most obvious practical applications, such encodings can be used to generate random terms for testing tools like compilers or source-to-source program transformers. At the same time, as our encodings are "size-proportionate", they provide a compact serialization mechanism for lambda terms. To derive a bijection to ℕ that is size-proportionate we will first extract a "Catalan skeleton" abstracting away the recursive structure of the compressed de Bruijn term, then implement a bijection from it to ℕ. The "content" fleshing out the term, represented as a list of natural numbers, will have its own bijection to ℕ by using a generalized Cantor tupling/untupling function, that will also help pairing/unpairing the code of the skeleton and the code of the content of the term.

The paper is organized as follows. Section 2 introduces the compressed de Bruijn terms and bijective transformations from them to standard lambda terms. Section 3 describes mappings from lambda terms to Catalan families of combinatorial objects, with focus on binary trees representing their inferred types and their applicative skeletons. These mappings lead in Sect. 4 to size-proportionate ranking and unranking algorithms for lambda terms and their inferred types. Section 5 discusses related work and Sect. 6 concludes the paper. The code in the paper has been tested with SWI-Prolog 6.6.6 and YAP 6.3.4. It is also available at http://www.cse.unt.edu/~tarau/research/2015/dbr.pro.

2   A Compressed de Bruijn Representation of Lambda Terms

We represent standard lambda terms [1] in Prolog using the constructors a/2 for applications and 1/2 for lambda abstractions. Variables bound by the lambdas as well as their occurrences are represented as *logic variables*. As an example, the lambda term $\lambda x0.(\lambda x1.(x0\ (x1\ x1))\ \lambda x2.(x0\ (x2\ x2)))$ will be represented as 1(A,a(1(B,a(A,a(B,B))),1(C,a(A,a(C,C))))).

2.1   De Bruijn Indices

De Bruijn indices [2] provide a *name-free* representation of lambda terms. All terms that can be transformed by a renaming of variables (α-conversion) will share a unique representation. Variables following lambda abstractions are omitted and their occurrences are marked with positive integers *counting the number of lambdas until the one binding them* is found on the way up to the root of the term. We represent them using the constructor a/2 for application, 1/1 for lambda abstractions (that we will call shortly *binders*) and v/1 for marking the integers corresponding to the de Bruijn indices.

For instance, the term 1(A,a(1(B,a(A,a(B,B))),1(C,a(A,a(C,C))))) is represented as 1(a(1(a(v(1),a(v(0),v(0)))),1(a(v(1),a(v(0),v(0)))))), corresponding to the fact that v(1) is bound by the outermost lambda (two steps away, counting from 0) and the occurrences of v(0) are bound each by the closest lambda, represented by the constructor 1/1.

From de Bruijn to Lambda Terms with Canonical Names. The predicate b21 converts from the de Bruijn representation to lambda terms whose canonical names are provided by logic variables. We will call them terms in *standard notation*.

```
b21(DeBruijnTerm,LambdaTerm):-b21(DeBruijnTerm,LambdaTerm,_Vs).

b21(v(I),V,Vs):-nth0(I,Vs,V).
b21(a(A,B),a(X,Y),Vs):-b21(A,X,Vs),b21(B,Y,Vs).
b21(l(A),l(V,Y),Vs):-b21(A,Y,[V|Vs]).
```

Note the use of the built-in nth0/3 that associates to an index I a variable V on the list Vs. As we initialize in b21/2 the list of logic variables as a free variable _Vs, free variables in open terms, represented with indices larger than the number of available binders will also be consistently mapped to logic variables. By replacing _Vs with [] in the definition of b21/2, one could enforce that only closed terms (having no free variables) are accepted.

From Lambda Terms with Canonical Names to De Bruijn Terms. Logic variables provide canonical names for lambda variables. An easy way to manipulate them at meta-language level is to turn them into special "$VAR/1" terms - a mechanism provided by Prolog's built-in numbervars/3 predicate. Given that "$VAR/1" is distinct from the constructors lambda terms are built from (l/2 and a/2), this is a safe (and invertible) transformation. To avoid any side effect on the original term, in the predicate l2b/2 that inverts b21/2, we will uniformly rename its variables to fresh ones with Prolog's copy_term/2 built-in. We will adopt this technique through the paper each time our operations would mutate an input argument otherwise.

```
l2b(StandardTerm,DeBruijnTerm):-copy_term(StandardTerm,Copy),
   numbervars(Copy,0,_),l2b(Copy,DeBruijnTerm,_Vs).

l2b('$VAR'(V),v(I),Vs):-once(nth0(I,Vs,'$VAR'(V))).
l2b(a(X,Y),a(A,B),Vs):-l2b(X,A,Vs),l2b(Y,B,Vs).
l2b(l(V,Y),l(A),Vs):-l2b(Y,A,[V|Vs]).
```

Note the use of nth0/3, this time to locate the index I on the (open) list of variables _Vs. By replacing _Vs with [] in the call to l2b/3, one can enforce that only closed terms are accepted.

2.2 Going One Step Further: Compressing the Blocks of Lambdas

Iterated lambdas (represented as a block of constructors l/1 in the de Bruijn notation) can be seen as a successor arithmetic representation of a number that counts them. So it makes sense to represent that number more efficiently in the usual binary notation. Note that in de Bruijn notation blocks of lambdas can wrap either applications or variable occurrences represented as indices. This suggests using just two constructors: v/2 indicating in a term v(K,N) that we have K lambdas wrapped around variable v(N) and a/3, indicating in a term a(K,X,Y) that K lambdas are wrapped around the application a(X,Y).

We call the terms built this way with the constructors v/2 and a/3 *compressed de Bruijn terms*.

2.3 Converting Between Representations

We can make precise the definition of compressed deBruijn terms by providing a bijective transformation between them and the usual de Bruijn terms.

From de Bruijn to Compressed. The predicate b2c converts from the usual de Bruijn representation to the compressed one. It proceeds by case analysis on v/1, a/2, 1/1 and counts the binders 1/1 as it descends toward the leaves of the tree. Its steps are controlled by the predicate up/2 that increments the counts when crossing a binder.

```
b2c(v(X),v(0,X)).
b2c(a(X,Y),a(0,A,B)):-b2c(X,A),b2c(Y,B).
b2c(l(X),R):-b2c1(1,X,R).

b2c1(K,a(X,Y),a(K,A,B)):-b2c(X,A),b2c(Y,B).
b2c1(K, v(X),v(K,X)).
b2c1(K,l(X),R):-up(K,K1),b2c1(K1,X,R).

up(From,To):-From>=0,To is From+1.
```

From Compressed to de Bruijn. The predicate c2b converts from the compressed to the usual de Bruijn representation. It reverses the effect of b2c by expanding the K in v(K,N) and a(K,X,Y) into K 1/1 binders (no binders when K=0). The predicate iterLam/3 performs this operation in both cases, and the predicate down/2 computes the decrements at each step.

```
c2b(v(K,X),R):-X>=0,iterLam(K,v(X),R).
c2b(a(K,X,Y),R):-c2b(X,A),c2b(Y,B),iterLam(K,a(A,B),R).

iterLam(0,X,X).
iterLam(K,X,l(R)):-down(K,K1),iterLam(K1,X,R).

down(From,To):-From>0,To is From-1.
```

A convenient way to simplify defining chains of such conversions is by using Prolog's Definite Clause Grammars (DCGs), which transform a clause defined with "-->" like

```
a0 --> a1,a2,...,an.
```

into

```
a0(S0,Sn):-a1(S0,S1),a2(S1,S2),...,an(Sn-1,Sn).
```

For instance, the predicate c21/2 which can be seen as specifying a composition of two functions, expands to something like c21(X,Z):-c2b(X,Y),b21(Y,Z) The predicate converts from compressed de Bruijn terms and standard lambda terms using de Bruijn terms as an intermediate step, while 12c/2 works the other way around.

```
c21 --> c2b,b21.

12c --> 12b,b2c.
```

2.4 Open and Closed Terms

Lambda terms might contain *free variables* not associated to any binders. Such terms are called *open*. A *closed* term is such that each variable occurrence is associated to a binder. Closed terms can be easily identified by ensuring that the lambda binders on a given path from the root outnumber the de Bruijn index of a variable occurrence ending the path. The predicate isClosed does that for compressed de Bruijn terms.

```
isClosed(T):-isClosed(T,0).
isClosed(v(K,N),S):-N<S+K.
isClosed(a(K,X,Y),S1):-S2 is S1+K,isClosed(X,S2),isClosed(Y,S2).
```

3 Catalan Approximations of Lambda Terms

We can see our compressed de Bruijn terms as binary trees decorated with integer labels. The underlying binary trees provide a skeleton that describes the applicative structure of the terms.

The Catalan Family of Combinatorial Objects. Binary trees are among the most well-known members of the Catalan family of combinatorial objects [4], that has at least 58 structurally distinct members, covering several data structures, geometric objects and formal languages.

3.1 Type Inference with Logic Variables

Simple types - seen as binary trees built by the constructor "->/2" with empty leaves (representing the unique primitive type "o") can be seen as a "Catalan approximation" of lambda terms, centered around ensuring safe and terminating evaluation (strong normalization) of lambda terms.

While in a functional language inferring types requires implementing unification with occur check, as shown, for instance, in [7], this is readily available in Prolog. The predicate extractType/2 works by turning each logical variable X into a pair _:TX where TX is a fresh variable denoting its type. As logic variable bindings propagate between binders and occurrences, this ensures that types are consistently inferred.

```
extractType(_:TX,TX):-!.
extractType(l(_:TX,A),(TX->TA)):-
  extractType(A,TA).
extractType(a(A,B),TY):-
  extractType(A,(TX->TY)),
  extractType(B,TX).
```

Instead of using unification with occurs-check at each step, we ensure that at the end, our term is still acyclic, with the built-in ISO-Prolog predicate `acyclic_term/1`.

```
hasType(CTerm,ExtractedType):-
  c2l(CTerm,LTerm),
  extractType(LTerm,ExtractedType),
  acyclic_term(LTerm),
  bindType(ExtractedType).
```

At this point, most general types are inferred by `extractType` as fresh variables, similar to multi-parameter polymorphic types in functional languages. However, as we are only interested in simple types, we will bind uniformly the leaves of our type tree to the constant "o" representing our only primitive type, by using the predicate `bindType/1`.

```
bindType(o):-!.
bindType((A->B)):-bindType(A),bindType(B).
```

We can also define the predicate `typeable/1` that checks if a lambda term is well typed, by trying to infer and then ignore its inferred type.

```
typeable(Term):-hasType(Term,_Type).
```

Example 1. *Illustrates typability of the term corresponding to the S combinator* $\lambda x0.\ \lambda x1.\ \lambda x2.((x0\ x2)\ (x1\ x2))$ *and untypabilty of the term corresponding to the Y combinator* $\lambda x0.(\ \lambda x1.(x0\ (x1\ x1))\ \lambda x2.(x0\ (x2\ x2)))$, *in de Bruijn form.*

```
?- hasType(a(3,a(0,v(0,2),v(0,0)),a(0,v(0,1),v(0,0))),T).
T = ((o->o->o)-> (o->o)->o->o).
?- hasType(
   a(1,a(1,v(0,1),a(0,v(0,0),v(0,0))),a(1,v(0,1),a(0,v(0,0),v(0,0)))),T).
false.
```

3.2   Generating Closed Well-Typed Terms of a Given Size

One can derive, along the lines of the type inferrer `hasType`, a generator working directly on de Bruijn terms with a given number of internal nodes, by controlling their creation with the predicate `down/2`.

The predicate `genTypedB/5` relies on Prolog's DCG notation to thread together the steps controlled by the predicate `down`. Note also the nondeterministic use of the built-in `nth0` that enumerates values for both I and V ranging over the list of available variables `Vs`, as well as the use of `unify_with_occurs_check` to ensure that unification of candidate types does not create cycles.

```
genTypedB(v(I),V,Vs)-->
  {
    nth0(I,Vs,V0),
    unify_with_occurs_check(V,V0)
  }.
genTypedB(a(A,B),Y,Vs)-->down,
  genTypedB(A,(X->Y),Vs),
  genTypedB(B,X,Vs).
genTypedB(l(A),(X->Y),Vs)-->down,
  genTypedB(A,Y,[X|Vs]).
```

Two interfaces are offered: `genTypedB` that generates de Bruijn terms of with exactly L internal nodes and `genTypedBs` that generates terms with L internal nodes or less.

```
genTypedB(L,B,T):-genTypedB(B,T,[],L,0),bindType(T).
```

```
genTypedBs(L,B,T):-genTypedB(B,T,[],L,_),bindType(T).
```

As expected, the number of solutions of `genTypedB`, $1, 2, 9, 40, 238, 1564, \ldots$ for sizes $1, 2, 3, \ldots$, matches entry A220471 in [8].

Example 2. *Generation of all well-typed closed de Bruijn terms of size 2.*

```
?- genTypedB(2,Term,Type).
Term = l(l(v(0))),Type = (o->o->o);
Term = l(l(v(1))),Type = (o->o->o).
```

Coming with Prolog's unification and non-deterministic search, is the ability to make more specific queries by providing a type pattern, that selects only terms of a given type.

The predicate `queryTypedTerm` finds closed terms of a given type of size exactly L.

```
queryTypedB(L,Term,QueryType):-
  genTypedB(L,Term,Type),
  Type=QueryType.
```

Example 3. *Terms of type x>x of size 4.*

```
?- queryTypedB(4,Term,(o->o)).
Term = a(l(l(v(0))), l(v(0)));
Term = l(a(l(v(1)), l(v(0))));
Term = l(a(l(v(1)), l(v(1)))).

?- queryTypedB(10,Term,((o->o)->o)).
false.
```

Note that the last query, taking about half a minute, shows that no closed terms of type (o->o)->o exist up to size 10. Generating closed terms that match a specific type is likely to be useful for combinatorial testing.

We will explore next a mechanism for generating terms and types by defining a bijection to N.

3.3   Size-Proportionate Encodings

In the presence of a bijection between two, usually infinite sets of data objects, it is possible that representation sizes on one side or the other are exponentially larger that on the other. Well-known encodings like Ackerman's bijection for hereditarily finite sets to natural numbers fall in this category.

We will say that a bijection is *size-proportionate* if the representation sizes for corresponding terms on its two sides are "close enough" up to a constant factor multiplied with at most the logarithm of any of the sizes.

Definition 1. *Given a bijection between sets of terms of two datatypes denoted M and N, $f : M \rightarrow N$, let $m(x)$ be the representation size of a term $x \in M$ and $n(y)$ be the representation size of $y \in N$. Then f is called* size-proportionate *if $|m(x) - n(y)| \in O(log(max(m(x), n(y))))$.*

Informally we also assume that the constants involved are small enough such that the printed representation of two data objects connected by the bijections is about the same.

3.4   The Language of Balanced Parentheses

Binary trees are in a well-known size-proportionate bijection with the language of balanced parentheses [4], from which we will borrow an efficient ranking/ unranking bijection. The reversible predicate `catpar/2` transforms between binary trees and lists of balanced parentheses, with 0 denoting the open parentheses and 1 denoting the closing one.

```
catpar(T,Ps):-catpar(T,0,1,Ps,[]).
catpar(X,L,R) --> [L],catpars(X,L,R).
catpars(o,_,R) --> [R].
catpars((X->Xs),L,R)-->catpar(X,L,R),catpars(Xs,L,R).
```

Example 4. *Illustrates the work of the reversible predicate* `catpar/2`.

```
?- catpar(((o->o)->o->o),Ps),catpar(T,Ps).
Ps = [0, 0, 0, 1, 1, 0, 1, 1],T = ((o->o)->o->o).
```

Note the extra opening/closing parentheses, compared to the usual definition of Dyck words [4], that make the sequence self-delimiting.

3.5   A Bijection from the Language of Balanced Parenthesis Lists to ℕ

This algorithm follows closely the procedural implementation described in [5].

The code of the helper predicates called by `rankCatalan` and `unrankCatalan` is provided in http://www.cse.unt.edu/~tarau/research/2015/dbr.pro. The details of the algorithms for computing `localRank` and `localunRank` are described at http://www.cse.unt.edu/~tarau/research/2015/dbrApp.pdf.

The predicate `rankCatalan` uses the Catalan numbers computed by `cat` in `rankLoop` to shift the ranking over the ranks of smaller sequences, after calling `localRank`.

```
rankCatalan(Xs,R):-
  length(Xs,XL),XL>=2,
  L is XL-2, I is L // 2,
  localRank(I,Xs,N),
  S is 0, PI is I-1,
  rankLoop(PI,S,NewS),
  R is NewS+N.
```

The predicate unrankCatalan uses the Catalan numbers computed by cat in unrankLoop to shift over smaller sequences, before calling localUnrank.

```
unrankCatalan(R,Xs):-
  S is 0, I is 0,
  unrankLoop(R,S,I,NewS,NewI),
  LR is R-NewS,
  L is 2*NewI+1,
  length(As,L),
  localUnrank(NewI,LR,As),
  As=[_|Bs],
  append([0|Bs],[1],Xs).
```

The following example illustrates the ranking and unranking algorithms:

```
?- unrankCatalan(2015,Ps),rankCatalan(Ps,Rank).
Ps = [0,0,1,0,1,0,1,0,0,0,0,1,0,1,1,1,1,1],Rank = 2015
```

3.6   Ranking and Unranking Simple Types

After putting together the bijections between binary trees and balanced parentheses with the ranking/unranking of the later we obtain the size-proportionate ranking/unranking algorithms for simple types.

```
rankType(T,Code):-
  catpar(T,Ps),
  rankCatalan(Ps,Code).

unrankType(Code,Term):-
  unrankCatalan(Code,Ps),
  catpar(Term,Ps).
```

Example 5. *Illustrates the ranking and unranking of simple types.*

```
?- I=100, unrankType(I,T),rankType(T,R).
I = R, R = 100,T = (((o->o)-> (o->o->o)->o)->o).
```

As there are $O(\frac{4^n}{n^{\frac{3}{2}}})$ binary trees of size n corresponding to 2^n natural numbers of bitsize up to n and our ranking algorithm visits them in lexicographic order, it follows that:

Proposition 1. *The bijection between types and their ranks is size-proportionate.*

3.7 Catalan Skeletons of Compressed de Bruijn Terms

As compressed de Bruijn terms can be seen as binary trees with labels on their leaves and internal nodes, their "Catalan skeleton" is simply the underlying binary tree. The predicate `cskel/3` extracts this skeleton as well as the list of the labels, in depth-first order, as encountered in the process.

```
cskel(S,Vs, T):-cskel(T,S,Vs,[]).
```

```
cskel(v(K,N),o)-->[K,N].
cskel(a(K,X,Y),(A->B))-->[K],cskel(X,A),cskel(Y,B).
```

The predicates `toSkel` and `fromSkel` add conversion between this binary tree and lists of balanced parenthesis by using the (reversible) predicate `catpar`.

```
toSkel(T,Skel,Vs):-
  cskel(T,Cat,Vs,[]),
  catpar(Cat,Skel).
```

```
fromSkel(Skel,Vs, T):-
  catpar(Cat,Skel),
  cskel(T,Cat,Vs,[]).
```

Example 6. *Illustrates the Catalan skeleton* `Skel` *and the list of variable labels* `Vs` *extracted from a compressed de Bruijn term corresponding to the* S *combinator.*

```
?- T = a(3, a(0, v(0, 2), v(0, 0)), a(0, v(0, 1), v(0, 0))),
     toSkel(T,Skel,Vs),fromSkel(Skel,Vs,T1).
T = T1, T1 = a(3, a(0, v(0, 2), v(0, 0)), a(0, v(0, 1), v(0, 0))),
Skel = [0,0,0,1,1,0,1,1],Vs = [3,0,0,2,0,0,0,0,1,0,0].
```

3.8 The Generalized Cantor k-tupling Bijection

As we we have already solved the problem of ranking and unranking lists of balanced parentheses, the remaining problem is that of finding a bijection between the lists of labels collected from the nodes of a compressed de Bruijn term and natural numbers.

We will use the generalized Cantor bijection between $\mathbb{N}^n$ and $\mathbb{N}$ as the first step in defining this bijection. The formula, given in [9] p.4, looks as follows:

$$K_n(x_1,\ldots,x_n) = \sum_{k=1}^{n} \binom{k-1+s_k}{k} \ where \ s_k = \sum_{i=1}^{k} x_i \tag{1}$$

Note that $\binom{n}{k}$ represents the number of subsets of k elements of a set of n elements, that also corresponds to the binomial coefficient of x^k in the expansion of $(x+y)^n$, and $K_n(x_1,\ldots,x_n)$ denotes the natural number associated to the tuple $(x_1,\ldots,x_n)$. It is easy to see that the generalized Cantor n-tupling function defined by Eq. (1) is a polynomial of degree n in its arguments.

The Bijection Between Sets and Sequences of Natural Numbers. We recognize in the Eq. (1) the *prefix sums* s_k incremented with values of k starting at 0. It represents the "set side" of the bijection between sequences of n natural numbers and sets of n natural numbers described in [10]. It is implemented in the online Appendix as the bijection list2set together with its inverse set2list. For example, list2set transforms [2,0,1,5] to [2, 3, 5, 11] as 3=2+0+1, 5=3+1+1, 11=5+5+1 and set2list transforms it back by computing the differences between consecutive members, reduced by 1.

3.9   The $\mathbb{N}^n \to \mathbb{N}$ Bijection

The bijection $K_n : \mathbb{N}^n \to \mathbb{N}$ is basically just summing up a set of binomial coefficients. The predicate fromCantorTuple implements the the $\mathbb{N}^n \to \mathbb{N}$ bijection in Prolog, using the predicate fromKSet that sums up the binomials in formula 1 using the predicate untuplingLoop, as well as the sequence to set transformer list2set.

```
fromCantorTuple(Ns,N):-
  list2set(Ns,Xs),
  fromKSet(Xs,N).


fromKSet(Xs,N):-untuplingLoop(Xs,0,0,N).

untuplingLoop([],_L,B,B).
untuplingLoop([X|Xs],L1,B1,Bn):-L2 is L1+1,
  binomial(X,L2,B),B2 is B1+B,
  untuplingLoop(Xs,L2,B2,Bn).
```

3.10   The $\mathbb{N} \to \mathbb{N}^n$ Bijection

We split our problem in two simpler ones: inverting fromKSet and then applying set2list to get back from sets to lists.

We observe that the predicate untuplingLoop used by fromKSet implements the sum of the combinations $\binom{X_1}{1} + \binom{X_2}{2} + \ldots + \binom{X_K}{K} = N$, which is nothing but the representation of N in the *combinatorial number system of degree K* due to [11]. Fortunately, efficient conversion algorithms between the conventional and the combinatorial number system are well known, [12].

We are ready to implement the Prolog predicate toKSet(K,N,Ds), which, given the degree K, indicating the number of "combinatorial digits", finds and repeatedly subtracts the greatest binomial smaller than N. It calls the predicate combinatoriallDigits that returns these "digits" in increasing order, providing the canonical set representations that set2list needs.

```
toKSet(K,N,Ds):-combinatoriallDigits(K,N,[],Ds).


combinatoriallDigits(0,_,Ds,Ds).
combinatoriallDigits(K,N,Ds,NewDs):-K>0,K1 is K-1,
```

```
    upperBinomial(K,N,M),M1 is M-1,
    binomial(M1,K,BDigit),N1 is N-BDigit,
    combinatoriallDigits(K1,N1,[M1|Ds],NewDs).

upperBinomial(K,N,R):-S is N+K,
    roughLimit(K,S,K,M),L is M // 2,
    binarySearch(K,N,L,M,R).
```

The predicate `roughLimit` compares successive powers of 2 with binomials $\binom{I}{K}$ and finds the first I for which the binomial is between successive powers of 2.

```
roughLimit(K,N,I, L):-binomial(I,K,B),B>N, !,L=I.
roughLimit(K,N,I, L):-J is 2*I,
    roughLimit(K,N,J,L).
```

The predicate `binarySearch` finds the exact value of the combinatorial digit in the interval [L,M], narrowed down by `roughLimit`.

```
binarySearch(_K,_N,From,From,R):-!,R=From.
binarySearch(K,N,From,To,R):-Mid is (From+To) // 2,binomial(Mid,K,B),
    splitSearchOn(B,K,N,From,Mid,To,R).

splitSearchOn(B,K,N,From,Mid,_To,R):-B>N,!,
    binarySearch(K,N,From,Mid,R).
splitSearchOn(_B,K,N,_From,Mid,To,R):-Mid1 is Mid+1,
    binarySearch(K,N,Mid1,To,R).
```

The predicates `toKSet` and `fromKSet` implement inverse functions, mapping natural numbers to canonically represented sets of K natural numbers.

```
?- toKSet(5,2014,Set),fromKSet(Set,N).
Set = [0, 3, 4, 5, 14], N = 2014.
```

The efficient inverse of Cantor's N-tupling is now simply:

```
toCantorTuple(K,N,Ns):-
    toKSet(K,N,Ds),
    set2list(Ds,Ns).
```

Example 7. *Illustrates the work of the generalized cantor bijection, on some large numbers:*

```
?- K=1000,pow(2014,103,N),toCantorTuple(K,N,Ns),fromCantorTuple(Ns,N).
K = 1000, N = 20802954558570368848441985145954726483138165...567744,
Ns = [0, 0, 2, 0, 0, 0, 0, 0, 1|...].
```

As the image of a tuple is a polynomial of degree n it means that the its bitsize is within constant factor of the sum of the bitsizes of the members of the tuple, thus:

Proposition 2. *The bijection between $\mathbb{N}^n$ and $\mathbb{N}$ is size-proportionate.*

4   Ranking/Unranking of Compressed de Bruijn Terms

We will implement a size-proportionate bijective encoding of compressed de Bruijn terms following the technique described in [13]. The algorithm will split a lambda tree into its *Catalan skeleton* and the list of atomic objects labeling its nodes. In our case, the Catalan skeleton abstracts away the applicative structure of the term. It also provides the key for decoding unambiguously the integer labels in both the leaves (two integers) and internal nodes (one integer). Our ranking/unranking algorithms will rely on the encoding/decoding of the Catalan skeleton provided by the predicates `rankCatalan/2` and `unrankCatalan/2` as well as for the encoding/decoding of the labels, provided by the predicates `toCantorTuple/3` and `fromCantorTuple/2`.

The predicate `rankTerm/2` defines the bijective encoding of a (possibly open) compressed de Bruijn term.

```
rankTerm(Term,Code):-
  toSkel(Term,Ps,Ns),
  rankCatalan(Ps,CatCode),
  fromCantorTuple(Ns,VarsCode),
  fromCantorTuple([CatCode,VarsCode],Code).
```

The predicate `rankTerm/2` defines the bijective decoding of a natural number into a (possibly open) compressed de Bruijn term.

```
unrankTerm(Code,Term):-
  toCantorTuple(2,Code,[CatCode,VarsCode]),
  unrankCatalan(CatCode,Ps),
  length(Ps,L2),L is (L2-2) div 2, L3 is 3*L+2,
  toCantorTuple(L3,VarsCode,Ns),
  fromSkel(Ps,Ns,Term).
```

Note that given the unranking of `CatCode` as a list of balanced parentheses of length 2*L+2, we can determine the number L of internal nodes of the tree and the number L+1 of leaves. Then we have 2*(L+1) labels for the leaves and L labels for the internal nodes, for a total of 3L+2, value needed to decode the labels encoded as `VarsCode`.

It follows from Propositions 1 and 2 that:

Proposition 3. *A compressed de Bruijn terms is size-proportionate to its rank.*

Example 8. *Illustrates the "size-proportionate" encoding of the compressed de Bruijn terms corresponding to the combinators* S *and* Y.

```
?- T = a(3,a(0,v(0,2),v(0,0)),a(0,v(0, 1),v(0,0))),
     rankTerm(T,R),unrankTerm(R,T1).
T = T1,T1 = a(3,a(0,v(0,2),v(0,0)),a(0,v(0, 1),v(0,0))),
R = 56493141.

?- T=a(1,a(1,v(0,1),a(0,v(0,0),v(0,0))),a(1,v(0,1),a(0,v(0,0),v(0,0)))),
     rankTerm(T,R),unrankTerm(R,T1).
T=T1,T1=a(1,a(1,v(0,1),a(0,v(0,0),v(0,0))),a(1,v(0,1),a(0,v(0,0),v(0,0)))),
R = 261507060.
```

4.1   Generation of Lambda Terms via Unranking

While direct enumeration of terms constrained by number of nodes or depth is straightforward in Prolog, an unranking algorithm is also usable for term generation, including generation of random terms.

Generating Open Terms in Compressed de Bruijn Form. Open terms are generated simply by iterating over an initial segment of ℕ with the built-in between/3 and calling the predicate unrankTerm/2.

```
ogen(M,T):-between(0,M,I),unrankTerm(I,T).
```

Reusing unranking-based open term generators for more constrained families of lambda terms works when their asymptotic density is relatively high. The extensive quantitative analysis available in the literature [7,14,15] indicates that density of closed and typed terms decreasing very quickly with size, making generation by filtering impractical for very large terms.

The predicate cgen/2 generates closed terms by filtering the results of ogen/2 with the predicate isClosed and tgen generates typeable terms by filtering the results of cgen/2 with typeable/2.

```
cgen(M,IT):-ogen(M,IT),isClosed(IT).
```

```
tgen(M,IT):-cgen(M,IT),typeable(IT).
```

Generation of Random Lambda Terms. Generation of random lambda terms, resulting from the unranking of random integers of a give bit-size, is implemented by the predicate ranTerm/3, that applies the predicate Filter repeatedly until a term is found for which the predicate Filter holds.

```
ranTerm(Filter,Bits,T):-X is 2^Bits,N is X+random(X),M is N+X,
  between(N,M,I),
    unrankTerm(I,T),call(Filter,T),
  !.
```

Random open terms are generated by ranOpen/2, random closed terms by the predicate ranClosed, random typeable term by ranTyped and closed typeable terms by closedTypeable/2.

```
ranOpen(Bits,T):-ranTerm(=(_),Bits,T).
```

```
ranClosed(Bits,T):-ranTerm(isClosed,Bits,T).
```

```
ranTyped(Bits,T):-ranTerm(closedTypeable,Bits,T).
```

```
closedTypeable(T):-isClosed(T),typeable(T).
```

Open terms based on unranking random numbers of 3000 bits of size above 1000, closed terms of size above 55 for 150 bits and closed typeable terms of size above 13 for 30 bits can be generated within a few seconds. The limited scalability for closed and well-typed terms is a consequence of their low asymptotic density, as shown in [7,14]. We refer to [7] for algorithms supporting random generation of large lambda terms.

Example 9. *Illustrates generation of some closed and well-typed terms in compressed de Bruijn form.*

```
?- ranClosed(10,T).
T = a(1, a(0, v(0, 0), v(0, 0)), a(0, a(0, v(0, 0), v(0, 0)), v(1, 0))).

?- ranTyped(20,T).
T = a(3, v(3, 1), v(2, 0)).
```

5   Related Work

The classic reference for lambda calculus is [1]. Various instances of typed lambda calculi are overviewed in [16]. De Bruijn's notation for lambda terms is introduced in [2]. The compressed de Bruijn representation of lambda terms proposed in this paper is novel, to our best knowledge.

While Gödel numbering schemes for lambda terms have been studied in several theoretical papers on computability, we are not aware of any size proportionate bijective encoding as the one described in this paper.

The combinatorics and asymptotic behavior of various classes of lambda terms are extensively studied in [7,14,15]. Distribution and density properties of random lambda terms are described in [14].

6   Conclusions

The most significant contributions of this paper are the size-proportionate ranking/unranking algorithm for lambda terms and the compressed de Bruijn representation that facilitated it. The ability to encode lambda terms bijectively can be used as a "serialization" mechanism in functional programming languages and proof assistants using them as an intermediate language.

Besides the newly introduced compressed form of de Bruijn terms, we have used ordinary de Bruijn terms as well as a canonical representation of lambda terms relying on Prolog's logic variables. We have switched representation as needed, though bijective transformers working in time proportional to the size of the terms. Our techniques, combining unification of logic variables with Prolog's backtracking mechanism and Definite Clause Grammar notation, suggest that logic programming is a suitable meta-language for the manipulation of various families of lambda terms and the study of their combinatorial and computational properties.

Acknowledgement. We thank the anonymous referees of **Calculemus'15** for their constructive criticisms and valuable suggestions that have helped improving the paper. This research was supported by NSF research grant **1423324**.

References

1. Barendregt, H.P.: The Lambda Calculus Its Syntax and Semantics, vol. 103, Revised edn. Elsevier, North Holland (1984)
2. Bruijn, N.G.D.: Lambda calculus notation with nameless dummies, a tool for automatic formula manipulation, with application to the Church-Rosser Theorem. Indag. Mathematicae **34**, 381–392 (1972)
3. McBride, C.: I am not a number, I am a classy hack (2010). Blog entry: http://mazzo.li/epilogue/index.html
4. Stanley, R.P.: Enumerative Combinatorics. Wadsworth Publishing Co., Belmont (1986)
5. Kreher, D.L., Stinson, D.: Combinatorial Algorithms: Generation, Enumeration, and Search. The CRC Press Series on Discrete Mathematics and its Applications. CRC Press INC, US (1999)
6. Gödel, K.: Über formal unentscheidbare Sätze der Principia Mathematica und verwandter Systeme I. Monatshefte für Mathematik und Physik **38**, 173–198 (1931)
7. Grygiel, K., Lescanne, P.: Counting and generating lambda terms. J. Funct. Program. **23**(5), 594–628 (2013)
8. Sloane, N.J.A.: The on-line encyclopedia of integer sequences (2014). Published electronically at https://oeis.org/
9. Cegielski, P., Richard, D.: On arithmetical first-order theories allowing encoding and decoding of lists. Theor. Comput. Sci. **222**(1–2), 55–75 (1999)
10. Tarau, P.: An embedded declarative data transformation language. In: Proceedings of 11th International ACM SIGPLAN Symposium PPDP 2009, Coimbra, Portugal, September 2009, pp. 171–182. ACM (2009)
11. Lehmer, D.H.: The machine tools of combinatorics. In: Edwin, F., Beckenbach, R.E. (eds.) Krieger Applied combinatorial mathematics, pp. 5–30. Wiley, New York (1964)
12. Knuth, D.E.: The Art of Computer Programming, Volume 4, Fascicle 3: Generating All Combinations and Partitions. Addison-Wesley Professional, Boston (2005)
13. Tarau, P.: Compact serialization of prolog terms (with Catalan skeletons, Cantor tupling and Gödel numberings). Theory Prac. Logic Program. **13**(4–5), 847–861 (2013)
14. David, R., Raffalli, C., Theyssier, G., Grygiel, K., Kozik, J., Zaionc, M.: Some properties of random lambda terms. Logic. Methods Comput. Sci. **9**(1) (2009). http://citeseerx.ist.psu.edu/viewdoc/download?doi=10.1.1.244.5073&rep=rep1&type=pdf
15. David, R., Grygiel, K., Kozik, J., Raffalli, C., Theyssier, G., Zaionc, M.: Asymptotically almost all λ-terms are strongly normalizing (2010). Preprint: arXiv: math.LO/0903.5505v3
16. Barendregt, H.P.: Lambda calculi with types. In: Abramsky, S., Gabbay, D., Maibaum, T. (eds.) Handbook of Logic in Computer Science, vol. 2. Oxford University Press, New York (1991)

Digital Mathematics Libraries

A Flexiformal Model of Knowledge Dissemination and Aggregation in Mathematics

Mihnea Iancu[(✉)] and Michael Kohlhase

Computer Science, Jacobs University, Bremen, Germany
{m.iancu,m.kohlhase}@jacobs-university.de

Abstract. In the traditional knowledge dissemination process in mathematics and sciences, authors write semi-selfcontained articles which are then published in journals, conference proceedings, preprint archives, and/or given as talks. Other scientists read these, extract the new knowledge, integrate it into their personal mental model of the field, and use this as the basis for creating new knowledge which is disseminated in the same form.

Somewhat surprisingly, this process has not been modeled from a formal or content-based perspective even though it is at the heart of human MKM and DML.

In this paper we tackle this problem starting from the practice of beginning papers with a "recap", which briefly introduces context, terminology, and notations and thus ties the paper into the knowledge commons. We propose a flexiformal model for knowledge dissemination and its aggregation into a communal, shared knowledge commons based on theory graphs and the newly introduced realms.

1 Introduction

Global mathematical knowledge grows – at least – at a rate 120,000 published articles a year to a current crop of about 3.5 Million articles. Even though these articles are scattered over several thousand journals they – together with papers in conferences, preprints in online or local archives, and talks given in seminars – function as a coherent scientific commons of communal knowledge about the various domains of mathematics. Other scientists read these documents, extract the new knowledge, integrate this into their personal mental model of the domain, and use this as the basis for creating new knowledge. This, in turn, is disseminated again through articles, conference papers, preprints, and talks, itself contributing to the knowledge commons.

In this process of knowledge dissemination and aggregation, scientific documents (articles, papers, preprints, and talks) play a great role: they have evolved from printed pamphlets or books and from postal letters in which a mathematician described progress to a colleague – and were then passed around by the latter among colleagues. Documents are assembled into topical journals and conference proceedings volumes, which are in turn assembled into libraries (physical and virtual ones), which give researchers and practitioners access to the scientific document commons – modulo physical distribution methods like inter-library loan

© Springer International Publishing Switzerland 2015
M. Kerber et al. (Eds.): CICM 2015, LNAI 9150, pp. 137–152, 2015.
DOI: 10.1007/978-3-319-20615-8_9

and access right restrictions like membership or commercial constraints. Documents are even classified into a domain-based classification schemes like the Math Subject Classification (MSC), and disseminated in information systems like Math Reviews and Zentralblatt Math.

Today's mathematical documents have a specific conventionalized structure and metadata which not only supports the production/dissemination processes outlined above, but also – we claim – the individual and communal aggregation processes which turn the document collections into a (virtual) knowledge space which mathematicians can operate on to find and apply existing knowledge and create new insights and knowledge.

In formalized mathematics, the situation is very different. Even though collections of formalized mathematics call themselves "libraries", the concept of a "formal document" does not exist or degenerates to a "file" which contains the formal development and possibly includes other files. Explanations for humans are generally relegated to comments or the informal literature described above (publishing about formalizations).

Notable exceptions are the Mizar Mathematical Library [MizLib] and the Mizar-inspired ISAR format in Isabelle [Wen07]. Both of these contain enough information to generate conventionally structured documents for publication,. e.g. Mizar articles in the Journal of Formalized Mathematics [JFM]. Dissemination, quality control, and "marketing" of results and developments is usually ad-hoc in formalized mathematics. Aggregation of developments into a knowledge space is ephemeral and executed by loading files with formal developments into the memory of a theorem prover or proof checker.

On the other hand, libraries of formalized mathematics directly represent the structure of a mathematical knowledge commons, usually in graph of files and file inclusions or a graph of theories and theory morphisms (see [RK13a] for a survey). The respective graphs supply identifiers for knowledge items and detail their relations to each other.

It stands to reason that the two dissemination and aggregation approaches can profit from each other. The scientific publication process can profit from a more explicitly represented knowledge commons, which enables added-value services for finding, understanding, and applying relevant knowledge items – after all the document/knowledge space even in mathematics is much too large and complex for a single human to process. Of course a prerequisite for this is computer support in the aggregation of the knowledge space. Conversely, formal libraries can profit from a dissemination process based on the publication of self-contained documents to scale the secondary aspects (quality control, checkpointing, citation stability, persistence, attention management) of assembling large bodies of knowledge. Even though formal developments are machine-checkable, their authoring, maintenance, refactoring, ... are processes that need at least some human intervention.

To reap these benefits we need a joint generalization of the two approaches to dissemination and aggregation that combines their advantages. But before we design such a system, we need a content-oriented model of the informal publication process. Somewhat surprisingly, such a model does not exist, even though knowledge dissemination and next-generation publication systems are a the heart of MKM and DML.

In this paper we propose a content-oriented model for knowledge dissemination and its aggregation into a communal, shared knowledge commons. As we make use of our previous development of the *flexiformal* – i.e. supporting flexible degrees of formality [Koh13] – OMDoc format [Koh06], which can represent formal and informal mathematical documents and developments, we think of this as a flexiformal model.

We use the practice of starting mathematical documents with a "recap", which briefly introduces context, terminology, and notations and thus ties the paper into the knowledge commons as a starting point and model it based on OMDoc/MMT theory graphs and the newly introduced realms [CFK14].

In Sect. 2 we briefly review the structure of mathematical documents and build our intuitions about "recaps" by looking at some examples. We discuss how to represent them using theory graphs in Sect. 3. Section 4 concludes the paper and discusses future work.

2 Common Ground in Mathematical Documents

With **dissemination** we mean the process of assembling a mathematical document for the purpose of publication. We use the term **aggregation** for the process of an individual integrating the knowledge gained from reading or experiencing the respective document into their mental model of the domain. For now we will use these two concepts intuitively only, it is the purpose of this paper to propose a more rigorous model for them. As a first step, we will now have a closer look at the practices in formal and informal mathematics.

2.1 The Structure of Informal Mathematical Documents

Mathematical documents traditionally have:

1. A **front/backmatter** and **page margins**, which identify the scientific metadata: (i) author's names, affiliations, and addresses, (ii) publication venue, date, and fragment identifiers (e.g. page numbers), (iii) classification data, e.g. keywords or MSC codes, (iv) acknowledgements of contributions of other researchers or funding agencies. (v) access conditions, e.g. copyright, confidentiality designations, or licenses.
2. An **abstract** that gives an executive overview over the document.
3. An **introduction** that leads the reader into the topic, discusses the problems solved in the document and their relation to the "real world", and generally argues that reading the paper is worth the reader's attention.
4. A **preview**, which outlines the structure of, the contributions in, and methods used in the document.
5. A discussion of the **state of the art** on the topic of document.

6. The establishment of a **common ground** between the reader and the author, which (*i*) recapitulates or surveys concepts and results from the documents/knowledge commons to make the document self-contained (for its intended audience) (*ii*) identifies any assumptions and gives the ensuing contributions a sound terminological basis.

7. The **contributions** part, which contains the development of new knowledge in form of e.g. new insights, new interpretations of known concepts, new theorems, new proofs, new applications/examples or new techniques of achieving results.

8. An **evaluation** of the contributions in terms of applicability or usability.

9. A discussion of **related work** which reviews the contributions and their relation to existing approaches and results from the literature.

10. A **conclusion** which summarizes the contribution with the benefit of hindsight and relates it to the claims made in the introduction.

11. Literature references, an index, a glossary, etc. and possibly appendices that contain material deemed supplementary to the contributions.

Even though the form or order of the structural elements may vary over publication venues, and certain elements may be implicit or even missing altogether, the overall structure is generally stable.

It may be surprising that only one in eleven parts of a mathematical document – the "contributions" – arguably the largest – is fully dedicated to transporting the payload of the paper. All other contribute to either the dissemination[1], understanding[2] and aggregation processes. We will see that the latter is mainly driven by the common ground (point 6. above), which we will analyze in more detail next.

2.2   Common Ground/Recapitulation in Mathematical Research

To get an overview over recaps in the literature, we randomly selected 30 papers from the new submissions to http://arxiv.org/archive/math and analyzed their structure. All had a significant common ground section that recapitulates the central notions and fixes notations. We show two examples where the mathematics involved is relatively elementary.

Example 1. Reference [HK15] discusses covers of the multiplicative group of an algebraically closed field which are formally introduced in the beginning of the paper as follows:

> **Definition 1.1.** Let V be a vector space over Q and let F be an algebraically closed field of characteristic 0. A *cover of the multiplicative group* of F is a structure represented by an exact sequence $0 \to K \to V \to F \to 1$, where the map $V \to F^*$ is a surjective group homomorphism from $(V, +)$ onto $(F, \cdot)$ with kernel K. We will call this map *exp*. (1)

---

[1] 1. for referencing, 2. for determining interest.

[2] 3. and 10. for broader context, 5. and 9. for problem context, 4. for document navigation, 8. for assessment of value, and 11. for further reading.

However, later, the authors source the concept origin to an earlier paper ("[13]") and effectively import the terminology, definitions and theorems. For instance, when establishing results, [HK15] mentions "Moreover, with an additional axiom (in $L_{\omega_1 \omega}$) stating $K \cong Z$, the class is categorical in uncountable cardinalities. This was originally proved in [13] but an error was later found in the proof and corrected in [2]. Throughout this article, we will make the assumption $K \cong Z$.".

In the second example, the situation is a bit more complex, since the import of the terminology and definitions is not direct, but involves a choice.

Example 2. Reference [Bar15] studies the properties of multinets. In the preliminaries section they are introduced with the following definition:

> **Definition 2.1.** The union of all completely reducible fibers (with a fixed partition into fibers, also called blocks) of a Ceva pencil of degree d is called a $(k, d) - multinet$ where k is the number of the blocks. The base X of the pencil is determined by the multinet structure and called the base of the multinet. $\qquad(2)$

Later in that section some properties of multinets are introduced with the phrase "Several important properties of multinets are listed below which have been collected from [4,10,12]". The referenced papers all use slightly different definitions of multinets but they are assumed to be equivalent so that the properties hold. In fact, in this paper [Bar15] the assumption is made explicit – although not proved – from the start: "There are several equivalent ways to define multinets. Here we present them using pencils of plane curves."

The next example is not from our 30 examples, since we want to show an even more complex situation.

Example 3. Reference [CS09] studies the halting problem for accelerated Turing machines and starts off the discussion with an informal introduction of the topic.

> An accelerated Turing machine (sometimes called Zeno machine) is a Turing machine that takes 2^{-n} units of time (say seconds) to perform its nth step; we assume that steps are in some sense identical except for the time taken for their execution. $\qquad(3)$

This is a telegraphic version of the full definition, which is given in the literature. Actually [CS09] continues with an overview of the literature, citing no less than 12 papers, which address the topic of accelerated Turing machines. One of these supposedly contains the formal definition, which involves generalizing Turing machines to timed ones, introducing computational time structures, and singling out accelerating ones, e.g. using (4).

> **Definition 1.3.** An **accelerated Turing machine** is a Turing machine $M = \langle X, \Gamma, S, s_o, \Box, \delta \rangle$ working with with a computational time structure $T = \langle \{t_i\}_i, <, + \rangle$ with $T \subseteq \mathbb{Q}_+$ ($\mathbb{Q}_+$ is the set of non-negative rationals) such that $\sum_{i \in \mathbb{N}} t_i < \infty$. $\qquad(4)$

Note that the definition of an ATM [CS09] is an instance of Definition 1.3, which allows arbitrary time structures.

2.3   Secondary Literature: Education/Survey

A similar effect can be observed with educational materials or survey articles, whose concern is not to make an original contribution to the knowledge commons, but to prepare a document that helps an individual or group study or better understand a body of already established knowledge. Consider for instance, slides and background materials (lecture notes, text books, encyclopaedias), where the slides often have telegraphic versions of the real statements, which verbalize more rigorous definition.

This is illustrated in Example 4 which is inspired from the notes of a first year computer science course taught by the first author. The example is a simplified and self-contained version of the original which in itself is only one instance of a commonly occurring pattern in the course notes.

Example 4 (A Course grounded in a Formal Library). A course which introduces (naive) set theory informally, but grounds itself in a formal, modular definition. In the cited source, we have a careful introduction in the form of a modular theory graph starting at a theory that introduces membership relation and the axioms of existence, extensionality, and separation and defines the set constructor $\{\cdot|\cdot\}$ from these axioms. In the course notes we have a theory that "adopts" the symbols $\in$ and $\{\cdot|\cdot\}$ but not the associated axioms. Instead it "defines" them by alluding to the intuitions of the students. Then the course notes continue with introducing set operations ranging from set union to the power set.

We observe that course notes in Example 4 are self-contained in the sense that they can be understood without knowing about the formal development. This self-containedness is important intra-course didactics. But it also has the problem that the courses become insular; how are students going to communicate with mathematicians who have learned their maths from other courses? This is where alluding to the literature comes in, by connecting the course notes with it.

Example 5. The situation in mathematical textbooks is similar in structure to that in research papers –perhaps more pronounced. Consider the following passage from Rudin's classical introductory textbook to Functional Analysis [Rud73, p.6f].

> **1.5 Topological spaces** A *topological space* is a set S in which a
> collection τ of subsets (called *open sets*) has been specified, with
> the following properties: S is open, $\emptyset$ is open, [...] Such a collection   (5)
> is called a *topology* on S. [...]

This is continued later – vector spaces have been recapped earlier in Sect. 1.4 – with:

> **1.6 Topological vector spaces** Suppose τ is a topology on a vector
> space X such that
>
> (a) *every point of X is a closed set, and*
> (b) *the vector space operations are continuous with respect to τ*   (6)
>
> Under these conditions, τ is said to be a *vector topology* on X,
> and X is a *topological vector space.*

Note that Rudin does not directly cite the literature in these quotes, but in the preface he mentions the vast literature on function analysis and in Appendix B he cites the original literature for each chapter. The situation in textbooks is also different from research articles in that textbooks – like survey articles, and by their very nature – do not add new knowledge or new results, but aggregate and organize the already published ones, possibly reformulating them for a more uniform exposition. But still, one can distinguish recap parts – as the ones above – which are much more telegraphic in nature from the primary material presented in the textbook.

2.4   Common Ground in Formal Mathematics

Where applicable, common ground in formal mathematics is typically established via direct imports of symbols, theorems, notations, etc. Formal documents emphasize correctness and do not focus on human readability so they do not reintroduce concepts or provide, verbalizations of definitions.

For instance, In Isabelle and Coq knowledge is organized in *Theories* and *Modules* which are effectively named sets of declarations. The incremental development process is enabled via the IMPORTS and, respectively, REQUIRE IMPORT statements that effectively opens a library module by name and enables its declarations to be used in the current development.

In Mizar, formal documents (called *articles*) can be exported as PDF files in a human readable format. The narrative documents contain a part that verbalizes the imports from the source documents and the notation reservations which can be seen as a common ground section.

Example 6. The common ground part for [RK13b]

> The notation and terminology used in this paper have been introduced in the following papers: [4], [11], [12], [19], [9], [3], [5], [6], [21], [22], [1], [2], [7], [18], [20], [24], [25], [23], [16], [13], [14], [10], [15], and [8]. [...] In this paper T, U are non empty topological spaces, t is a point of T, and n is a natural number.   (7)

3   Publication and Dissemination in Theory Graphs

In this section we look more closely at the examples from Sect. 2 and how each can be represented using theory graphs. But first, we look at the aspects common to all examples to form an intuition of the theory graphs structures that are needed.

The examples in Sect. 2 are each slightly different but they have fundamental common aspects. First, each paper starts with establishing a common ground on which the results of the paper are built. This leverages the literature in two ways.

– Firstly, concepts from the literature are used to conveniently build up the local definitions. From the theory graphs perspective this functions as a (possibly partial) import.

– Secondly, properties of locally introduced concepts are *adopted* from the literature. Mathematically, this is justified by and (implicit or explicit) subsumption
between the local definition and that used by the referenced theorem. From
the theory graph perspective this function as a theory morphism that induces
the properties locally due to its truth-preserving semantics.

Therefore, a paper corresponds, not to a single theory, but to a theory pattern
that leads to a theory of the main contribution of the paper.

Secondly, the notion of "literature" and the existence of concepts beyond a
particular definition (so that equivalent definitions imply one is talking about
the same platonic concept) are common to all examples. We believe that what
happens in mathematical practice is that definition and foundational choices are
abstracted away as implementation details and the important concepts and their
properties are used as an interface to each theory (in the mathematical sense,
e.g. group theory). But this is precisely the situation that realms try to capture
in theory graphs. Therefore, we maintain that, from a theory graph perspective, informal mathematical papers refer (and contribute to) realms rather than
individual theories.

3.1 Realms

Intuitively, a realm [CFK14] is a theory structure in a theory graph G (i.e. a
subgraph of G) that abstracts from the development and provides practitioners
with the useful symbols and theorems via an *interface theory*.

We briefly introduce realms and the background concepts below and refer to
[CFK14] for details.

First, in the following, *theories* are named sets of declarations (i.e. symbols,
axioms or theorems). Additionally, *theory morphisms* (or *views*) are truth-
preserving mappings from a source theory to a target theory and formalize
inheritance and applicability of theorems. Theories can access and use declarations from other theories by importing them, either directly (*plain includes*),
or via a translation (*structures*).

An important concept for realms is
that of a *conservative extension* which
usually occurs when a theory includes
another and contains only theorems and
derived symbols (i.e. adds no axioms or
primitive symbols). An essential property
of conservative extensions is that if S' is a
conservative extension of S then there is
view v between T and S iff there is a view
between T and S' in the same direction. In
fact, we will often talk about views *modulo conservativity* below. Figure 1 shows a
prototypical realm with F as its interface
theory (also called a *face*) and n pillars

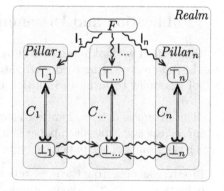

Fig. 1. The architecture of a realm

each representing a different (yet equivalent) development of the concepts in the face. Common examples are the different ways to define natural or real numbers. Each pillar is a conservative development in the sense that all theories in a pillar are conservative extensions of a bottom theory (denoted with $\perp$). A top theory (denoted with $\top$) aggregates all symbols, axioms and theorems declared within the pillar. The view pairs at the bottom establish the equivalence of the pillars and the n views I_k capture the relation of interface-implementation between the face and each pillar.

3.2   Realms as a Model for Dissemination and Aggregation

Figure 2 shows the general case for the representation of a paper as part of a theory graph. The "literature" for the mathematical theory to which the paper contributes is represented as a realm with a face and several pillars. The paper references a document within the field, that is naturally part of a pillar and grounds the recap theory. The contribution of the paper is a theory in itself that includes the recap theory and is a conservative extension of it. Again, the fact that we are representing the contribution in a single theory is a simplification for presentational simplicity which does not lead to a loss of generality. The view v ensures that the paper can make use of concepts and theorems from the realm, as they can be accessed via v.

In our analysis we first restrict ourselves to the case where there is a single recap for simplicity and expositional clarity. This already covers the majority of research papers we have analyzed; they mainly build on one earlier paper and extend it. Indeed, all three examples from Sect. 2.2 fall into this category, they import the definitions and terminology from a central cited paper, but call on others from the same realm for results, context, and support.

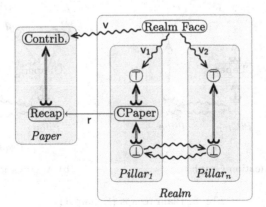

Fig. 2. General case for recaps

We recognize four special cases for (single) recaps based on the nature of r and discuss each individually below. First we have to decide the home theory of the symbols that the recap introduces. If the home is the cited theory then r is an import and we have a *plain recap* (Sect. 3.3). Otherwise, we have new symbols in the recap theory that are somehow related with the ones in the cited one. In that situation we have three sub-cases depending on the relation between the recap and cited theory: *equivalence recap* (Sect. 3.4), *specialization recap* (Sect. 3.5) and, in the informal case, *postulated recap* (Sect. 3.6).

Finally, we have the case where the paper builds on several others and, therefore, has *multiple recaps* (Sect. 3.7).

3.3   Special Case: Plain Recaps

One situation is that of plain recaps where the relation r is an inclusion into the recap from the cited paper. Typically the include r is a conservative extension of the cited paper. For instance the "covers of the multiplicative group" from Example 1 directly uses the concept from the cited paper (CPaper), but gives a concise verbalization of its definition. This allows it to make use of the results in two other papers higher up in the pillar of the cited paper. The situation is shown in Fig. 3a. Note that, if r is conservative, then we have a **pillar extension** for the realm which justifies the new paper becoming part of the realm's literature (see Fig. 3b). It also makes v exist as induced by v_1 modulo conservativity.

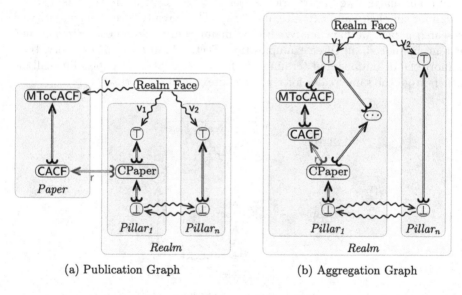

(a) Publication Graph   (b) Aggregation Graph

Fig. 3. Plain recaps (Example 1)

Plain recaps can also model the formal examples (e.g. Example 6) but in that situation it is not too interesting as we have the degenerate case for the realm itself.

3.4 Special Case: Equivalence Recap

Another common situation is that of equivalence recaps where the relation r is an equivalence (isomorphism) between the two theories. We can represent the relation r, in this case, as two views v_{to} and v_{from}, one in each direction between the recap and the cited paper that ensure their isomorphism. Then, the view v is induced by $v_{from} \circ v_1$ modulo conservativity. Moreover, the contribution of the paper carries over to the realm via the view v_{to}.

This occurs, for instance, in Example 2 where this intuition is explicitly written down in the paper as "There are several equivalent ways to define multinets." (although not proved). In fact it is the most common situation in the sample papers we studied.

Note that adding an equivalent definition corresponds to a **realm extension**, where the face is fixed, and the view from the face to the current theory can be postulated. Therefore, in Fig. 4a the paper effectively extends the realm (or the current pillar) as introduced in Sect. 3.1. This corresponds to the mathematical practice of "contributing to" a field (or mathematical theory). This resulting realm after knowledge aggregation is shown in Fig. 4b, where the new paper contributes a new pillar to the realm. The equivalence is ensured by v_{from} and v_{to} as we take into account conservativity to reduce them to the $\perp$ theory.

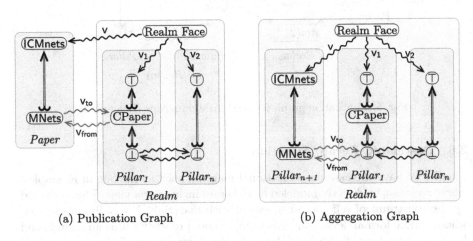

(a) Publication Graph   (b) Aggregation Graph

Fig. 4. Equivalence recaps (Example 2)

3.5 Special Case: Specialization Recap

Thirdly, we have the case where r is a specialization relation that can be represented as a view v_{from} from the cited theory to the recap. Same as in the previous case, this ensures the existence of v as $v_{from} \circ v_1$ modulo conservativity. However it does not directly contribute the results of the paper back to the (same) realm as they concern only a special case of the concepts in the realm.

This is the case in Example 3 where the definition from the paper is a specialization of the one in the literature. In [CS09], the definition of the accelerated Turing machine involves a concrete step size (2^{-n}), whereas the definition it recaps allows arbitrary sequences of step sizes as long as their sum remains finite. Thus we have the situation in Fig. 5. Theory ATM contains the (opaque) sentence (3), but there cannot be a view from ATM to atm that is more general. But we do have a view to atm(2^{-n}), which naturally arises in treatments of accelerated Turing machines as an example. That special case can form a realm of its own, namely the realm of accelerated Turing machines with step size 2^n. Then we can talk about aggregation with that realm (via the view v_{to}) but we omit that here for simplicity – the aggregation is similar as for equivalence recaps, except with the specialization realm.

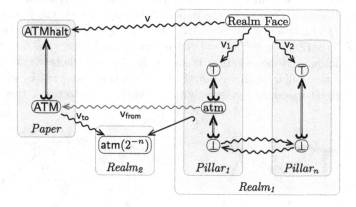

Fig. 5. Publication graph for specialization recaps (Example 3)

3.6   Postulated Recap/Adoption

Finally, we have the case for educational material such as the one in Example 4 where r cannot be directly modeled as either an include or a view. This is caused by the constraint of self-containedness of such materials. Normally, in the case where a more formal development is used we could represent it as an include and be in the case for plain recaps. However, the home theory of the new symbols must be the current development in order for it to be self-contained, so we cannot use an include. Instead we envision a special kind of import that *adopts* the included symbols effectively changing their home theory to the current one. But, then the view v is not justified so we must also assert its existence. In that case we call v a *postulated* view and the relation r is an *adoption* (see Fig. 6). We leave working out the precise details of postulated views and adoptions in flexiformal theory graphs for future work.

This is the situation in Example 4 where the recap theory SET includes only the symbols $\in$ and $\{\cdot, \cdot\}$ from the formal development ZFset, but not their axioms.

Instead the symbols are "defined" by alluding to the literature (common knowledge). We claim this verbalization effectively postulates the existence of v, by implying that the semantics of the two symbols is compatible with that given in the literature (which we represent as a realm).

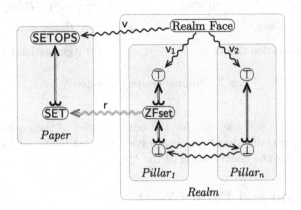

Fig. 6. Publication graph for Generalization/Unspecified recaps (Example 4)

Note that we omit the aggregation part for this case as the purpose of such educational or survey material is typically to provide a concise overview of a realm rather than to contribute to it.

3.7 Multiple Recaps

Up to now we have only treated cases with single recaps to ease the exposition. But papers and especially textbooks often recap from different realms and base the rest of the exposition on them.

This is the situation in Fig. 7a inspired by Example 5 from Rudin's book. Note that the two recaps import directly from the faces rather than from a specific pillar or paper. This is intended and covers the typical case of recaps in textbooks and survey articles. For instance, as mentioned in Sect. 2.3, Rudin does not directly cite literature in the recaps, but aggregates the vast literature in an appendix.

For the aggregation phase the multiple recaps situation begs the question of where the contribution should be placed. In the recap in Example 5 we have separate recaps of vector spaces and topological spaces (5), and we analyze them as theory morphisms from their respective realms. In this case, there is the realm of topological vector spaces (6) which imports from both realms, this is the natural place for the contributions. In the case such a realm does not exist yet, the paper can be used as the natural starting point for (first pillar of) the realm. Actually, the "union realm" concept in Fig. 7b is a bit simplified. The contribution of the paper will usually add some conditions – like conditions

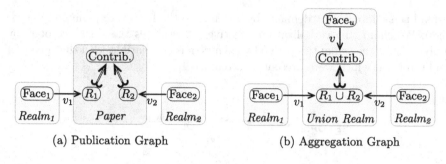

(a) Publication Graph   (b) Aggregation Graph

Fig. 7. Multiple recaps (Example 5)

(a) and (b) in (6) – and use that for the base theories of the realms. This does not invalidate our claim that there is always a natural realm – which may have to be created – for the contribution of the "paper".

4   Conclusion and Future Work

We have presented a flexiformal model of the mechanics of paper-based dissemination of research results and their aggregation into a structured knowledge commons. We model the latter as an underlying theory graph structured by inclusions and views that is further structured into a graph of realms to abstract from details of the particular low-level developments of the mathematical domains.

We identify the recap+contribution structure in mathematical papers as the mechanism by which papers can at the same time be made self-contained for human readers and by which the contribution can be integrated into the knowledge commons: the recap anchors the contribution in the commons. It is the realms structure with its equivalent pillars and abstraction capabilities that gives the recaps the necessary flexibility to adequately model the variety of anchors we see in mathematical documents.

We have validated our model by identifying the recaps and their types in 30 recent papers randomly selected from a preprint archive. To obtain a more scientific evaluation of the model, we need a much larger and more varied sample. We are currently developing an annotation ontology for realms and recaps for the KAT annotator [Dum+14] as a basis for a more principled and sustainable analysis. This will also give us the data to develop our model further.

In the future we want to look into the communication-enabling partial isomorphisms postulated in Sect. 3.6 and see whether [KRSC11] is directly applicable.

We believe that the realms-based model can be extended to handle recaps from multiple realms in one document. For the document model, this is not a problem, since we would just have multiple bases for the conservative development. For the aggregation things become more complex. Intuitively, the contribution must be integrated into a realm that is the "union" of the realms, and if that does not exist yet, the realm can be initialized with the paper at hand.

An implementation of realms in the MMT API [Rab13] is under way, this will allow us to validate the model proposed in this paper from the synthetic direction: If we have a realm-structured knowledge commons, then we may be able to auto-generate recaps and common ground sections to obtain narrative presentations of fragments that are more self-contained and readable to the human reader. This is particularly interesting for the concept of "guided tours" in content-based eLearning systems: auto-generated explanatory narratives leading to a given mathematical concept by topologically sorting the dependency relation given by the theory graph in the content commons. For the "early" parts on the border to the estimated common ground, recaps might be more suitable than direct copies of the definitions.

Acknowledgements.. This work has been supported by the Leibniz Association under grant SAW-2012-FIZ_KA-2 and the German Research Foundation (DFG) under grant KO 2428/13-1.

References

[Bar15]   Bartz, J.: Induced and Complete Multinets. In: ArXiv e-prints (February 2015). arXiv:1502.02059 [math.AG]

[CFK14]   Carette, J., Farmer, W.M., Kohlhase, M.: Realms: a structure for consolidating knowledge about mathematical theories. In: Watt, S.M., Davenport, J.H., Sexton, A.P., Sojka, P., Urban, J. (eds.) CICM 2014. LNCS, vol. 8543, pp. 252–266. Springer, Heidelberg (2014)

[CS09]   Calude, C., Staiger, L.: A note on accelerated turing machines. CDMTCS Research Report 350. Centre for Discrete Mathematics and Theoretical Computer Science, Auckland University (2009). http://www.cs.auckland.ac.nz/CDMTCS/researchreports/350cris.pdf

[Dum+14] Dumitru et al. M.A.: System description: KAT an annotation tool for STEM documents (2014). http://kwarc.info/kohlhase/submit/cicm14-kat.pdf

[HK15]   Hyttinen, T., Kangas, K.: On model theory of covers of algebraically closed fields (2015) (visited on 16 February 2015). http://arxiv.org/pdf/1502.01042.pdf

[JFM]   J. Formalized Math. JFM (visited on 27 September 2012). http://www.mizar.org/

[Koh06]   Kohlhase, Michael: A development graph for elementary algebra. In: Kohlhase, Michael (ed.) OMDoc – An Open Markup Format for Mathematical Documents [version 1.2]. LNCS (LNAI), vol. 4180, pp. 59–63. Springer, Heidelberg (2006)

[Koh13]   Kohlhase, M.: The flexiformalist manifesto. In: Voronkov, A. et al. (eds.) 14th International Workshop on Symbolic and Numeric Algorithms for Scientific Computing (SYNASC 2012), Timisoara, Romania, pp. 30–36. IEEE Press, California (2013). http://kwarc.info/kohlhase/papers/synasc13.pdf

[KRSC11] Rabe, F., Kohlhase, M., Sacerdoti Coen, C.: A foundational view on integration problems. In: Davenport, J.H., Farmer, W.M., Urban, J., Rabe, F. (eds.) MKM 2011 and Calculemus 2011. LNCS, vol. 6824, pp. 107–122. Springer, Heidelberg (2011)

[MizLib] Mizar mathematical library (visited on 27 September 2012). http://www.mizar.org/library

[Rab13] Rabe, F.: The MMT API: a generic MKM system. In: Carette, J., Aspinall, D., Lange, C., Sojka, P., Windsteiger, W. (eds.) CICM 2013. LNCS, vol. 7961, pp. 339–343. Springer, Heidelberg (2013)

[RK13a] Rabe, F., Kohlhase, M.: A scalable module system. In: Information and Computation 0.230, pp. 1–54 (2013). http://kwarc.info/frabe/Research/mmt.pdf

[RK13b] Riccardi, M., Kornilowicz, A.: Fundamental group of n-sphere for $n \geq 2$. Formalized Math. **20**(2), 97–104 (2013). doi:10.2478/v10037-012-0013-1

[Rud73] Rudin, W.: Functional Analysis. McGraw Hill (1973)

[Wen07] Wenzel, M.: Isabelle/Isar - a generic framework for human- readable proof documents. In: Matuszewski, R., Zalewska, A. (eds.) From Insight to Proof: Festschrift in Honour of Andrzej Trybulec, vol. 10(23). Studies in Logic, Grammar and Rhetoric. University of Bia lystok, pp. 277–298 (2007). http://mizar.org/trybulec65/

Mathematical Knowledge Management

Structure Formation in Large Theories

Serge Autexier and Dieter Hutter[✉]

German Research Center for Artificial Intelligence,
Bibliothekstr. 1, 28359 Bremen, Germany
{serge.autexier,dieter.hutter}@dfki.de

Abstract. Structuring theories is one of the main approaches to reduce the combinatorial explosion associated with reasoning and exploring large theories. In the past we developed the notion of development graphs as a means to represent and maintain structured theories. In this paper we present a methodology and a resulting implementation to reveal the hidden structure of flat theories by transforming them into detailed development graphs. We review our approach using plain TSTP-representations of MIZAR articles obtaining more structured and also more concise theories.

1 Introduction

It has been long recognized that the modularity of specifications is an indispensable prerequisite for an efficient reasoning in complex domains. Algebraic specification techniques provide appropriate frameworks for structuring complex specifications and the authors introduced the notion of an development graph [1,5,6] as a technical means to work with and reason about such structured specifications. While its use presupposes the development of theories having the intended structures already in mind, there are various applications of Formal Methods in which theories are automatically generated in an entirely unstructured representation. Thus, there is a need for a computer-aided structure formation for large theories, which allows for an efficient reasoning in such theories.

In this paper we present an initial approach to support structure formations in large unstructured specifications. The idea is to provide a calculus and a corresponding methodology to crystalize intrinsic structures hidden in a specification and represent them explicitly in terms of development graphs. Step by step, the specification is split into different nodes resulting in increasingly richer development graphs. On the opposite, common concepts that are scattered in different specifications are identified and unified in a common theory.

We start with a discussion on syntactical properties to measure the appropriateness of a structuring and specify invariants underlying a structure formation process. Based on this general framework we present a calculus (and heuristics to guide this calculus) to transform development graphs in order to enrich the explicitly given structure. We review our framework with the help of the Mizar Mathematical Library (http://www.mizar.org/) providing hundreds of articles which are subject to our structure formation process.

© Springer International Publishing Switzerland 2015
M. Kerber et al. (Eds.): CICM 2015, LNAI 9150, pp. 155–170, 2015.
DOI: 10.1007/978-3-319-20615-8_10

2   Development Graphs for Structure Formation

We base our framework on the notions of development graphs (and thus on the notion of institutions [4]) to specify and reason about structured specifications. Development graphs $\mathcal{D}$ are acyclic, directed graphs $\langle \mathcal{N}, \mathcal{L} \rangle$, the nodes $\mathcal{N}$ denote individual theories and the links $\mathcal{L}$ indicate theory inclusions with respect to signature morphisms attached to the links. Each node $N \in \mathcal{N}$ of the graph is a tuple (sig^N, ax^N, lem^N) such that sig^N is called the *local signature* of N, ax^N a set of *local axioms* of N, and lem^N a set of *local lemmas* of N. $\mathcal{L}$ is a set of global definition links $M \overset{\sigma}{\Longrightarrow} N$. Each link imports the mapped theory of M (by the signature morphism σ) as part of the theory of N. A node N is globally reachable from a node M via a signature morphism σ, $\mathcal{D} \vdash M \overset{\sigma}{\Longrightarrow} N$ for short, iff 1. either $M = N$ and $\sigma = id$, or 2. $M \overset{\sigma'}{\Longrightarrow} K \in \mathcal{L}$, and $\mathcal{D} \vdash K \overset{\sigma''}{\Longrightarrow} N$, with $\sigma = \sigma'' \circ \sigma'$. The global signature (global axioms and global lemmata, respectively) of a node $N \in \mathcal{N}$ is the union of its local signature (local axioms and local lemmata) and the mapped global signatures of all nodes from which N is globally reachable. A node is valid if all signature symbols occurring in its global axioms and lemmata are declared in its global signature. A development graph is well-defined, if all its nodes are valid.

The *maximal nodes* (root nodes) $\lceil \mathcal{D} \rceil$ of a graph $\mathcal{D}$ are all nodes without outgoing links. $Dom_{\mathcal{D}}(N) := Sig_{\mathcal{D}}(N) \cup Ax_{\mathcal{D}}(N) \cup Lem_{\mathcal{D}}(N)$ is the set of all signature symbols, axioms and lemmata visible in a node N. The *local domain* of N, $dom^N := sig^N \cup ax^N \cup lem^N$ is the set of all local signature symbols, axioms and lemmata of N. The *imported domain* $Imports_{\mathcal{D}}(N)$ of N in $\mathcal{D}$ is the set of all signature symbols, axioms and lemmata imported via incoming definition links. $Dom_{\mathcal{D}} = \bigcup_{N \in \mathcal{N}} Dom_{\mathcal{D}}(N)$ is the set of all signature symbols, axioms and lemmata occurring in $\mathcal{D}$. Analogously we define $Sig_{\mathcal{D}}$, $Ax_{\mathcal{D}}$, and $Lem_{\mathcal{D}}$. $Dom_{\lceil \mathcal{D} \rceil} = \bigcup_{N \in \lceil \mathcal{D} \rceil} Dom_{\mathcal{D}}(N)$ is the set of all signature symbols, axioms and lemmata occurring in the maximal nodes of $\mathcal{D}$.

Given a node $N \in \mathcal{N}$ its associated class $\mathbf{Mod}^{\mathcal{D}}(N)$ of models (or N-models for short) consists of those $Sig_{\mathcal{D}}(N)$-models n for which (i) n satisfies the local axioms ax^N, and (ii) for each $K \overset{\sigma}{\Longrightarrow} N \in \mathcal{S}$, $n|_{\sigma}$ is a K-model. In the following we denote the class of Σ-models that fulfill the Σ-sentences Ψ by $\mathbf{Mod}_{\Sigma}(\Psi)$.

Given a signature Σ and $Ax, Lem \subseteq \mathbf{Sen}(\Sigma)$, a *support mapping Supp for Ax and Lem* assigns each lemma $\varphi \in Lem$ a subset $H \subseteq Ax \cup Lem$ such that (i) $\mathbf{Mod}_{\langle sym(H) \cup sym(\varphi) \rangle_{\Sigma}}(H) \models \varphi^1$ (ii) The relation $\sqsubset \subseteq (Ax \cup Lem) \times Lem$ with $\Phi \sqsubset \varphi \Leftrightarrow (\Phi \in Supp(\varphi) \lor \exists \psi. \Phi \in Supp(\psi) \land \psi \sqsubset \varphi)$ is a well-founded strict partial order. If $\mathcal{D}$ is a development graph, then a support mapping $Supp$ is a *support mapping for $\mathcal{D}$* iff for all $N \in \mathcal{D}$ $Supp$ is a support mapping for $Ax_{\mathcal{D}}(N)$ and $Lem_{\mathcal{D}}(N)$.

We will now formalize the requirements on development graphs that reflect our intuition of an appropriate structuring for formal specifications in the following principles.

---

[1] Where $\langle S \rangle_{\Sigma}$ denotes the smallest valid sub-signature of Σ containing S.

The first principle is *semantic appropriateness*, saying that the structure of the development graph should be a syntactical reflection of the relations between the various concepts in our specification. This means that different basic specifications are located in different nodes of the graph and the links of the graph reflect the logical relations between these specifications. The second principle is *closure* saying, for instance, that deduced knowledge should be located close to the axioms guaranteeing the proofs. Also the specification defined by the theory of an individual node of a development graph should have a meaning of its own and provide some source of deduced knowledge. The third principle is *minimality* saying that each concept (or part of it) is only represented once in the graph. When splitting a monolithic theory into different theories common foundations for these theories should be (syntactically) shared between them by being located at a unique node of the graph.

We now translate these principles into syntactical criteria on development graphs and into procedures of how to transform or refactor development graphs. In a first step we formalize technical requirements to enforce the minimality-principle in terms of development graphs. Technically, we demand that each signature symbol, each axiom and each lemma has a unique location in the development graph. When we enrich a development graph with more structure we forbid to have multiple copies of the same definition in different nodes. We therefore require that we can identify for a given signature entry, axiom or lemma a *minimal theory* in a development graph and that this minimal theory is unique. We define:

Definition 1 (Providing Nodes). *Let $\langle \mathcal{N}, \mathcal{L} \rangle$ be a development graph. An entity e is* provided *in $N \in \mathcal{N}$ iff $e \in Dom_{\langle \mathcal{N}, \mathcal{L} \rangle}(N)$ and $\forall M \overset{\sigma}{\Longrightarrow} N.\ e \notin Dom_{\langle \mathcal{N}, \mathcal{L} \rangle}(M)$. Furthermore,*

1. *e is* locally provided *in N iff additionally $e \in dom^N$ holds.*
2. *e is* provided *by a link $l : M \overset{\sigma}{\Longrightarrow} N$ iff e is not locally provide in N and $\exists e' \in Dom_{\langle \mathcal{N}, \mathcal{L} \rangle}(M).\ \sigma(e') = e$ holds. In this case we say that l provides e from e'. e is* exclusively provided *by l iff e is not provided by any other link $l' \in \mathcal{L}$.*

The closure-principle demands that there are no spurious nodes in the graph not contributing anything new. We combine these requirements into the notion of location mappings:

Definition 2 (Location Mappings). *Let $\mathcal{D} = \langle \mathcal{N}, \mathcal{L} \rangle$ be a development graph. A mapping $loc_\mathcal{D} : Dom_\mathcal{D} \to \mathcal{N}$ is a* location *mapping for $\mathcal{D}$ iff*

1. *$loc_\mathcal{D}$ is surjective (closure)*
2. *$\forall N \in \mathcal{N}.\ \forall e \in dom^N.\ loc_\mathcal{D}(e) = N$*
3. *$\forall e \in Dom_\mathcal{D}.\ loc_\mathcal{D}(e)$ is the only node providing e (minimality)*

For a given $loc_\mathcal{D}$ we define $loc_\mathcal{D}^{-1} : \mathcal{N} \to 2^{Dom_\mathcal{D}}$ by

$$loc_\mathcal{D}^{-1}(N) := \{e \in Dom_\mathcal{D} | loc_\mathcal{D}(e) = N\}.$$

We write loc and loc^{-1} instead of $loc_\mathcal{D}$ and $loc_\mathcal{D}^{-1}$ if $\mathcal{D}$ is clear from the context.

Based on the notion of location mappings we formalize our intuition of a *structuring*. The idea is that the notion of being a structuring constitutes the invariant of the structure formation process and guarantees both, requirements imposed by the minimality-principle as well as basic conditions on a development graph to reflect a given formal specification.

Definition 3 (Structuring). *Let* $\mathcal{D} = \langle \mathcal{N}, \mathcal{L} \rangle$ *be a valid development graph,* $loc : Dom_\mathcal{D} \to \mathcal{N}$, $\Sigma \in |\mathbf{Sign}|$, $Ax, Lem \subseteq \mathbf{Sen}(\Sigma)$ *and Supp be a support mapping for* $\mathcal{D}$. *Then* $(\mathcal{D}, loc, Supp)$ *is a* structuring *of* (Σ, Ax, Lem) *iff*

1. *loc is a location mapping for* $\mathcal{D}$.
2. *let* $Dom_{\lceil \mathcal{D} \rceil} = \Sigma' \cup Ax' \cup Lem'$ *then* $\Sigma = \Sigma'$, $Ax = Ax'$ *and* $Lem \subseteq Lem'$.
3. $\forall \phi \in Lem_\mathcal{D} . \forall \psi \in Supp(\phi). \exists \sigma. loc(\psi) \overset{\sigma}{=\!\!=\!\!\Rightarrow} loc(\phi) \wedge \sigma(\psi) = \psi$.

3   Refactoring Rules

In the following we present the transformation rules on development graphs that transform a structuring again into a structuring. Using these rules we are able to structure the initially trivial development graph consisting of exactly one node that comprises all given concepts step by step. This initial development graph consisting of exactly one node satisfies the condition of a structuring provided that we have an appropriate support mapping at hand.

We define four types of structuring-invariant transformations: (i) horizontal splitting and merging of development graph nodes, (ii) vertical splitting and merging of development graph nodes, (iii) factorization and multiplication of development graph nodes, and (iv) removal and insertion of specific links. Splitting and merging as well as factorization and multiplication are dual operations. For lack of space and because we are mainly interested in rules increasing the structure of a development graph we will omit the formal specification of the merging and multiplication rules here.

Horizontal Split. The first refactoring rule aims at the separation of specifications in independent theories. In terms of the development graph a node is replaced by a series of independent nodes; each of them contains a distinct part from a partitioning of the specification of the original node. In order to ensure a valid new development graph, each of the new nodes imports the same theories as the old node and contributes to the same theories as the old node did. To formalize this rule we need constraints on how to split a specification in different chunks such that local lemmata are always located in a node which provides also the necessary axioms and lemmata to prove it (Fig. 1).

Definition 4. *Let* $\mathcal{S} = (\mathcal{D}, loc, Supp)$ *be a structuring of* (Σ, Ax, Lem) *and* $N \in \mathcal{N}_\mathcal{D}$. *A partitioning* $\mathcal{P}$ *for* N *is a set* $\{N_1, \ldots, N_k\}$ *with* $k > 1$ *such that 1.* $sig^N = sig^{N_1} \uplus \ldots \uplus sig^{N_k}$, $ax^N = ax^{N_1} \uplus \ldots \uplus ax^{N_k}$, $lem^N = lem^{N_1} \uplus \ldots \uplus lem^{N_k}$ *2.* $sig^{N_i} \cup ax^{N_i} \cup lem^{N_i} \neq \emptyset$ *for* $i = 1, \ldots, k$. *A node* $N_i \in \mathcal{P}$ *is* lemma independent *iff* $Supp(\psi) \cap (ax^N \cup lem^N) \subseteq (ax^{N_i} \cup lem^{N_i})$ *for all* $\psi \in lem^{N_i}$.

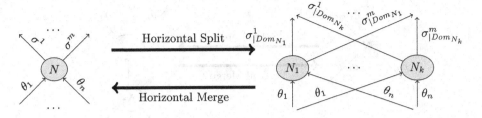

Fig. 1. Horizontal split and Merge

Definition 5 (Horizontal Split). *Let* $S = (\langle \mathcal{N}, \mathcal{L} \rangle, loc, Supp)$ *be a structuring of* (Σ, Ax, Lem), $\mathcal{P} = \{N_1, \ldots, N_k\}$ *be a partitioning for some node* $N \in \mathcal{N}$ *such that each* $N_i \in \mathcal{P}$ *is lemma independent and* $loc^{-1}(N) = dom^N$. *The horizontal split of* S *wrt.* N *and* $\mathcal{P}$ *is* $S' = (\mathcal{D}', loc', Supp)$ *with* $\mathcal{D}' = \langle \mathcal{N}', \mathcal{L}' \rangle$ *where*

1. $\mathcal{N}' := \{N_1, \ldots, N_k\} \uplus (\mathcal{N} \setminus N)$
2. $\mathcal{L}' := \{M \overset{\sigma}{\Longrightarrow} M' \in \mathcal{L} | M \neq N \wedge M' \neq N\}$

$$\cup \{M \overset{\theta}{\Longrightarrow} N_i | M \overset{\theta}{\Longrightarrow} N \in \mathcal{L}, i \in \{1, \ldots, k\}\}$$

$$\cup \{N_i \overset{\tau|Dom_{N_i}}{\Longrightarrow} M | N \overset{\tau}{\Longrightarrow} M \in \mathcal{L}, i \in \{1, \ldots, k\}\}$$

3. $loc'(e) := N_i$ *if* $e \in dom^{N_i}$ *for some* $i \in \{1, \ldots, k\}$ *and* $loc'(e) := loc(e)$ *otherwise.*

such that $Sig_{\mathcal{D}'}(N_i)$ *are valid signatures and* $ax_i, lem_i \subseteq \mathbf{Sen}(Sig_{\mathcal{D}'}(N_i))$ *for* $i = 1, \ldots, k$.

Vertical Split. Similar to a horizontal split we introduce a vertical split which divides a node into two nodes and locates one node on top of the other. While all outgoing links start at the top node, we are free to reallocate incoming links to either node (Fig. 2).

Definition 6 (Vertical Split). *Let* $S = (\langle \mathcal{N}, \mathcal{L} \rangle, loc, Supp)$ *be a structuring of* (Σ, Ax, Lem) *and* $\mathcal{P} = \{N_1, N_2\}$ *be a partitioning for some* $N \in \mathcal{N}$ *such that* N_1 *is lemma independent. Then, the vertical split* S *wrt.* N *and* $\mathcal{P}$ *is* $S' = (\mathcal{D}', loc', Supp)$ *with* $\mathcal{D}' = \langle \mathcal{N}', \mathcal{L}' \rangle$ *where*

$$\mathcal{N}' := \{N_1, N_2\} \uplus (\mathcal{N} \setminus N)$$

$$\mathcal{L}' := \{M \overset{\sigma}{\Longrightarrow} M' \in \mathcal{L} | M \neq N \wedge M' \neq N\} \cup \{N_1 \overset{id}{\Longrightarrow} N_2\}$$

$$\cup \{M \overset{\sigma}{\Longrightarrow} N_1 | M \overset{\sigma}{\Longrightarrow} N \in \mathcal{L}\} \cup \{N_2 \overset{\sigma}{\Longrightarrow} M | N \overset{\sigma}{\Longrightarrow} M \in \mathcal{L}\}$$

$$loc'(e) = \begin{cases} N_2 & \text{if } loc(e) = N \text{ and } e \in Dom_{\mathcal{D}'}(N_2) \\ N_1 & \text{if } loc(e) = N \text{ and } e \notin Dom_{\mathcal{D}'}(N_2) \\ loc(e) & \text{otherwise} \end{cases}$$

such that $Sig_{\mathcal{D}'}(N_i), i = 1, 2$, *are valid signatures and* $ax_i, lem_i \subseteq \mathbf{Sen}(Sig_{\mathcal{D}'}(N_i))$, $i = 1, 2$. *Conversely,* S *is a vertical merge of* N_1 *and* N_2 *in* S'.

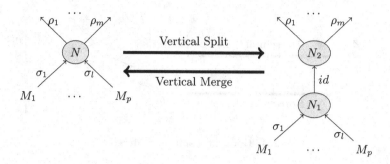

Fig. 2. Vertical split and Merge

Example 1. We illustrate the horizontal and vertical split rules by considering a single theory axiomatizing a Field with binary operations $+$ and $\times$ consisting of a Distributivity axiom ($\Phi_D := \forall x, y, z.x \times (y + z) = x \times y + x \times z$) and the axioms of an Abelian Group for $+$ and $\times$, respectively ($\Phi_{AG}^+ := \forall x, y, z \,.\, x+(y+z) = (x+y)+z, \forall x, y \,.\, x+y = y+x, \forall x \,.\, x+0 = x, \forall x \,.\, x+-(x) = 0$ and $\Phi_{AG}^\times := \forall x, y, z \,.\, x \times (y \times z) = (x \times y) \times z, \forall x, y \,.\, x \times y = y \times x, \forall x \,.\, x \times 1 = x, \forall x \,.\, x \times inv(x) = 1$). Assume axioms are contained in a single node Field, which forms a trivial structuring. In a first step we can split that node vertically by separating the distributivity axiom from the other axioms. In a second step we can separate the Abelian Group axioms for $+$ and $\times$ by a horizontal split. This is shown in the following Figure:

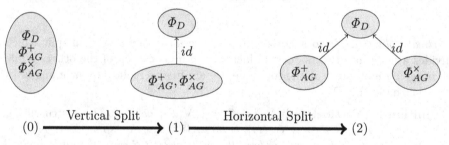

Factorization. The factorization rule allows one to merge equivalent specifications into a single generalized specification and then to represent the individual ones as instantiations of the generalized specification. A precondition of this rule is that all individual specifications inherit the same (underlying) theories (Fig. 3).

Definition 7 (Factorization). *Let $\mathcal{S} = (\langle \mathcal{N}, \mathcal{L} \rangle, loc, Supp)$ be a structuring of (Σ, Ax, Lem). Let $K_1, \ldots, K_n, M_1, \ldots, M_p \in \mathcal{N}$ with $p > 1$ such that $sig^{M_j} \cup ax^{M_j} \neq \emptyset$ and $\exists \sigma_{i,j}. K_i \xrightarrow{\sigma_{i,j}} M_j \in \mathcal{L}$ for $i = 1, \ldots, n, j = 1, \ldots, p$.*

Suppose there are sets sig, ax and lem with $(sig \cup ax \cup lem) \cap Dom_{\mathcal{D}} = \emptyset$ and signature morphisms $\theta_1, \ldots, \theta_p$ and $\sigma_1, \ldots, \sigma_n$ such that

– $\forall e \in Dom_{\mathcal{D}}(K_i). \; \theta_j(\sigma_i(e)) = \sigma_{i,j}(e)$ and $\sigma_{i,j}(e) = e \lor \sigma_{i,j}(e) \notin Dom_{\mathcal{D}}$

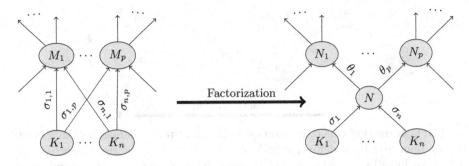

Fig. 3. Factorization (with $\sigma_{i,j} := \theta_j \circ \sigma_i$)

- $sig^{M_j} \subseteq \theta_j(sig) \subseteq Dom_{\mathcal{D}}(M_j)$, $ax^{M_j} \subseteq \theta_j(ax) \subseteq Dom_{\mathcal{D}}(M_j)$
- $\forall e \in lem$ holds $\exists l \in \{1, \ldots p\}$. $\theta_l(e) \in lem^{M_l}$, $\theta_i(e) = \theta_j(e)$ implies $i = j$ and $\theta_j(e) \in Dom_{\mathcal{D}}$ implies $loc(\theta_j(e)) \in M_j$
- there is a support mapping $Supp_N$ for $ax \cup \bigcup_{i=1,\ldots,n} \sigma_i(Dom_{\mathcal{D}}(K_i))$ and lem.

Then $\mathcal{S}' = (\langle \mathcal{N}', \mathcal{L}' \rangle, loc', Supp')$ is a factorization of $\mathcal{S}$ wrt. $M_1, \ldots, M_p$ and $Supp_N$ iff

$$\mathcal{N}' := \{N\} \cup \{N_j | j \in \{1, \ldots p\}\} \cup \mathcal{N} \setminus \{M_1, \ldots M_p\}$$

$$\text{with } N = \langle sig, ax, lem \rangle, N_j = \langle \emptyset, \emptyset, lem^{M_j} \setminus \theta_j(lem) \rangle$$

$$\mathcal{L}' := \{K \overset{\sigma}{\Longrightarrow} K' \in \mathcal{L} | K, K' \notin \{M_1, \ldots M_p\}$$

$$\cup \{K_i \overset{\sigma_i}{\Longrightarrow} N | K_i \overset{\sigma_{i,j}}{\Longrightarrow} M_j, j \in \{1, \ldots p\}, i \in \{1, \ldots n\}\}$$

$$\cup \{N \overset{\theta_j}{\Longrightarrow} N_j | j \in \{1, \ldots p\}\}$$

$$\cup \{K \overset{\tau}{\Longrightarrow} N_j | K \overset{\tau}{\Longrightarrow} M_j \wedge (\forall i \in \{1, \ldots n\}.K \neq K_i \wedge \tau \neq \sigma_{i,j})$$

$$\cup \{N_j \overset{\tau}{\Longrightarrow} K | M_j \overset{\tau}{\Longrightarrow} K \in \mathcal{L}, j \in \{1, \ldots p\}\}$$

$$loc'(x) := \begin{cases} N & \text{if } x \in Dom_{\mathcal{D}'}(N) \setminus \bigcup_{i=1,\ldots,n} Dom_{\mathcal{D}'}(K_i) \\ N_j & \text{if } x \in Dom_{\mathcal{D}}(N_j) \text{ and } \forall K \overset{\sigma}{\Longrightarrow} N_j. x \notin Dom_{\mathcal{D}'}(K) \\ loc(x) & \text{otherwise.} \end{cases}$$

$$Supp' := Supp \cup Supp_N.$$

Example 2. Consider again our example a Field axioms, which we have transformed into the structuring (3) (p. 6). On the last structuring (3) we can apply the factorization rule to extract the general abelian group axioms ($\Phi_{AG}^\circ :=$ $\forall x, y, z. x \circ (y \circ z) = (x \circ y) \circ z, \forall x, y. x \circ y = y \circ x, \forall x. x \circ e = x, \forall x. x \circ i(x) = e$) and obtain the respective axioms for $+$ and $\times$ by morphisms $\sigma_1 := \circ \mapsto +, e \mapsto 0, i \mapsto -$ and $\sigma_2 := \circ \mapsto \times, e \mapsto 1, i \mapsto$ inv. This is illustrated in the following diagram and the final structuring contains 5 axioms and the initial structuring contained 9 axioms.

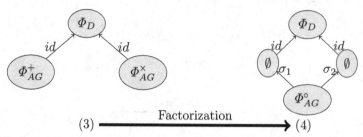

$$(3) \xrightarrow{\text{Factorization}} (4)$$

The factorization rule only covers a sufficient criterion demanding that each theory imported by a definition link to one specification is also imported via definition links by all other specifications. The more complex case in which a theory is imported via a path of links can be handled by allowing one to shortcut a path in a single global link. This results in the following rule.

Definition 8 (Transitive Enrichment). *Let* $S = (\langle \mathcal{N}, \mathcal{L} \rangle, loc, Supp)$ *be a structuring of* (Σ, Ax, Lem), $K, N \in \mathcal{N}$ *and there is a path* $K \overset{\sigma}{\Longrightarrow} N$ *between both. Then,* $S' = (\langle \mathcal{N}, \mathcal{L} \cup \{ K \overset{\sigma}{\Longrightarrow} N \} \rangle, loc, Supp)$ *is a transitive enrichment of* $\mathcal{D}$.

Definition links in a development graph can be redundant, if there are alternatives paths which have the same morphisms or if they are not used in any reachable node of the target. We formalize these notions as follows:

Definition 9 (Removable Link). *Let* $S = (\mathcal{D}, loc, Supp)$ $(\mathcal{D} = \langle \mathcal{N}, \mathcal{L} \rangle)$ *be a structuring of* (Σ, Ax, Lem). *Let* $l \in \mathcal{L}$ *and* $\mathcal{D}' = \langle \mathcal{N}, \mathcal{L} \setminus \{l\} \rangle$. l *is removable from* S *and* $S' = (\mathcal{D}', loc, Supp)$ *is a reduction of* S *iff*

1. $\forall l' : M \overset{\sigma}{\Longrightarrow} N.$ *if* l' *provides exclusively* $\sigma(e)$ *from some* $e \in Dom_{\mathcal{D}}(M)$ *then* $e \in Dom_{\mathcal{D}'}(N)$ *and* $l \neq l'$;
2. $\forall e \in Dom_{\mathcal{D}}.\forall M \in \lceil \mathcal{D} \rceil.$ *if* $loc(e) \overset{\sigma}{\Longrightarrow} M$ *then there exists* $M' \in \lceil \mathcal{D}' \rceil$ *such that* $loc(e) \overset{\sigma}{\Longrightarrow} M'$;
3. $\forall \phi \in Lem_{\mathcal{D}}.$ $Supp(\phi) \subseteq Dom_{\mathcal{D}'}(N)$ *and* $\forall Sig_{\mathcal{D}}^{loc}(N) \subseteq Dom_{\mathcal{D}'}(N)$.

Theorem 1 (Structuring Preservation). *Let* $S := (\mathcal{D}, loc, Supp)$ $(\mathcal{D} = \langle \mathcal{N}, \mathcal{L} \rangle)$ *be a structuring of* (Σ, Ax, Lem). *Then*

1. *every horizontal split of* S *wrt. some* $N \in \mathcal{N}$ *and partitioning* $\mathcal{P}$ *of* N,
2. *every vertical split of* S *wrt. some* $N \in \mathcal{N}$ *and partitioning* $\mathcal{P}$ *of* N,
3. *every factorization of* S *wrt. nodes* $M_1, \ldots M_p \in \mathcal{N}$,
4. *every transitive enrichment of* S, *and*
5. *every reduction of* S

is a structuring of (Σ, Ax, Lem).

The theorem follows from the soundness proofs for each rule given in Appendix 6.

4   Refactoring Process

In order to evaluate the refactoring rules on real theories we have implemented the development graphs and the rules in Scala[2] and added support to read formulas in TSTP format [9] using the Java parser from [8]. The support mapping is given as an extra datastructure representing the information which formula has been used in the proof of a theorem. In the case of TSTP we extract that information from the files by using the names of the formulas. Since the TSTP format does not include signature declarations, we add declarations for all occurring symbols in a TSTP file in an initialization step. We used the untyped part of TSTP and hence the declarations only contain arity information but no types.

The refactoring rules are parameterized over the theories and possibly the subsets of the local signature, axioms and lemmata to split over. To compute the parametric information we provided some basic heuristic tactics. Using the support mapping, we define that an axiom (resp. lemma) depends on a symbol declaration, if the symbol occurs in the axiom (resp. lemma) and a lemma depends on another axiom or lemma, if the latter is in its support mapping. A symbol declaration is always independent. This dependency relation induces a partial order on the local domain of each node in a development graph.

Tactic for Horizontal Split. This rule requires the partitioning of the local signature, axioms and lemmas for a given theory into independent parts such that given the same imports than the original node, each part is a valid theory and lemma independent of the other part. We implemented a heuristic that given a local domain of some node, searches for a largest subset which has a non-empty intersection of its occurring symbols and supporting axioms and lemmata. If such a set exists, the largest such set is used to split the theory horizontally into that set and the rest.

Tactics for Vertical Split. The rule requires to find a subset of the local domain, which is independent of the rest and use it as the content of the lower theory. We implemented two heuristics to search for this subset. First, we consider all maximal elements wrt. the dependency relation and use that as content for the new upper theory constructed by vertical split. Second, we consider all minimal elements and use it as content for the lower theory constructed by vertical split. These two tactics allow one to incrementally split a theory into layered slices of the dependency relation.

Tactic for Factorization. This rule requires to find isomorphic subsets in two different theories to factorize over. The notion of isomorphism between formulas is very strict, as we only search for renamings. Furthermore, we extended the isomorphism to the support mapping such that lemmata can only be identified with isomorphic lemmata which supporting axioms and lemmata are also isomorphic wrt. the same renaming. Thus, an axiom can never be factorized with a lemma and vice-versa. Even with that strict notion, computation of such subsets is already expensive. If the entire local domain of a given node is isomorphic to

---

[2] http://www.scala-lang.org/.

the local domain of the second node, both nodes are factorized according the definition of the factorization rule. If the identified subset in the first node does not cover the complete second node, we first try to split the second node to isolate the subset. To this end we first try to split the second node horizontally using the identified subset. If that fails, we first try to split vertically using the subset for the upper part and finally as the lower part. If one of these splittings was successful, the factorization is applied on the isolated part. Otherwise the factorization fails.

In addition to these main tactics, we have implemented the tactics to delete superfluous links as well as deletion of empty nodes which technically corresponds to vertically merging the empty node with their importing theories.

Automatic Procedure. In order to automate the theory formation process we have implemented the usual tacticals to describe more complex search behaviors. The tactic language is defined as follows starting from the basic tactics described above:

$$T ::= SplitHorizontal \mid SplitVerticallyMaximal \mid SplitVerticallyMinimal$$
$$\mid \ Factorize \mid RemoveSuperfluousEmptyTheories$$
$$\mid \ T* \mid T+ \mid T;T \mid T \ onfail \ T$$

The tactics take as argument a structuring and if they could be applied, return a new structuring and otherwise fail. The tacticals for as many as possible iteration ($*$), as many as possible but at least one ($+$) and sequencing (;) are standard. The tactical *onfail* executes the second tactic expression only if the first failed. Using this language we have implemented the following automatic procedure. The goal of the procedure is starting from an unstructured graph, i.e. a single theory containing all declarations, axioms and lemmata, to search for possibilities to factorize common patterns. Factorization is only possible if at least one application of the horizontal split rule was possible, which in turn may require the application of a preparatory vertical split. Following that initial part, we try to split further vertically using the maximal elements of the theory and finally removing the superfluous links and empty theories. Hence, the initial phase of the automation consists of

$$inittac \triangleq ((SplitVerticallyMinimalEntries+; SplitHorizontally*)$$
$$onfail \ SplitHorizontally+);$$
$$SplitVerticallyMaximalEntries*;$$
$$RemoveSuperfluousEmptyTheories*$$

That initialization tactic succeeds only if at least one vertical split or one horizontal split could be done. Following that, we start to factorize. If at least one factorization was possible, we first clean up the structuring by removing superfluous links and empty theories before trying again to split vertically. The overall tactic is thus

$$inittac; (Factorize+; RemoveSuperfluousEmptyTheories*;$$
$$SplitVerticallyMinimalEntries*)*$$

Article	Axioms	Theorems	Reduction	Timeout
binop_2.top.rated	21 / 19	28 / 28	5%	yes
bintree1.top.rated	62 / 61	16 / 16	2%	no
cfuncdom.top.rated	25 / 24	40 / 40	2%	no
ff_siec.top.rated	52 / 51	32 / 32	2%	no
finsub_1.top.rated	38 / 37	16 / 16	2%	no
heine.top.rated	96 / 95	13 / 13	1%	no
membered.top.rated	17 / 17	36 / 16	38%	no
mssubfam.top.rated	84 / 83	55 / 55	1%	no
msualg_1.top.rated	49 / 48	13 / 13	2%	no
power.top.rated	103 / 102	61 / 61	1%	yes
qc_lang1.top.rated	86 / 85	23 / 23	1%	no
rsspace.top.rated	46 / 45	20 / 20	2%	no
setfam_1.top.rated	51 / 48	44 / 44	4%	no

Fig. 4. Factorization results on TSTP versions of the Mizar articles

5   Evaluation

We have applied the factorization procedure presented in the previous section
to TSTP versions of the Mizar library articles www.mizar.org, which have been
created by Joseph Urban and are available at http://www.cs.miami.edu/~tptp/
MizarTPTP/TPTPArticles/. This is a collection of 922 files in TSTP format
(www.cs.miami.edu/~tptp/TSTP) where theorems are annotated by information which theorems and axioms have been used in their proofs. The files consist
of the axioms and theorems of each article including all directly included articles,
but without transitive expansion of all inclusions. Hence, the knowledge in each
file is already quite tailored to the knowledge necessary to define the additional
mathematical concepts and to enable the proofs of the theorems. We have run
the procedure on all examples with a timeout of 5 min each. The environment
was a virtual machine with 4 virtual CPUs, 16 GB RAM, under openSuSE 12.2
64-bit, running on a host with 2 Intel Xeon Westmere E5620 QuadCore CPUs,
2,4 GHz, 96 GB RAM and VMware ESXi 4.1.

For most articles no factorization has been found. However, there are 13
articles where factorization was possible, which are presented in the table Fig. 4.
The results are summarized in the following format: for each file we indicate
in the **Axioms** column the number of axioms in the initial development graph
and the final development graph. Analogously, the **Theorems** column indicates
the number of theorems respectively in the initial and the final development
graph. The **Reduction** column indicates how much the factorization reduced
the overall number of axioms and theorems. The last column indicates if the
automatic procedure had terminated within the 5 min time frame or timeout
was reached.

While reducing the number of axioms by factorization is already interesting in order to reduce the search space for automatic provers, reducing the number of theorems is more interesting as it means less theorems to prove. For all but one file where factorizations have been found, only axiom factorization have been found. However, in the article **membered. top.rated** obtained from the Mizar article [10] "On the Sets Inhabited by Numbers" we could factorize 36 theorems into 16 theorems. On closer inspection this is not surprising because it concerned theorems about sets of reals, sets of rationals, sets of integers, sets of naturals and sets of complex numbers, all defined and proved according to the same schema. The resulting development graph is shown on the right side of Fig. 4, and the factor theory containing the 5 theorems, from which all others are obtained by renaming, is node 9 in gray/orange. The factorization is visible via the 5 outgoing edges towards node 11 which are annotated with the respective morphisms (Fig. 5).

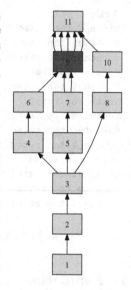

Fig. 5. Resulting DG

6   Related Work and Conclusion

Related to the structuring of theories, there is a large work on anti-unification, i.e. computing common generalizations of different formuala or theories (e.g. [2,3,7]). The resulting structuring approach is primarily bottom-up and driven by the pure existence of anti-unifiers. In contrast, our approach is top-down as it introduces measures for the intended structuring (i.e. semantic appropriateness, closure and minimality) to guide the formation process. For example, we split up theories in smaller ones but that are still self-contained in the sense that each theorem of the original theory can be proven in one of the new (smaller) ones. Anti-unification is an important technique to test the applicability of the factorization rule, for instance, but applicability of a rule is not the driving force of the formation process.

In this paper we were concerned with trying to reveal shared definitions, axiomatizations and theorems in a given formal theory. Based on *structurings* which extend development graphs with notions to exclude redundancies and include dependency information, we presented a set of rules on structurings. We implemented the rules with simple heuristics to detect isomorphic subsets which are sufficient to find simple factorization and applied it to the TSTP formulations of the Mizar articles. Not surprisingly, not many factorizations could be found, which is due to Mizar's non-transitive reuse principle of other articles and the fact that these were chosen carefully by the authors of the Mizar article. Moreover, the heuristics to compute isomorphic axioms and theorems was very restricted. However, a few factorizations could be found, and especially one were the number of theorems could be halved. This indicates that adding theory morphisms to the

Mizar language may be useful, but that needs to be confirmed by further analysis of larger subsets. On the other hand the non-transitive import mechanisms of Mizar already seems to allow for a good organization of the knowledge. That kind of mechanism is typically not implemented in specification languages, but exists in development graphs in form of local axiom links.

Future work will consist of analyzing larger subsets of the whole Mizar library, i.e. sets of Mizar articles, for possible factorizations. We also plan to apply it to libraries of other proof assistants assuming we can get the dependency information which axioms/theorems have been used in which proof. Also other automation tactics and especially heuristics to identify isomorphic formulas need to be explored, as well as heuristics to identify subsets for horizontal and vertical splits. On a more theoretical level, we will investigate how axioms and theorems could be identified, in order to allow to factorize alternative axiomatizations of the same theory without losing information, such as, e.g., alternative forms to axiomatize groups. Finally, the whole system can be applied to any untyped first-order subset of TPTP theories to search for redundancies. However, the resulting development graphs cannot be saved as TPTP theories, as it does not support renaming. Hence, we propose to extend the TPTP language in that respect.

Proof of Theorem 1 (Structure Preservation)

Horizontal Split

It holds trivially that $Dom_D = Dom_{D'}$.

- loc' is surjective because by construction each N_i, $i = 1, \ldots, k$ has a local entity. Furthermore, for each N_i and each $e \in dom^{N_i}$ holds $loc'(e) = N_i$ by construction. Furthermore, since $loc^{-1}(N) = dom^N$, none of the incoming links into N provided any entity, and consequently none of the incoming links into $N_1, \ldots, N_k$ do. Hence, $loc'^{-1}(N_i) = dom^{N_i}$, $i = 1, 2$ and since $dom^N := dom^{N_1} \uplus \ldots \uplus dom^{N_k}$, $loc'(e)$ is unique for $e \in dom^N$.
- If N is not a top-level node in D, then $Dom_{\lceil D' \rceil} = Dom_{\lceil D \rceil} = \Sigma \uplus Ax \uplus Lem$ because the domains of nodes reachable from N are not affected by the horizontal split. If N is a top-level node, then all N_i with $1 \le i \le k$ are top-level nodes. Since $dom^N = dom^{N_1} \uplus \ldots \uplus dom^{N_k}$ and $Imports_D(N) = Imports_{D'}(N_1) = \ldots = Imports_{D'}(N_k)$, it holds

$$Dom_D(N) = dom^N \cup Imports_D(N) = dom^{N_1} \cup \ldots dom^{N_k} \cup Imports_D(N)$$
$$= dom^{N_1} \cup \ldots dom^{N_k} \cup Imports_{D'}(N_1) \cup \ldots \cup Imports_{D'}(N_k)$$
$$= dom^{N_1} \cup Imports_{D'}(N_1) \cup \ldots \cup dom^{N_k} \cup Imports_{D'}(N_k)$$
$$= Dom_{D'}(N_1) \cup \ldots \cup Dom_{D'}(N_k)$$

Thus, $Dom_{\lceil D' \rceil} = Dom_{\lceil D \rceil} = \Sigma \uplus Ax \uplus Lem$.

- Assume $\phi \in Lem_{\mathcal{D}}$ and $\psi \in Supp(\phi)$. If $loc_{\mathcal{D}}(\psi) \neq N$ and $loc_{\mathcal{D}}(\phi) \neq N$, then both $loc_{\mathcal{D}}(\psi), loc_{\mathcal{D}}(\phi)$ are in $\mathcal{D}'$ and we consider $p : loc_{\mathcal{D}}(\psi) \overset{\sigma}{\Longrightarrow} loc_{\mathcal{D}}(\phi)$. If $N \in p$ then $p := [p_1, M \overset{\theta}{\Longrightarrow} N \overset{\tau}{\Longrightarrow} M', p_2]$ and by construction the path $[p_1, M \overset{\theta}{\Longrightarrow} N_i \overset{\tau_{|Dom_{N_i}}}{\Longrightarrow} M', p_2]$ are in $\mathcal{D}'$ for $1 \leq i \leq k$. Since $loc_{\mathcal{D}}(\psi) \neq N$, each $\tau_{|Dom_{N_i}}$ behaves equivalently on the image of ψ imported in N_i and hence $loc_{\mathcal{D}'}(\psi) \overset{\sigma'}{\Longrightarrow} loc_{\mathcal{D}'}(\phi)$ for some σ' such that $\sigma'(\psi) = \sigma(\psi)$. If $N \notin p$, then p is also a path in $\mathcal{D}'$ and $loc_{\mathcal{D}'}(\psi) \overset{\sigma}{\Longrightarrow} loc_{\mathcal{D}'}(\phi)$ holds trivially.

  If $loc_{\mathcal{D}}(\phi) = N$ then since all N_i are mutually lemma independent, without loss of generality we can assume $\phi \in ax^{N_1} \cup lem^{N_1}$ and this $loc'_{\mathcal{D}'}(\phi) = N_1$. If $loc_{\mathcal{D}}(\psi) = N$, then $\psi' \in ax^{N_1} \cup lem^{N_1}$ because N_1 is lemma independent. Thus, $loc'_{\mathcal{D}'}(\psi) = N_1$ and $loc'_{\mathcal{D}'}(\psi) = N_1 \overset{id}{\Longrightarrow} N_1 = loc'_{\mathcal{D}'}(\phi)$ holds trivially. Otherwise, $loc_{\mathcal{D}}(\psi) = loc'_{\mathcal{D}'}(\psi)$ and since N was reachable from $loc_{\mathcal{D}}(\psi)$ by construction N_1 is also reachable from $loc'_{\mathcal{D}'}(\psi)$.

Vertical Split

- First, we have to prove that loc' is a location mapping. loc' is surjective because by construction each node N_i (with $i = 1, 2$) has some local entity $e \in dom^{N_i}$. Thus $loc'(e) = N_i$ and N_i is in the range of loc'. Furthermore, $\forall e \in dom^{N_i}$. $loc'(e) = N_i$ holds by definition. Finally, let $e \in Dom_{\mathcal{D}'} = Dom_{\mathcal{D}}$: $loc'(e) = N_i$ implies $loc(e) = N$ and therefore there is no node in $\mathcal{N} \setminus \{N\}$ which provides e. Furthermore, since $N_1 \overset{id}{\Longrightarrow} N_2 \in \mathcal{L}'$, N_1 and N_2 cannot provide the same entity e.
- By definition $\forall e \in dom^{N_i}$ implies $loc'(e) = N_i$ for $i = 1, 2$ in $\mathcal{D}'$. For all other nodes in $\mathcal{D}' \setminus \{N_1, N_2\}$ the property is inherited by $(\mathcal{D}, loc, Supp)$ being a structuring and $loc(e) = loc'(e)$ if $loc(e) \neq N$.
- Since $Dom_{\mathcal{D}}(N) = Dom_{\mathcal{D}'}(N_2)$ and $N \overset{\sigma}{\Longrightarrow} M \in \mathcal{D}$ iff $N_2 \overset{\sigma}{\Longrightarrow} M \in \mathcal{D}'$ $Dom_{\lceil \mathcal{D} \rceil} = Dom_{\lceil \mathcal{D}' \rceil}$.
- Suppose $\phi \in Lem_{\mathcal{D}}, \psi \in Supp(\phi)$ with $loc(\phi) = M$ and $loc(\psi) = M'$. If $N \notin \{M, M'\}$ then $loc'(\phi) = M, loc'(\psi) = M'$ and $M \overset{\sigma}{\Longrightarrow} M'$ in $\mathcal{D}'$ trivially. If $M = N$ and $M' \neq N$ then $loc'(\phi) \in \{N_1, N_2\}$, and again $N_i \overset{\sigma}{\Longrightarrow} M'$ in $\mathcal{D}'$. The case of $M \neq N$ and $M' = N$ is proven analogously. We are left with the case of $M = M' = N$.

  Since N_1 is independent of N_2 , it holds that for all $\phi' \in ax^{N_1} \cup lem^{N_1}$. $Supp(\phi) \cap (ax^{N_2} \cup lem^{N_2}) = \emptyset$.

  Thus $\phi \in ax^{N_1} \cup lem^{N_1}$ implies that $\psi \in ax^{N_1} \cup lem^{N_1}$ as well and $N_1 \overset{id}{\Longrightarrow} N_1$ holds trivially. $\qquad \square$

Factorization

- We have to prove that loc' is a location mapping. First, we prove that loc' is surjective. For any node $K \in \mathcal{N}' \setminus \{N, N_1, \ldots N_p\}$ $loc^{-1}(K) = loc^{-1}(K)$

holds. Since $sig^N \cup ax^N \neq \emptyset$ but $(sig^N \cup ax^N) \cap Dom_{\mathcal{D}} = \emptyset$ it holds that $sig^N \cup ax^N \subseteq loc'^{-1}(N)$. Furthermore, $sig^{M_j} \cup ax^{M_j} \subseteq loc'^{-1}(N_j)$ since $sig^{M_j} \cup ax^{M_j} \subseteq \theta_j(sig^N \cup ax^N)$ and $\theta_j(sig^N \cup ax^N) \cap (sig^N \cup ax^N) = \emptyset$.

Second we have to prove $\forall K \in \mathcal{N}'. \forall e \in dom^K. loc'(e) = K$ holds. If $K \notin \{N, N_1, \ldots N_p\}$ then $loc'(e) = loc(e) = K$. If $K = N$ then $dom^N \in Dom_{\mathcal{D}'}(N)$ and $dom^N \nsubseteq Dom_{\mathcal{D}}(K_i)$ for $i = 1, \ldots, n$ because $dom^N \cap Dom_{\mathcal{D}} = \emptyset$. Thus $\forall e \in dom^N. loc'_{\mathcal{D}'}(e) = N$. Finally, if $K = N_j$ then $dom^{N_j} = lem^{M_j} \setminus \theta_j(lem)$ In particular, $dom^{N_j} \cap Dom_{\mathcal{D}'}(N) = \emptyset$ implying that $loc'_{\mathcal{D}'}(e) = N_j$ for all $e \in dom^{N_j}$.

Third, we prove that all $e \in Dom_{DG'}$ are provided by a unique node. The only interesting case is that e is provided by N or some N_j. In case of N both dom^N and also entries provided by some link from K_i are by definition not in $Dom_{\mathcal{D}}$ and thus not provided by any node already in $\mathcal{D}$ but by definition also not provided by N_j. It remains the case that an entry e is provided by two nodes N_i and N_j. Since all $e \in Dom_{DG}$ were provided by a unique node, this implies that e has to be a mapped lemma of N but that violates the precondition that each θ_i has to map e into a different entity.

- Next we prove that $\mathcal{D}$ and $\mathcal{D}'$ coincide in the entities they provide at their maximal nodes. Since N is not a maximal node, it is sufficient to prove that N_j and M_j coincide in their provided entities:

$$Dom_{\mathcal{D}'}(N_j) = lem^{M_j} \setminus \theta_j(lem) \cup \bigcup\{\sigma(Dom_{\mathcal{D}'}(K)) \mid K \overset{\sigma}{\Longrightarrow} N_j\}$$
$$= lem^{M_j} \setminus \theta_j(lem) \cup \bigcup\{\sigma(Dom_{\mathcal{D}'}(K)) \mid K \overset{\sigma}{\Longrightarrow} N_j, K \neq N\}$$
$$\cup \; \theta_j(sig) \cup \theta_j(ax) \cup \theta_j(lem) \cup \bigcup\{\sigma_{i,j}(Dom_{\mathcal{D}}(K_{i,j})) \mid i = 1 \ldots n\}$$
$$= lem^{M_j} \cup sig^{M_j} \cup ax^{M_j}$$
$$\cup \bigcup\{\sigma(Dom_{\mathcal{D}}(K)) \mid K \overset{\sigma}{\Longrightarrow} M_j, K \neq K_i, \sigma \neq \sigma_{i,j}\}$$
$$\cup \bigcup\{\sigma_{i,j}(Dom_{\mathcal{D}}(K_{i,j})) \mid i = 1 \ldots n\} \cup \theta_j(lem)$$
$$= Dom_{\mathcal{D}}(M_j) \cup \theta_j(lem).$$

- Suppose $\phi \in Lem_{\mathcal{D}'}$ and $\psi \in Supp_{\mathcal{D}'}(\phi)$. If $loc'(\phi), loc'(\psi) \notin \{N, N_1, \ldots N_p\}$ then $loc'(\phi) = loc(\phi)$ and $loc'(\psi) = loc(\psi)$ and therefore, $\exists \sigma. loc(\psi) \overset{\sigma}{\rightarrowtail} loc(\phi)$ with $\sigma(\psi) = \psi$ in $\mathcal{D}$. Since $\mathcal{D}'$ inherits all links away from $M_1, \ldots M_p$ and paths travesing some K_i and M_j can be mapped to paths traversing K_i, N, and N_j. $\exists \sigma. loc'(\psi) \overset{\sigma}{\rightarrowtail} loc'(\phi)$ with $\sigma(\psi) = \psi$ also in $\mathcal{D}'$- Next, let $loc'(\phi) = N_j$: by definition we know that $\phi \in M_j$ and $Supp(\phi) \subseteq Dom_{\mathcal{D}}(M_j)$. Since $Dom_{\mathcal{D}}(M_j) \subseteq Dom_{\mathcal{D}'}(N_j)$ we know that $Supp'(\phi) = Supp(\phi) \subseteq Dom_{\mathcal{D}'}(N_j)$ and thus $\forall \psi \in Supp'(\phi). loc'(\psi) \overset{\sigma}{\rightarrowtail} N_j$ with $\sigma(\psi) = \psi$. Finally, let $loc'(\phi) = N$. Then $Supp_N \subseteq Supp'$ is a support mapping for ϕ in particular.

Transitive Enrichment

Obviously, the inclusion of the global link does not affect the visibility (e.g. Dom) of any node in $\mathcal{N}$ nor the local entities provided by the individual nodes (i.e. dom). Hence, all properties of a structuring are trivially forwarded to the enriched structuring.

Removable Link

- We have to prove that loc is also a location mapping for $\mathcal{D}'$. It holds that $\forall N \in \mathcal{N}.\ loc_{\mathcal{D}}(N) = loc_{\mathcal{D}'}(N)$ since $dom(N)$ remains unchanged and also all $e \in loc_{\mathcal{D}}(N)$ that are exclusively provided by some link in $\mathcal{D}$ are still provided exclusively in $\mathcal{D}'$. Thus, loc is also surjective in $\mathcal{D}'$, also $\forall N \in \mathcal{N}.\forall e \in dom^N.\ loc\mathcal{D}'(e) = loc\mathcal{D}(e) = N$ and $\forall e \in Dom_{\mathcal{D}'}.\ loc\mathcal{D}'(e)$ is the only node providing e.
- $\mathcal{D}'$ and $\mathcal{D}'$ coincide in the entities they provide at their maximal nodes, which is an immediate consequence of condition (2) of Definition 9.
- Also $\forall \phi \in Lem_{\mathcal{D}'}.\ \forall \psi \in Supp(\phi).\ \exists \sigma.\ loc(\psi) \overset{\sigma}{=\!=\!\Rightarrow} loc(\phi) \wedge \sigma(\psi) = \psi$ is implied by condition (3) of Definition 9.   □

References

1. Autexier, S., Hutter, D.: Mind the gap - maintaining formal developments in MAYA. In: Siekmann, J.H. (ed.) LNCS 2605. Springer, Heidelberg (2005)
2. Frisch, A.M., David Page Jr., C.: Generalization with taxonomic information. In: 8th National Conference on Artificial Intelligence, pp. 775–761. AAAI-Press (1990)
3. Gauthier, T., Kaliszyk, C.: Matching concepts across HOL libraries. In: Watt, S.M., Davenport, J.H., Sexton, A.P., Sojka, P., Urban, J. (eds.) CICM 2014. LNCS, vol. 8543, pp. 267–281. Springer, Heidelberg (2014)
4. Goguen, J.A., Burstall, R.M.: Institutions: abstract model theory for specification and programming. J. Assoc. Comput. Mach. **39**(221–256), 95–146 (1992). Predecessor in: LNCS **164**(221–256) (1984)
5. Hutter, D.: Management of change in verification systems. In: Proceedings 15th IEEE International Conference on Automated Software Engineering, ASE-2000, pp. 23–34. IEEE Computer Society (2000)
6. Mossakowski, T., Autexier, S., Hutter, D.: Development graphs - proof management for structured specifications. J. Logic Algebr. Program. Special Issue Algebr. Specif. Dev. Tech. **67**(1–2), 114–145 (2006)
7. Normann, I., Kohlhase, M.: Extended formula normalization for ε-retrieval and sharing of mathematical knowledge. In: Kauers, M., Kerber, M., Miner, R., Windsteiger, W. (eds.) MKM/CALCULEMUS 2007. LNCS (LNAI), vol. 4573, pp. 356–370. Springer, Heidelberg (2007)
8. Riazanov, A., Tchaltsev, A.: Reusable TPTP parser in java (2007). http://www.freewebs.com/andrei_ch/TPTP_2007.01.30.tgz
9. Sutcliffe, G.: The TPTP world – infrastructure for automated reasoning. In: Clarke, E.M., Voronkov, A. (eds.) LPAR-16 2010. LNCS, vol. 6355, pp. 1–12. Springer, Heidelberg (2010)
10. Trybulec, A.: On the sets inhabited by numbers. J. Formaliz. Math. **15** (2003). http://mizar.uwb.edu.pl/JFM/Vol15/membered.html

Formal Logic Definitions for Interchange Languages

Fulya Horozal$^{(\boxtimes)}$ and Florian Rabe

Jacobs University Bremen, Bremen, Germany
fulyahorozal@gmail.com

Abstract. System integration often requires standardized interchange languages, via which systems can exchange mathematical knowledge. Major examples are the MathML-based markup languages and TPTP. However, these languages standardize only the syntax of the exchanged knowledge, which is insufficient when the involved logics are complex or numerous. Logical frameworks, on the other hand, allow representing the logics themselves (and are thus aware of the semantics), but they abstract from the concrete syntax.

Maybe surprisingly, until recently, state-of-the-art logical frameworks were not quite able to adequately represent logics commonly used in formal systems. Using a recent extension of the logical framework LF, we show how to give concise formal definitions of the logics used in TPTP. We can also formally define translations and combinations between the various TPTP logics. This allows us to build semantics-aware tool support such as type-checking TPTP content.

While our presentation focuses on the current TPTP logics, our approach can be easily extended to other logics and interchange languages. In particular, our logic representations can be used with both TPTP and MathML. Thus, a single definition of the semantics can be used with either interchange syntax.

1 Introduction

Interchange Languages. System integration is one of the biggest challenges concerning formal systems. One major line of research has been the development of interchange languages that allow exchanging mathematical knowledge in standardized formats.

Two languages have been particularly successful: the MathML family (which we use here to group together MathML [1], OpenMath [4], and OMDoc [13]) developed in the CICM community and the TPTP family [21] developed in the deduction community.

It may be surprising to some readers that we consider these two families to be very similar. Indeed, TPTP was originally introduced for benchmark problems for first-order logic (FOL) provers, whereas MathML is meant to support all mathematical content. But TPTP syntax is becoming more and more expressive, including typing [22], higher-order logic (HOL), [2], polymorphism [3] and

© Springer International Publishing Switzerland 2015
M. Kerber et al. (Eds.): CICM 2015, LNAI 9150, pp. 171–186, 2015.
DOI: 10.1007/978-3-319-20615-8_11

arithmetic [22]. As a consequence, the two languages share a central property: Both are primarily defined by an extremely expressive context-free grammar and use only informal descriptions of the fragment of well-formed, meaningful content. MathML is broader, allowing users to define and document well-formed fragments in content dictionaries. TPTP is deeper, slowly specifying individual fragments in an ongoing series of papers (which is barely keeping up with user demand for specifying more expressive languages – the most recent extension proposal [15] appears in the same volume as our present paper). But in both cases, a formal description (e.g., with a context-sensitive inference system) is lacking.

Maybe provocatively, we can say that both MathML and TPTP standardize the syntax but not the semantics. Because successful system integration usually requires a common understanding of the semantics, and because the number and complexity of formal systems is growing, this is becoming an important issue for interface languages.

Moreover, it is not always obvious what the relations are between the various well-formed fragments used in different applications. For example, there are intuitive sublanguage relations between TPTP's untyped FOL, typed FOL, and HOL. But these can be difficult to specify precisely, e.g., when the larger language introduces new concepts and then recovers the smaller language as a special case. Moreover, it is desirable to specify language fragments modularly so that they can be combined flexibly. For example, arithmetic should be combinable with any typed logic, and the extension of FOL with polymorphism should be consistent with a future extension of higher-order logic with polymorphism.

Logical Frameworks. Logical frameworks [17] such as Twelf [18] or Isabelle [16], in some sense, suffer from the opposite problem: They allow very concise definitions of the well-formed fragments (often with tool-support to decide which expressions are meaningful) but have little-to-no bearing on the actual syntax used in communication between systems. In fact, logics specifications in logical frameworks tend to be idealized and not in sync with those implicitly used in the above interchange languages.

Moreover, state-of-the-art logical frameworks are surprisingly unable to fully specify standard logics: They allow specifying the well-formed objects but not the well-formed theories. For example, it is not possible to specify in LF that a first-order theory should not declare higher-order function symbols. Similarly, Isabelle cannot specify the shape of HOL type definitions. This problem was remedied recently in [10], which develops a logical framework that extends LF with declaration patterns [11], which specify the shape of declarations in well-formed theories.

Contribution. We apply the framework of [10] to give concise, fully formal, human- and machine-readable specifications of a set of logics commonly used with interchange languages. We focus on the logics in the TPTP family, whose practical value is well-documented, but our approach is applicable to most other logics. Because our specifications are modular, we can combine features conveniently. We demonstrate this by defining TPTP's polymorphic HOL (which has not been described previously) by taking a pushout of HOL and polymorphic FOL.

Our specifications are tightly integrated with both TPTP and MathML. Firstly, the framework's syntax can match the TPTP syntax almost exactly (the necessary minor transformation is maintained by Geoff Sutcliffe and the second author). In fact, our specifications have quasi-official status because they are now used by Sutcliffe to type-check the TPTP library. Secondly, the logical framework we use is implemented within MMT [20]. Thus, our specifications double as content dictionaries (which MMT internally maintains in OMDoc syntax), which MMT uses to check MathML content.

Thus MathML and TPTP become alternative concrete encodings, and a single logic specification defines well-formed MathML and TPTP content at the same time.

Overview. We summarize the logical framework we use in Sect. 2. Then we specify the logics and their relations in Sects. 3 and 4, respectively. We describe implementation aspects in Sect. 5 and conclude in Sect. 6.

2   The Logical Framework

Technically, our logical framework arises in a rather complex way using four different features: We use *(i)* the foundation-independent module system MMT [20] *(ii)* extended with the declaration patterns of [10] and *(iii)* instantiated with a version of LF [8] with *(iv)* sequences [10,12]. We will refer to the resulting framework as LFS. Below we give the fragment of LFS that is sufficient for our purposes as a self-contained grammar in Fig. 1:

Modules	M	$::=$ $\texttt{theory } T = \{\Sigma\} \mid \texttt{view } v : T_1 \rightarrow T_2 = \{\sigma\}$
Theories	Σ	$::=$ $c : E \mid \texttt{include } T \mid \texttt{pattern } p = P$
Views	σ	$::=$ $c := E \mid \texttt{include } v \mid \texttt{pattern } p := P$
Expressions	E	$::=$ $c \mid x \mid \{x : E\} E \mid [x : E] E \mid E E$
		$\mid \quad \cdot \mid E, E \mid [E]_{i=1}^{E} \mid E_E \mid \texttt{nat} \mid 0 \mid succ(E) \mid \texttt{type}^E$
Patterns	P	$::=$ $p \mid \{\Sigma\} \mid [x : E] P \mid P E$

Fig. 1. Grammar

Module System. Modules are the toplevel declarations. Their semantics is defined in terms of the category of MMT *theories* and theory morphisms. We call the latter *views*.

A non-modular theory Σ declares a list of typed constants c. Correspondingly, views from a theory T_1 to a theory T_2 consist of assignments $c := E$, which map T_1-constants to T_2-expressions. Views extends homomorphically to a (type-preserving) map of T_1-expressions to T_2-expressions.

To this, the module system adds the ability for theories and views to *include* other theories and views, respectively.

LF. Expressions are formed from constants c, bound variables x, dependent function types (Π-types) $\{x : E\} E$, λ-abstraction $[x : E]E$, and application $E\,E$. As usual, we write $E_1 \rightarrow E_2$ instead of $\{x : E_1\} E_2$ if x does not occur in E_2.

As an example, consider the following declarations of the theory *Forms*, which we will use for our representations in Sect. 3. It declares an LF-type $o of propositions and an $o-indexed type family ⊢. This type family exemplifies how logic encodings in LF follow the Curry-Howard correspondence to represent judgments as types and proofs as terms: Terms of type ⊢ F represent derivations of the judgment "F is true". Furthermore, *Forms* declares all propositional connectives, among which we give & as an example. It also has one declaration pattern *axiom*, which we explain below.

```
theory Forms = {
    $o : type
    ⊢ : $o → type
    & : $o → $o → $o
    ⋮
    pattern axiom = [F : $o] {
        m : ⊢ F
    }
}
```

Declaration Patterns. Declaration patterns formalize the shape of well-formed theories of a logic L: In a well-formed L-theory, each declaration must match one of the L-patterns. For example, in a well-formed first-order theory, each symbol declaration must match one of these patterns: *(i)* n-ary first-order function symbols, *(ii)* n-ary first-order predicates symbols and *(iii)* axioms.

But in LF, nothing stops users from writing theories that contain non-first-order declarations such as $f : (\$i \rightarrow \$i) \rightarrow \$i$ or $g : \$o \rightarrow \$i$ (where $i is the LF-type of first-order terms). These are still well-typed in LF even in the context of first-order logic. But in LFS, we can formalize the above three declaration patterns, and LFS will reject declarations that do not match one of the patterns.

Declaration patterns P are formed from pattern constants p, theories $\{\Sigma\}$, λ-abstractions $[x : E]\, P$, and applications of patterns P to expressions E. Further details on declaration patterns are given in [10].

Example 1. The declaration pattern *axiom* in *Forms* formalizes the shape of axiom declarations. These must be of the form $m : ⊢ F$ for some proposition F, which matches the declaration pattern *axiom* F.

Adequacy. Representations of logics in LF are well-known to be adequate. However, technically, that is only true for the representation of *expressions*. Representations in LF are not actually adequate with respect to *theories* because the LF type theory cannot rule out declarations that do not match a pattern. The declaration patterns of LFS are exactly what is needed to obtain an adequate representation of the theories of a logic.

Moreover, for all logics L we consider, all well-formed L-theories in LFS simplify to theories in the LF fragment of LFS. Therefore, we can inherit the adequacy for expressions from LF.

Sequences. Even though most logics do not use sequences, it turns out that sequences are usually necessary to write down declaration patterns. For example, in theories of typed first-order logic, function symbol declarations use a sequence of types – the argument types of the function symbol. Therefore, our language also uses *expression sequences* and *natural numbers*. These are formed by the underlined productions for expressions:

- for the empty sequence,
- E_1, E_2 for the concatenation of two sequences,
- E_n for the n-th element of E,
- $[E(x)]_{x=1}^n$ for the sequence $E(1), \ldots, E(n)$ where n has type **nat** and $E(x)$ denotes an expression E with a free variable x : **nat**; we write this sequence as E^n if x does not occur free in E,
- **nat** for the type of natural numbers,
- 0 and $succ(n)$ for zero and the successor of a given natural number n,
- **type**^{n} for the kind of a sequence expression of length n.

We avoid giving the type system for this extension of LF and refer to [10] for the details. Intuitively, natural numbers and sequences occur only in pattern expressions, and fully applied closed pattern expressions normalize to expressions of the form $\{\Sigma\}$ where Σ is an LF theory. We will give examples below when we introduce specific declaration patterns.

A powerful feature of our sequences is that we can elegantly extend the primitives of LF to flexary operators. In particular, for a sequence A of types that normalizes to $A_1, \ldots, A_n$ and for a type B, the type $A \to B$ normalizes to $A_1 \to \ldots \to A_n \to B$. Correspondingly, for a function f of that type and a sequence E that normalizes to $E_1, \ldots, E_n$, the expression $f E$ normalizes to $(\ldots (f E_1) \ldots E_n)$.

Finally, we also extend views from T_1 to T_2 to map pattern constants to pattern expressions. The semantics of such a view is a functor mapping well-formed T_1-theories to well-formed T_2-theories. We will give examples when we introduce specific views.

3 Representing Logics

In this section, we represent the TPTP languages in our logical framework. Specifically, we present the untyped first-order (*FOF*), the typed first-order (*TF0*) and its extension with arithmetic (*TFA*), the polymorphic first-order (*TF1*) and the typed higher-order (*TH0*) languages of TPTP.

Our representations form a diagram of LFS-theories as shown in Fig. 2, where $\hookrightarrow$ denotes inclusion. Where compatible with Twelf's concrete syntax, we will use the same symbol names as TPTP.

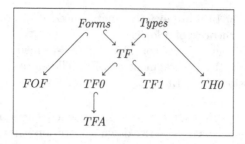

Fig. 2. TPTP logics

Untyped First-Order Language. The theory *FOF* is given on the right. It includes the auxiliary theory *Forms* from Sect. 2 and adds the LFS-type $i : type for the universe of first-order individuals. Moreover, it declares the first-order universal (!) and existential (?) quantifiers using higher-order abstract syntax, and the binary predicate symbol == for equality of individuals.

FOF contains two declaration patterns, *fun* and *pred*. These allow n-ary function and predicate symbols in *FOF*-theories, respectively. Recall that here i^n abbreviates the sequence $i,\ldots,$i of length n and that

```
theory FOF = {
   include Forms
   $i  : type
   !   : ($i → $o) → $o
   ?   : ($i → $o) → $o
   == : $i → $i → $o

   pattern fun = [n : nat] {
      f : $i^n → $i
   }
   pattern pred = [n : nat] {
      p : $i^n → $o
   }
}
```

$($i,\ldots,$i) \to i normalizes to $i \to \ldots \to $i \to $i. This includes the case $n = 0$ of constant declarations. *FOF* additionally has the declaration pattern *axiom*, which is inherited from *Forms*.

Typed First-Order Languages. To maximize reuse, we use two additional auxiliary LFS-theories, *Types* and *TF*, which contain the respective shared components of the typed first-order languages of TPTP.

Types is a base theory for all the typed TPTP languages. It declares an LFS-type

```
theory Types = {
   $tType : type
   $i     : $tType
   $tm    : $tType → type
}
```

$tType that represents the universe of all TPTP types. It also declares a distinguished base type $i : $tType. We also use an LFS-type family $tm, which is an artifact of our Church-style, intrinsically typed representation and does not have an analog in TPTP. $tm assigns to each TPTP-type A an LFS-type $tm A which contains the TPTP-terms of A. For example, the TPTP-terms of type $i are represented as LFS-terms of LFS-type $tm $i.

TF contains all the shared components of $TF0$ and $TF1$. Besides typing and propositions, which are included from $Types$ and $Forms$, respectively, it declares the logical symbols that are polymorphic over all TPTP types. These are the typed quantifiers ! (universal) and ? (existential) and typed equality ==. These cannot be declared in $Forms$,

```
theory TF = {
   include Types
   include Forms
   !  : ($tm A → $o) → $o
   ?  : ($tm A → $o) → $o
   == : $tm A → $tm A → $o
}
```

because they take a type argument $A : \$tType$. Note that here we make A an implicit argument in the style of Twelf that is automatically inferred from the context.

We extend TF to obtain the languages $TF0$ and $TF1$. TF already declares all logical symbols of $TF0$ so that we only have to add the three declaration patterns:

```
theory TF0 = {
   include TF
   pattern baseType = {
      t : $tType
   }
   pattern typedFun = [n : nat] [A : $tType^n] [B : $tType] {
      f : [$tm A_i]_{i=1}^n → $tm B
   }
   pattern typedPred = [n : nat] [A : $tType^n] {
      p : [$tm A_i]_{i=1}^n → $o
   }
}
```

These patterns specify the form of the declarations of non-logical symbols that are allowed in $TF0$-theories:

- $baseType$ allows the declaration of $TF0$-types t,
- $typedFun$ allows the declaration of typed function symbols f that take arguments of $TF0$-type $A_1, \dots, A_n$ and return an expression of type B,
- $typedPred$ allows the declaration of typed predicate symbols p with arguments of $TF0$-types $A_1, \dots, A_n$.

Note that our representation of $TF0$ uses an LFS-type $\$o$: type in order to distinguish formulas from terms. This is different from the description in [22], where a TPTP-type $\$o : \$tType$ is used. Our representation has the advantage that we do not need case distinctions in order to avoid $\$o$ as an argument of a function or predicate symbol or of a quantifier.

Example 2 (TF0-Theories). Assume that a base type $nat : \$tType$ has already been declared (using the pattern $baseType$). Then the declaration of a binary function symbol on nat matches the pattern expression *typed Fun* $2\,(nat, nat)\,nat$. The latter β-reduces to $\{f : [\$tm\,(nat, nat)_i]_{i=1}^2 \to \$tm\,nat\}$, which can be simplified to $\{f : (\$tm\,(nat, nat)_1, \$tm\,(nat, nat)_2) \to \$tm\,nat\}$ and eventually normalizes to $\{f : \$tm\,nat \to \$tm\,nat \to \$tm\,nat\}$.

Interestingly, the logical symbols of *TF1* are almost the same as those of *TF0*. It only adds the universal (!°) and existential (?°) quantifiers over types. In the TPTP syntax, these are identified with ! and ?, but in LFS their types are different so that they must be distinguished.

The crucial difference between the representations of *TF0* and *TF1* is in the legal declarations: *TF1*-theories may declare n-ary type operators and polymorphic function and predicate symbols. This shows the importance of declaration patterns in our framework as this difference could not be captured in LF.

```
theory TF1 = {
    include TF
    !° : ($tType → $o) → $o
    ?° : ($tType → $o) → $o

    pattern typeOp = [n : nat] {
        t : $tType^n → $tType
    }
    pattern polyFun = [m : nat] [n : nat] [A : ($tType^m → $tType)^n]
                      [B : $tType^m → $tType] {
        f : {a : $tType^m} [$tm (A_i a)]^n_{i=1} → $tm (B a)
    }
    pattern polyPred = [m : nat] [n : nat] [A : ($tType^m → $tType)^n] {
        p : {a : $tType^m} [$tm (A_i a)]^n_{i=1} → $o
    }
}
```

The pattern *typeOp* in *TF1* describes type operators t of arity n.

polyFun describes polymorphic function symbols f, which take m type arguments $a_1, \ldots, a_m$ and then n term arguments of types $A_1(a_1, \ldots, a_m), \ldots, A_n(a_1, \ldots, a_m)$ and return an expression of type $B(a_1, \ldots, a_m)$. Note that we use higher-order abstract syntax in the style of LF to represent expressions of type A : $tType with m free variables of type $tType as terms of type $tType^m → $tType.

Finally, *polyPred* describes polymorphic predicate symbols p, which take m type arguments $a_1, \ldots, a_m$ and then n term arguments of types $A_1(a_1, \ldots, a_m), \ldots, A_n(a_1, \ldots, a_m)$.

Note that via the inclusion of *TF*, *TF1* inherits the declaration pattern *axiom* of *Forms* that allows axioms of LFS-type ⊢ F for some *TF1*-formula F.

Example 3 (TF1-Theories). Consider a unary type operator *list* has already been declared (using the pattern expression *typeOp* 1). Then the declaration of the *cons* operation on lists matches the pattern expression

$$polyFun\, 1\, 2 \left(([a : $tType^1]a), ([a : $tType^1]list\, a) \right) ([a : $tType^1]list\, a)$$

which normalizes to $\{f : \{a : $tType\} a → list\, a → list\, a\}$.

Higher-Order Language. *TH0* is based on the one in [2]. Like *TF0* and *TF1*, it is based on the theory *Types*. It adds the logical symbols > for function type formation, ˆ for λ-abstraction, and @ for application. As usual, we will write > and @ as right and left-associative infix operators, respectively.

TH0 is not based on the theory *Forms*, which introduced the LFS-type $o : type of formulas. Instead, it treats formulas as terms using a TPTP type $o : $tType. Consequently, the logical connectives and quantifiers and the truth judgment are declared based on $o. Here we give only some example declarations.

Finally, *TH0* uses three declaration patterns:

- *baseType* allows the declaration of *TH0*-base types *t*,
- *typedCon* allows the declaration of typed constants *c* of type *A* for some *TH0*-type *A*,
- *axiom* allows the declaration of axioms *F* for some *TH0*-formula *F*.

```
theory TH0 = {
    include Types
    >  : $tType → $tType → $tType
    ˆ  : ($tm A → $tm B) → $tm (A > B)
    @  : $tm (A > B) → $tm A → $tm B
    $o : $tType
    &  : $tm ($o > $o > $o)
    !  : $tm ((A > $o) > $o)
    :
    ⊢  : $tm $o → type
    pattern baseType = {
        t : $tType
    }
    pattern typedCon = [A : $tType] {
        c : $tm A
    }
    pattern axiom = [F : $tm $o] {
        m : ⊢ F
    }
}
```

Using the declaration patterns in *TH0*, we are able to define the theories of *TH0* precisely. This is important because a number of different definitions are plausible. For example, the original higher-order logic of [5] arises if we drop the declaration pattern *baseType*. Another option is to use the pattern

```
pattern neBaseType = {t : $tType, noempty :⊢ ? @ (ˆ [x : $tm t] $true)}
```

so that every base type *t* is nonempty. Type definitions in the style of [7] can be obtained similarly.

Arithmetic. To add arithmetic as described in [22], we extend *TF0* with arithmetic operations in the theory *TFA* below. We only give a representative fragment of the encoding. Arithmetic domains are added as elements of the type $adom, and $atype includes the arithmetic domains into the universe $tType of types. This indirection is useful to quantify over exactly the arithmetic domains: It permits declaring the polymorphic operations $sum, $less, etc. for an arbitrary arithmetic domain *D*.

```
theory TFA = {
    include TF0
    $adom   : type
    $atype : $adom → $tType
    $int    : $adom
    $rat    : $adom
    $real   : $adom
    $sum    : $tm ($atype D) → $tm ($atype D) → $tm ($atype D)
    $less   : $tm ($atype D) → $tm ($atype D) → $o
    ⋮
}
```

Other Syntactic Features. There is a variety of further syntactic variants, which can be seen as orthogonal features that can be added to a logic on demand. These include product types, choice operators, conditional terms, let-expressions, etc. We separate these into individual modules to gain fine-grained control over the strength of a logic. Akin to [6], we call this the *little logics* approach.

Semantic Variants. The above has specified *well-formed* formulas. But we can also specify the *valid* formulas by formalizing appropriate calculi. For each logical operator, we formalize natural deduction proof rules. These rules are straightforward, and we refer to our formalizations in [14].

We explicitly mention only some axioms, which are important because they distinguish semantic variants of the same syntax. As a guiding principle, we make the base calculi as weak as possible and define axioms in separate modules, which can be included on demand.

For first-order logics, only one semantic variant is of major importance: the one between classical and intuitionistic logic. Therefore, we use the module on the right for the axiom of excluded middle.

```
theory ExclMid = {
    include Forms
    em : ⊢ (A | (∼ A))
}
```

The situation is more complicated for higher-order logic. The theory *MinHOL* defines the minimal proof theory of HOL using only (i) β-conversion for λ-abstraction and (ii) congruence rules for equality. All further rules are added in separate modules that extend *MinHOL* as introduced in Fig. 3: *PropExt* (propositional extensionality) identifies equality on booleans with logical equivalence. *Xi* provides the ξ-rule, a weak form of functional extensionality that can also be seen as a congruence rule for λ-abstraction. *Eta* provides η-conversion. *FuncExt* and *BoolExt* are functional and boolean extensionality. *ExclMidHOL* states excluded middle. *NonEmptyTypes* makes all types non-empty.

By combining these and if necessary other extensions of *MinHOL*, we obtain the various incarnations of higher-order logics. Of particular importance is the logic *BaseHOL*, which combines *MinHOL*, *Xi*, and *PropExt*: It is the weakest

Theory	Added axioms/rules
PropExt	$(\vdash F \to \vdash G) \to (\vdash G \to \vdash F) \to \vdash F == G$
Xi	$(\{x : \$\mathtt{tm}\, A\} \vdash (S\,x) == (T\,x)) \to \vdash \hat{\ }[x]\,(S\,x) == \hat{\ }[x]\,(T\,x)$
FuncExt	$(\{x : \$\mathtt{tm}\, A\} \vdash (S\,@\,x) == (T\,@\,x)) \to \vdash S == T$
Eta	$\vdash \hat{\ }[x]\,(F\,@\,x) == F$
BaseHOL	*PropExt, Xi*
BoolExt	$BaseHOL, \vdash F\,\$\mathtt{true} \to \vdash F\,\$\mathtt{false} \to \vdash !\,@\,F$
ExclMidHOL	$BaseHOL, \vdash |\,@\,F\,@\,(\sim @\,F)$
NonEmptyTypes	$BaseHOL, \vdash ?\,@\,(\hat{\ }[x : \$\mathtt{tm}\, A]\,\$\mathtt{true})$

Fig. 3. Modules extending *MinHOL*

reasonable variant of HOL that is strong enough to define all first-order connectives and quantifiers and their (intuitionistic) natural deduction rules. We use it as a base logic for all *BoolExt*, *ExclMidHOL*, and *NonEmptyTypes*, which can only be formulated in the presence of some logical connectives.

Note that *Eta* and *FuncExt* are equivalent, and so are *BoolExt* and *Excl MidHOL*. These relations can be formalized concisely as views between the respective signatures; these are given in [14].

4 Translating and Combining Logics

We will now relate the logics from the previous section to each other using views and combine them to create a new logic *TH1* (polymorphic higher-order logic), resulting in the diagram in Fig. 4. Because each view induces a theory translation functor, this permits moving theories between the TPTP logics. We will focus on the most important views representing sublanguage relations; there are also (possibly partial) translations in the opposite directions, but they are substantially more complicated to formalize.

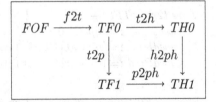

Fig. 4. Translating TPTP logics

Notably, users working with implementations of higher-order logic have already started using ad hoc variants of *TH1* (independent of our work) in expectation of an eventual adoption as an official TPTP logic. This shows that our work offers an efficient way for TPTP to keep up with the growing demand for reference definitions of logics.

Translating Logics. The **view** *f2t* from *FOF* to *TF0* is given below. Since *FOF* and *TF0* share the theory *Forms*, the view implicitly includes the identity translation of *Forms*.

The main characteristic of the translation is that the individuals of *FOF* are interpreted as the individuals of *TF0* of the distinguished base type $i. This is expressed as an assignment of the type $i of *FOF*-individuals to the type $tm $i in *TF0*. Correspondingly, *FOF*-quantifiers and

```
view f2t : FOF → TF0 = {
    $i    := $tm $i
    !     := [f] ! [x : $tm $i] (f x)
    ?     := [f] ? [x : $tm $i] (f x)
    ==    := [x : $tm $i] [y : $tm $i] x == y
    fun   := [n : nat] {f : ($tm $i)^n → $tm $i}
    pred  := [n : nat] {p : ($tm $i)^n → $o}
}
```

equality are interpreted as the *TF0*-quantifiers and equality on the type $tm $i. Therefore, we map, for instance, ! to the expression $[f] ! [x : \$tm \$i](f\,x)$ that takes a *TF0*-formula f with a free variable $x : \$tm \$i$ and returns the universally quantified *TF0*-formula $! [x : \$tm \$i](f\,x)$.

For each *FOF*-pattern p, the pattern translation maps every *FOF* declaration that matches p to a *TF0* declaration. This is defined by two assignments to the *FOF*-patterns *fun* and *pred*. For instance, *fun* is mapped to the pattern expression $[n : \mathrm{nat}]\{f : (\$tm \$i)^n → \$tm \$i\}$ so that every n-ary *FOF*-function symbol declaration is translated to the *TF0*-declaration of an n-ary function on $tm $i.

Note that our framework enforces that all views preserve typing. For example, the *FOF*-symbol ==: $i → $i → $o must be mapped to a *TF0*-expression of type $tm $i → $tm $i → $o. Similarly, the *FOF*-pattern *fun*, which takes a natural number and returns a theory, must be mapped to a *TF0*-pattern expression, which takes a natural number and returns a theory.

We give the **view** *t2p* **from** *TF0* **to** *TF1* below. Since *TF0* and *TF1* share *TF* and *TF0* only adds declaration patterns, the view only consists of declaration pattern assignments.

```
view t2p : TF0 → TF1 = {
    baseType  := typeOp 0
    typedFun  := [n : nat] [A : $tType^n] [B : $tType] polyFun 0 n A B
    typedPred := [n : nat] [A : $tType^n] polyPred 0 n A
}
```

Every *TF0*-type is interpreted as a nullary type operator in *TF1*. This is given as an assignment of the pattern *baseType* of *TF0* to the pattern *typeOp* supplied with 0 as the argument for the number of type arguments. Note that β-reducing *typeOp* 0 results in the pattern expression $\{t : \$tType^0 → \$tType\}$, where $\$tType^0$ normalizes to the empty sequence so that the whole type normalizes to $tType.

Every n-ary typed function symbol of *TF0* is interpreted as an n-ary polymorphic function symbol that does not take type arguments. This is given as an assignment from the pattern *typedFun* to the pattern expression $[n : \mathrm{nat}][A : \$tType^n][B : \$tType]polyFun\,0\,n\,A\,B$, which takes the arity n of the function symbol, the sequence A of argument types, and the return type B and returns the corresponding monomorphic *TF1*-declaration. Note that after η-contraction, this pattern expression is equal to $polyPred\,0$. The pattern $typedPred$ is translated accordingly.

Example 4 (Translating Theories). Consider a *TF0*-theory T containing the two declarations from Example 2. Applying the view $t2p$ to it yields a *TF1*-theory T'. Due to the assignment to $baesType$, $t2p$ translates the T-type nat to a T'-type of the same name. Due to the assignment to $typedFun$, $t2p$ translates the pattern expression $typedFun\,2\,(nat, nat)\,nat$ to

$$([n : \mathbf{nat}][A : \$\mathtt{tType}^n][B : \$\mathtt{tType}]polyFun\,0\,n\,A\,B)\,2\,(nat, nat)\,nat$$

which simplifies to $\{f : \{x : \mathbf{type}^0\}\,\$\mathtt{tm}\,nat \to \$\mathtt{tm}\,nat \to \$\mathtt{tm}\,nat\}$. Here $\$\mathtt{tType}^0$ normalizes to the empty sequence of types so that the binding $\{x : \mathbf{type}^0\}$ binds no variables and disappears, yielding the expected declaration.

The **view** $t2h$ **from** *TF0* **to** *TH0* interprets the *TF0* type $\$\mathtt{o} : \mathbf{type}$ in terms of the *TH0* constant $\$\mathtt{o} : \$\mathtt{tType}$ and translates the connectives to their higher-order analogues. Function and predicate symbols declared in terms of $\to$ over *TF0* are translated to the respective declarations in terms of $>$ over *TH0*. The translation of axioms is straightforward.

$$
\begin{aligned}
\mathbf{view}\ t2h\ :\ &TF0\ \to\ TH0\ =\ \{\\
\$\mathtt{o} \quad &:= \$\mathtt{tm}\,\$\mathtt{o}\\
\vdash \quad &:= [F : \$\mathtt{tm}\,\$\mathtt{o}] \vdash F\\
\& \quad &:= [A : \$\mathtt{tm}\,\$\mathtt{o}]\,[B : \$\mathtt{tm}\,\$\mathtt{o}]\,\&\,@\,A\,@\,B\\
! \quad &:= [f : \$\mathtt{tm}\,\$\mathtt{i} \to \$\mathtt{tm}\,\$\mathtt{o}]\,!\,@\,(\hat{}\,f)\\
\vdots \quad &\\
typedFun \quad &:= [n : \mathbf{nat}]\,[A : \$\mathtt{tType}^n]\,[B : \$\mathtt{tType}]\,\{f' : \$\mathtt{tm}\,(A >^* B)\}\\
typedPred \quad &:= [n : \mathbf{nat}]\,[A : \$\mathtt{tType}^n]\,\{p' : \$\mathtt{tm}\,(A >^* \$\mathtt{o})\}\\
axiom \quad &:= [F : \$\mathtt{tm}\,\$\mathtt{o}]\,axiom\,F\\
\}&
\end{aligned}
$$

Note that $>^* : \$\mathtt{tType}^n \to \$\mathtt{tType} \to \mathbf{type}$ is the flexary operator derived from $>$ (in LFS, the flexary version of a binary operator is definable, see [12]).

Combining Logics. A particular strength of a logical framework like ours is the ability to combine logics using colimits as studied in [9]. In the simplest case, this is just taking the union of two logics. More generally, we can use pushouts in the category of theories. We will give two interesting examples how our framework guides the design of new TPTP logics.

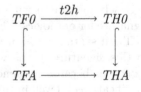

Fig. 5. TPTP logics with arithmetic

Firstly, we obtain *THA*, the extension of *TH0* with arithmetic, by applying the theory translation functor induced by the view $t2h$. It maps the *TF0*-theory *TFA* to the corresponding *TH0*-theory. This construction is obtained automatically from our framework and results in the commuting diagram in Fig. 5.

Secondly, we combine *TF1* and *TH0* into a new logic: polymorphic higher-order logic *TH1*. This construction uses the commuting diagram in Fig. 6.

If we ignore the patterns and consider only the underlying LF-theories, this diagram is obtained automatically as a pushout. However, we have to add the patterns of *TH1* – which merge the patterns of *TH1* and *TH0* in a non-trivial way – manually. The relevant fragments of the LFS-theory for *TH1* is given below where *TH0'* represents *TH0* without its patterns. We omit the straightforward views *p2ph* and *h2ph*.

$$
\begin{array}{ccc}
TF0 & \xrightarrow{\ t2h\ } & TH0 \\
{\scriptstyle t2p}\downarrow & & \downarrow{\scriptstyle h2ph} \\
TF1 & \xrightarrow[\ p2ph\ ]{} & TH1
\end{array}
$$

Fig. 6. Constructing *TH1*

```
theory TH1 = {
    include TH0'
    !° : ($tType → $tm $o) → $tm $o
    ?° : ($tType → $tm $o) → $tm $o

    pattern typeOp = [n : nat] {
        t : $tType^n → $tType
    }
    pattern typedPolyCon = [m : nat] [A : $tType^m → $tType] {
        c : {a : $tType^m} $tm (A a)
    }
}
```

5 Practical Aspects

Processing Content. Declaration patterns are implemented as a part of MMT [19]. Our specifications are written in MMT instantiated with LF and sequences. MMT can check the logic specifications and generates content dictionaries (in the form of OMDOC theories) from them. As the namespaces for these content dictionaries, we use URIs derived from http://www.tptp.org/.

If theories or objects of these logics are given in OMDOC/MathML syntax, they can be read and type-checked natively by MMT.

If theories or objects are given in TPTP syntax, we use the fact that MMT is closely integrated with the Twelf tool [18]. In particular, we obtain a Twelf signature for our logics. The TPTP distribution includes a converter from TPTP to Twelf syntax, and Sutcliffe uses Twelf to type-check TPTP content relative to this signature. (This works even though Twelf only supports LF and not sequences because sequences never occur in TPTP theories, only in the logic specifications.) Twelf in turn can export its input as OMDOC, which can be used for further processing by MMT or other tools. Twelf can also act as a reference proof checker if systems produce proofs.

Logic Ascription in TPTP Content. MMT and OMDOC require content to reference the logic it is written in, but TPTP does not. Indeed, users can mix and match language features in the same TPTP file. Therefore, the above-mentioned converter actually translates all content into the largest logic, i.e., polymorphic HOL with arithmetic.

We propose adding a value to the header of a TPTP theory, which is a list of strings and defines the target logic to be used during type checking. The meaning of the value L1 ... Ln would be that the theory is formed over the union of the logics L1, ..., Ln. Incidentally, it could be used by problem authors and system implementers to determine whether an ATP system is applicable to a specific problem. For example, this information could be included in the value of the existing SPC header field.

Our proposal is also the best solution to the problem of semantic variants: The status of a problem (i.e., whether it is a theorem) may depend on the chosen logic. So far, TPTP has side-stepped this issue because it mostly occurred in the form of intuitionistic vs. classical FOL. (The official TPTP policy is that intuitionistic provers are welcome but incomplete.) But with higher-order provers becoming more sophisticated, it is likely to become necessary to record which logic a problem is supposedly provable in.

6   Conclusion

We observed that interchange languages commonly used for system integration – like MathML or TPTP – focus on standardizing the context-free syntax. But they do not formalize the context-sensitive language fragments corresponding to the well-formed expressions of individual logics. By formalizing these logics in a logical framework, it becomes possible to concisely specify these fragments.

We systematically applied this approach to obtain a suite of formal specifications of logics commonly used in formal systems. We focused on the TPTP logics, the quasi standard for automated deduction systems, but further logics can be defined easily, possibly reusing existing ones. We applied this modular design to obtain a new TPTP-style logic for polymorphic higher-order logic.

Our specifications are both human- and machine-readable. And they are tightly integrated with the concrete syntax and tool support of the MathML and TPTP interchange languages, inducing type checkers and serving as content dictionaries. Therefore, we propose them as reference definitions of these logics. In fact, TPTP has effectively adopted our proposal already by using our specifications for type-checking.

References

1. Ausbrooks, R., Buswell, S., Carlisle, D., Dalmas, S., Devitt, S., Diaz, A., Froumentin, M., Hunter, R., Ion, P., Kohlhase, M., Miner, R., Poppelier, N., Smith, B., Soiffer, N., Sutor, R., Watt, S.: Mathematical Markup Language (MathML) Version 2.0 (2nd edn.) (2003). See http://www.w3.org/TR/MathML2
2. Benzmüller, C.E., Rabe, F., Sutcliffe, G.: THF0 – the core of the TPTP language for higher-order logic. In: Armando, A., Baumgartner, P., Dowek, G. (eds.) IJCAR 2008. LNCS (LNAI), vol. 5195, pp. 491–506. Springer, Heidelberg (2008)

3. Blanchette, J.C., Paskevich, A.: TFF1: the TPTP typed first-order form with rank-1 polymorphism. In: Bonacina, M.P. (ed.) CADE 2013. LNCS, vol. 7898, pp. 414–420. Springer, Heidelberg (2013)
4. Buswell, S., Caprotti, O., Carlisle, D., Dewar, M., Gaetano, M., Kohlhase, M.: The Open Math Standard, Version 2.0. Technical report, The Open Math Society (2004). See http://www.openmath.org/standard/om20
5. Church, A.: A formulation of the simple theory of types. J. Symbolic Logic 5(1), 56–68 (1940)
6. Farmer, W., Guttman, J., Thayer, F.: Little theories. In: Kapur, D. (ed.) Conference on Automated Deduction, pp. 467–581. Saratoga Spings, NY (1992)
7. Gordon, M., Pitts, A.: The HOL logic. In: Gordon, M., Melham, T. (eds.) Introduction to HOL, Part III, pp. 191–232. Cambridge University Press, New York (1993)
8. Harper, R., Honsell, F., Plotkin, G.: A framework for defining logics. J. Assoc. Comput. Mach. 40(1), 143–184 (1993)
9. Harper, R., Sannella, D., Tarlecki, A.: Structured presentations and logic representations. Ann. Pure Appl. Logic 67, 113–160 (1994)
10. Horozal, F.: A Framework for Defining Declarative Languages. Ph.D. thesis. Jacobs University Bremen (2014)
11. Horozal, F., Kohlhase, M., Rabe, F.: Extending MKM formats at the statement level. In: Campbell, J.A., Jeuring, J., Carette, J., Dos Reis, G., Sojka, P., Wenzel, M., Sorge, V. (eds.) CICM 2012. LNCS, vol. 7362, pp. 65–80. Springer, Heidelberg (2012)
12. Horozal, F., Rabe, F., Kohlhase, M.: Flexary operators for formalized mathematics. In: Watt, S.M., Davenport, J.H., Sexton, A.P., Sojka, P., Urban, J. (eds.) CICM 2014. LNCS, vol. 8543, pp. 312–327. Springer, Heidelberg (2014)
13. Kohlhase, M.: OMDoc: An Open Markup Format for Mathematical Documents (Version 1.2). Lecture Notes in Artificial Intelligence, vol. 4180. Springer Heidelberg (2006)
14. Kohlhase, M., Mossakowski, T., Rabe, F.: The LATIN Project (2009). see https://trac.omdoc.org/LATIN/
15. Kotelnikov, E., Kovacs, L., Voronkov, A.: A first class boolean sort in first-order theorem proving and TPTP. In: Kerber, M., Carette, J., Kaliszyk, C., Rabe, F., Sorge, V. (eds.) Intelligent Computer Mathematics. Springer, Stockholm (2015)
16. Paulson, L.: Isabelle: A Generic Theorem Prover. Lecture Notes in Computer Science, vol. 828. Springer, Heidelberg (1994)
17. Pfenning, F.: Logical frameworks. In: Robinson, J., Voronkov, A. (eds.) Handbook of Automated Reasoning, pp. 1063–1147. Elsevier, The Netherlands (2001)
18. Pfenning, F., Schürmann, C.: System description: Twelf - a meta-logical framework for deductive systems. In: Ganzinger, H. (ed.) CADE 1999. LNCS (LNAI), vol. 1632, pp. 202–206. Springer, Heidelberg (1999)
19. Rabe, F.: The MMT API: a generic MKM system. In: Carette, J., Aspinall, D., Lange, C., Sojka, P., Windsteiger, W. (eds.) CICM 2013. LNCS, vol. 7961, pp. 339–343. Springer, Heidelberg (2013)
20. Rabe, F., Kohlhase, M.: A scalable module system. Inf. Comput. 230(1), 1–54 (2013)
21. Sutcliffe, G.: The TPTP problem library and associated infrastructure. The FOF and CNF parts, v3.5.0. J. Autom. Reasoning 43(4), 337–362 (2009)
22. Sutcliffe, G., Schulz, S., Claessen, K., Baumgartner, P.: The TPTP typed first-order form with arithmetic. In: Bjørner, N., Voronkov, A. (eds.) LPAR-18 2012. LNCS, vol. 7180, pp. 406–419. Springer, Heidelberg (2012)

Math Literate Knowledge Management
via Induced Material

Mihnea Iancu[✉] and Michael Kohlhase

Computer Science, Jacobs University, Bremen, Germany
{m.iancu,m.kohlhase}@jacobs-university.de

Abstract. Mathematicians integrate acquired knowledge into a mental model. For trained mathematicians, the mental model seems to include not just the bare facts, but various induced forms of knowledge, and the amount of this and the ability to perform all reasoning and knowledge operations taking that into account can be seen as a measure of mathematical training and literacy. Current MKM systems only act on the bare facts given to them; we contend that they – their users actually – would profit from a good dose of mathematical literacy so that they can better complement the abilities of human mathematicians and thus enhance their productivity.

In this paper we discuss how we can model induced knowledge naturally in highly modular, theory-graph based, mathematical libraries and establish how to access it to make it available for applications, creating a form of mathematical literacy. We show two examples of math-literate MKM systems – searching for induced statements and accessing a knowledge via induced theories – to show the utility of the approach.

1 Introduction

There is an interesting duality between the forms and extents of mathematical knowledge that is verbally expressed (published in articles, scribbled on blackboards, or presented in talks/discussions) and the forms that are needed to successfully extend mathematical knowledge and/or apply it. To "do mathematics", we need to extract the relevant knowledge structures from documents and reconcile them with the context of our existing knowledge – recognizing parts as already known and identifying those that are new to us. In this process we may abstract from syntactic differences, chain together known and acquired facts, and even employ interpretations via non-trivial mappings as long as they are meaning-preserving. We will call the ability to do all of this relatively effortlessly **mathematical literacy** as it is a prerequisite for doing mathematics effectively. Mathematical literacy is a distinguishing characteristic of a trained mathematician.

Current MKM systems are essentially illiterate mathematically as they only act on the bare facts given to them; this may be one of the reasons why they are not routinely used to support mathematics: mathematicians expect math literacy in their discussion partners. For a query of "binomial coefficient" a math-literate

© Springer International Publishing Switzerland 2015
M. Kerber et al. (Eds.): CICM 2015, LNAI 9150, pp. 187–202, 2015.
DOI: 10.1007/978-3-319-20615-8_12

search engine would also find formulae of the form $\binom{3}{5!}$ and $C(n, k)$, instances of the formula $\frac{n!}{(n-k)!}$, and even "choice without repetition" or "Pascal's triangle". A math-literate proof checker would try to recognize an idempotent monoid as Abelian and extend its repertoire of applicable theorems accordingly. And finally, a math-literate eLearning system would pose the same exercises, but generate different explanations for students who know groups as axiomatized via an associative composition operation $\circ$ that admits units and inverses and defined a division operation $x/y := x \circ y^{-1}$ and students to whom groups were introduced by axioms for a division operation and composition, unit, and inverses were defined from them. As these examples show, mathematical literacy would make the interaction with MKM systems more natural and effective (Fig. 1).

We contend that a large part of mathematical literacy is a function of having at our disposal – or being able to generate on demand – a large space of knowledge that is induced in some way by the explicitly represented knowledge we have acquired previously. We call this the **Mathematical Knowledge Space** (MKS).

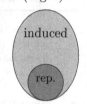

Fig. 1. MKS

In this paper, we will show two ways of how knowledge items can systematically be induced from existing representations to arrive at more mathematically literate services. The first is based on the mathematical practice of viewing an object of class A as one of class B – which we call **framing**. Following [KK09] we model framing via theory morphisms in modular theory graphs – which act as the MKS – and extend our MATHWEBSEARCH engine [KMP12] so that it answer queries "modulo framing". The second case study takes up the notion of realms that structure theory graphs in a more human-oriented fashion. The various user roles identified in [CFK14] allow us to induce special versions of the underlying theory graph for different roles. Indeed, we show that the realm faces – which were assumed to be hand-curated in [CFK14] – can be induced from the developments in the realms. We regard these two systems as initial case studies that show what math literate MKM could look like only; more case studies are certainly needed and more forms of induced knowledge need to be identified.

We will introduce the two forms of induced knowledge in the next two sections: Sect. 2 interprets the knowledge space as a MMT [Rab08, RK13] theory graph, the induced statements are computed by flattening (see Sect. 3). The realms case study is presented in Sect. 4, where we discuss how realm faces (induced theories) can be generated and pillars can be opened for inspection from the faces. Section 5 shows an application of induced material: we can use flattening to make a math search engine literate, and correspondingly, Sect. 6 discusses how realm-based access can be used in a library of formalizations. Section 7 concludes the paper.

2   Induced Statements in Theory Graphs

To build math literate MKM services as defined in Sect. 1 we need to first address the issues of generating (part of) the mathematical knowledge space and then accessing the induced knowledge in order to make it available for MKM applications.

We use the (theory-graph enabled) MMT language and system as a basis of discussion and we briefly introduce it below. MMT [RK13] is a generic, formal module system for mathematical knowledge. We will only give a brief introduction to MMT here and then discuss the concepts using examples in the following sections. We refer to [RK13] for further details.

The central notion is that of a **theory graph** containing *theories* and *views*. Theories S are formed from a set of *constant* declarations which have a name and an optional type and definition. Due to the Curry-Howard isomorphism, MMT constants can be used to declare not only symbols but also axioms and theorems describing their properties. Views $v : S \rightarrow T$ are structure-preserving mappings (morphisms) from the source to the target theory which are also truth-preserving in the sense that they map axioms of the source theory to theorems in the target theory. These properties ensure that all theorems of the source theory induce theorems in the target theory. In addition to views, the module level structure in MMT theory graphs is given by theory inheritance. The most general kind of inheritance in MMT is represented by *structures* which are (possibly partial) named imports (and defined using theory morphisms). We will use the term *includes* to refer to the trivial structures which are unnamed and total.

Every MMT declaration is identified by a canonical, globally unique URI. Theories and views can be referenced relative to the URI G of the theory graph (document) that contains them by $G?\langle\!\langle\text{theory-name}\rangle\!\rangle$ and $G?\langle\!\langle\text{view-name}\rangle\!\rangle$, respectively. Constant declarations can be referenced relative to the URI of their containing theory T by $T?\langle\!\langle\text{constant-name}\rangle\!\rangle$. Similarly, assignment declarations can be referenced relative to the URI of their containing view v by $v?\langle\!\langle\text{constant-name}\rangle\!\rangle$.

Note that the names of constants, theories and views can have multiple /-separated fragments and are of the general form $f_1/\ldots/f_n$. This makes MMT URIs much more expressive and, in particular, allows the following additional access methods:

- if theory T contains structure s_1, (the target of) s_1 contains structure s_2, ..., and s_n contains constant const then we can use the constant name $s_1/\ldots/s_n/$const, to refer to const (as translated over the assignments from the structures) from T.
- if there is a view $v_1 : T_1 \rightarrow T$ and a view $v_2 : T_2 \rightarrow T_1$, ..., and a view $v_n : T_n \rightarrow T_{n-1}$ where T_n contains constant const then we can use the constant name $[G?v_1]/\ldots/[G?v_n]/$const, to refer to const (as translated over the assignments from the views) from T.
- if T_1 is a nested module in T, ..., and T_n is a nested module in T_{n-1} we can use the theory name $T/T_1/\ldots/T_n$ to refer to theory T_n.

The MMT system provides an API to the MMT data structures described above and the MMT implementation [Rab08,RK13] provides a Scala-based [OSV07] open source implementation of the MMT API.

Generating Induced Knowledge. In the context of theory graphs we model the process of generating the knowledge space as an operation on theory graphs. Specifically, one that takes a theory graph G and return an enriched graph $\overline{G}$ where a new part of the mathematical knowledge space is explicitly represented. We call $\overline{G}$ the *induced theory graph*.

Accessing Induced Knowledge. A key aspect of MMT is that it's URI language is expressive enough to produce URIs for the induced statements that are not only unique but also informative. Specifically, we can compute the induced knowledge entities from the induced theory graph by their URI and the original graph alone, and furthermore, we can generate explanations for the existence of each induced statement in terms of the original theory graph. We call this property of MMT URIs *information completeness.*

3   Flattening Theory Graphs

To better understand the concept of framing in modular libraries, consider the theory graph U in Fig. 2. The right side of the graph introduces the elementary algebraic hierarchy building up algebraic structures step by step up to rings; the left side contains a construction of integer arithmetics. In this graph, the nodes are theories[1], the solid edges are structures (imports) and the wavy edges are views.

As discussed in Sect. 2, MMT structures can carry a name, and inherited constants can be disambiguated by the name of the structure that induced them. An application of this is in the definition of the ring theory, which inherits all of its operators (and their axioms) via the two structures m (for the multiplicative operations) and a (for the additive operations). To complete the ring we only need to add the two distributivity axioms in the inherited operators m/∘ and a/∘.

Furthermore, since structures are defined using morphisms (just like views) they can carry an assignment which maps symbols and axioms from the source theory to terms in the target theory. We see this in the view e from Monoid to NatArith, which assigns N to the base set G, multiplication ($\cdot$) to ∘ and the number 1 to the unit e. To satisfy the obligations of the theory morphism property, e also contains proofs for all Monoid axioms in NatArith. It is a special feature of MMT that assignments can also map morphisms into the source theory to morphisms into the target theory. We use this to specify the morphism c modularly (in particular, we can re-use the proofs from e and c).

Note that already in this small graph, there are a lot of induced statements. For instance, the associativity axiom is inherited in seven times (via inclusions; twice into Ring) and induced four times (via views; twice each into NatArith and IntArith). All in all, we have more than an hundred induced statements from the axioms alone. If we assume just 5 theorems proven per theory (a rather

---

[1] We have left out the quantifiers for the variables x, y, and z from the axioms to reduce visual complexity. The always range over the respective base set. Furthermore, all axioms are named; but we only state the names we actually use in the examples.

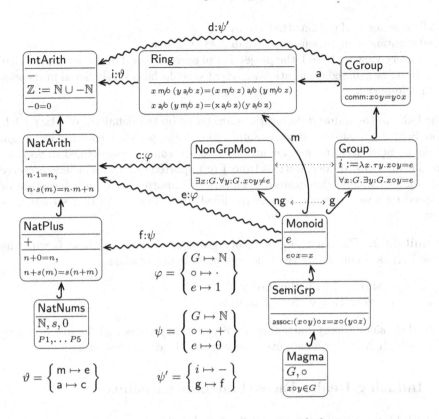

Fig. 2. A MMT graph for elementary algebra

conservative estimation), then we obtain a number of induced statements that is an order of magnitude higher.

Another important property of MMT is that, as discussed in Sect. 2, MMT URIs are expressive enough to supply names for all induced statements. In fact, we can already access the induced statements in Fig. 2 in MMT. For example, the statement $\forall x, y, z : \mathbb{Z}.(x + y) + z = x + (y + z)$ induced by the view c in IntArith has the MMT URI U?IntArith?c/g/assoc. Still, for external applications, it is essential to have the induced statements explicitly represented.

Generating Induced Statements. We already hinted in Sect. 2 that generating the induced statements is a *theory graph operation* i.e. it takes a theory graph as input and returns a different one, specifically the induced theory graph. Below, we define theory graph flattening as an instance of such a generation procedure that produces those statements induced by framing.

Definition 1. Given a theory graph G the flattening of a theory T in G is a theory $\overline{\mathsf{T}}$ with the same URI as T containing:

– all constant declarations that are in T.
– all constant declarations that are imported into T.
– for every view v : S → T the projection of every S-based declaration over view
   v. Here, by S-based declaration we refer to the declarations in S and in theories
   that import S.

The URIs of the induced declarations are based on the definition of MMT URIs
from Sect. 2 (see also the assoc example above) and permit recovering the origin
of the induced declarations (i.e. are *information complete* as defined in Sect. 2).
Specifically, constant declarations from T or imported in T by an include preserve
their name. Meanwhile constant declarations imported in T by a structure or
induced by a view are additionally qualified with the name of that structure or
view.

Definition 2. The flattening of theory graph G is a theory graph $\overline{G}$ with the
same URI as G containing the following module declarations:

– for every theory T in G the theory $\overline{T}$
– a copy of every view v : S → T from G

Note that, since theory flattening preserves theory URIs and doesn't add new
axioms (only new theorems), every view v in G is also a view in $\overline{G}$.

4   Inducing Realm Faces/Flattening Realms

In [CFK14] we introduce the concept of **realms** to con-
solidate knowledge about mathematical theories. This is
motivated by the intuition that users of a knowledge collection can have different roles and therefore want to see
different kinds of materials. In a nutshell, a realm – pic-
tured schematically on the right – is super-structure of a
fragment of a theory graph that singles out a set of conservative developments called **pillars** and extends them
with a theory called the **face** of the realm that abstracts
from all development details in the pillars, which are
required to be linked by a chain of views that makes them isomorphic.

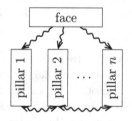

Fig. 3. Realm schema

The idea of [CFK14] is that practitioners only need access to the face that
supplies all the useful facts about a mathematical domain, whereas the student
also wants to know how these are established and needs to access the "development history" in one (or more) pillar. The developer finally wants to develop the
knowledge about the domain by extending one (or more) pillar, the development
of the face is just regarded as a side effect.

In Fig. 4, we have refined the schematic of Fig. 3 by graying out the parts of
the realms the users in their respective roles will not be able to see. We can see the
restricted theory/realm graphs as *induced material* adapted to particular users:
(i) the face graph – the faces inherit the graph structure of the developments –
for the practitioners, *(ii)* the graph of faces with selectively opened developments

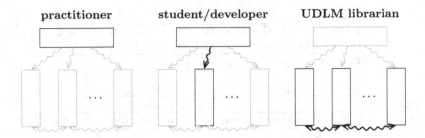

Fig. 4. Realms and user roles

for the student *(iii)* the developments with their faces for the developers, and *(iv)* the original theory graph for the knowledge librarian – a "user" of the library who maintains the library, e.g. refactoring theories, renaming theorems to avoid name clashes, etc. For the developer and the librarian, the realm faces are secondary objects which have to be maintained without being the primary object of study. For them it would be very convenient to let them be computed from the developments automatically as induced material. Indeed this can be done, as we will show in the rest of this section.

4.1   Generating Realm Faces as Induced Theories

For realms we consider their face as inducible from the pillars and discuss in the following how that can be mechanized.

Firstly, if faces are induced, we can define a realm by just giving it a **name**, specifying the pillars, and listing the cycle of views that shows that the pillars are equivalent. Figure 5 on the right gives the general form of the specification.

For the induced face, the URI g of the file with the realm specification induce the MMT

$$\text{realm } r = \{$$
$$\text{pillar } p_1 = \{t_1^{p_1}, t_2^{p_1}, \ldots, t_{k_1}^{p_1}\}$$
$$\text{pillar } p_2 = \{t_1^{p_2}, t_2^{p_2}, \ldots, t_{k_2}^{p_2}\}$$
$$\vdots$$
$$\text{pillar } p_n = \{t_1^{p_n}, t_2^{p_n}, \ldots, t_{k_n}^{p_n}\}$$
$$\text{equivcyc} = \{v_1, \ldots, v_n\}\}$$

Fig. 5. A realm specification

URI of the face, and the MMT URIs of the symbols inside are induced from the pillar names. To induce the face we need to solve three main issues:

1. select which symbols from each pillar should be in the face
2. merge equivalent symbols (such as the e in Fig. 6)
3. resolve naming conflicts (equivalent symbols with different names and distinguishable symbols with same name)

Before we formalize this in Definition 4 below, let us adapt the groups realm from [CFK14] to the MMT setting for intuitions.

Example 1. Figure 6 shows a realm with two pillars for the equivalent group definitions based on composition ∘ and, respectively, division/in the usual way. The corresponding realm specification is in Fig. 7. We have added the unit e

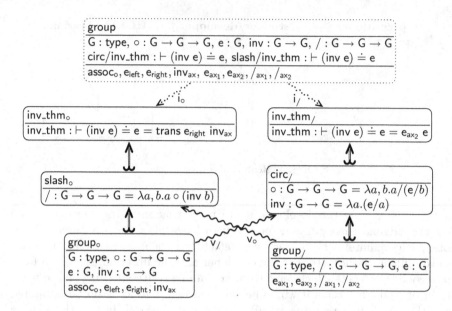

Fig. 6. A realm of groups

with its axiom e_{ax_1}, e_{ax_2} to group$_/$ even though it is mathematically redundant –
$e = x/x$ for all x in G – because that allows us to show all aspects of the face
generation algorithm below.

Theories slash$_\circ$ and inv_thm$_\circ$ as well as circ$_/$
and inv_thm$_/$are each conservative developments
of group$_\circ$ and group$_/$ respectively as they only
introduce defined symbols. The views $v_\circ$ and $v_/$
ensure the equivalence of the two pillars with the
obvious assignments. The proofs that the axioms
hold (i.e. the assignments for them in the views)

```
realm group = {
  pillar circ = {group₀, slash₀}
  pillar slash = {group/, circ/}
  equivcyc = {v₀,v/}}
```

Fig. 7. A group realm spec.

are all straightforward and omitted for simplicity. The **face** for the realm is
shown in theory group and has the intuitive shape, containing all the important
concepts as primitive symbols and all their properties as axioms. We explain
how the face was produced below.

Definition 3. We define the following partial ordering on symbols in a realm
r. Let t and s be symbols in theories T and S respectively such T and S are in
different pillars. If there is an assignment s := t in one of the pillar equivalence
views then we write $s \leq_r t$. If there is also an assignment t := s in such a view
then we write. $s =_r t$, otherwise $s <_r t$. We call a symbol **essential** in r if it is
$<_r$-minimal.

The intuition behind Definition 3 is that if there is an assignment s := t then
there is an equivalence between s and t at the realm level and we need to decide
which should appear in the face. Then, the ordering captures the fact that s

is a primitive concept in its realm while t is derived, possibly only to give the equivalence view. In Fig. 6 the symbols inv and ∘ from theory circ$_/$ are derived.

Definition 4. Let r be a realm as specified in Fig. 5. Then, we generate a face for r by adding copies of all essential symbols (with type but no definition – following the definition of a realm face from [CFK14]) with the following provisions:

1. If two essential symbols in different pillars have the same name we prefix them with the pillar name (to ensure unique URIs)
2. If n essential symbols $s_1, \ldots, s_n$ are equal (with respect to $=_r$) we add the later ones as aliases of the first (order is irrelevant but must be consistent). An exception occurs when (some) of the equal symbols have the same name in which case we only add them once and effectively merge them (instead of prefixing the names with the pillar name as usual).

Example 2. The face group from Fig. 6 is generated following Definition 4. The essential symbols (omitting axioms for simplicity) are G, ∘, e, inv, inv_thm for the first pillar and G, /, e, inv_thm for the second. We have two name clashing pairs: group$_∘$?e and group$_/$?e as well as inv_thm$_∘$?inv_thm and inv_thm$_/$?inv_thm. For the first pair (e) we have an equality since v$_∘$ and v$_/$ assign them to each other so we merge. Then, for the second pair (inv_thm) we prefix with the realm name producing circ/inv_thm and slash/inv_thm to obtain the face shown in Fig. 6.

4.2   Curating Realms Through Alignments

The problem with Definition 4 is that we can have duplicate symbols that are actually equivalent but appear as different because of slightly different formalizations. A common example is theorems with different proofs as is the case of inv_thm in Example 2.

In [CFK14] realm faces are meant to be manually generated and curated to avoid such issues. However, we propose an alternative method of curating faces by giving *alignments* between pillar theories to establish symbols as being equivalent. This idea is inspired from [KRSC11], but has not been made formal before.

Definition 5 (Alignment). An **alignment** is a view pair $v_1 : \widehat{S} \to T$ and $v_2 : \widehat{T} \to S$, where $\widehat{S}$ is the **abstraction** of S: $\widehat{S}$ omits all definitions in S.

The abstraction operation $\widehat{}$ is needed to allow us to assign a new interpretation for defined symbols which would otherwise be translated via definition expansion. We will concentrate on the case where S and T are in different pillars here.

For instance, take the situation in Fig. 6 where the (trivial) theorem inv_thm proving that e is its own inverse appears in each pillar. Still, the proofs are different over the translation so that both symbols appear different at the MMT level. However, we can fix the problem by giving an alignment between the two theories containing the theorem. Listing 1.1 below shows the alignment and the resulting, curated face.

Listing 1.1. Alignment Example

view $a_/$: $\widehat{\text{inv\_thm}_o} \to$ inv_thm$_/$ = {inv_thm := inv_thm}
view a_o : $\widehat{\text{inv\_thm}_/} \to$ inv_thm$_o$ = {inv_thm := inv_thm}
theory group = {G : type, e : G, ..., / : G→G→G ..., inv_thm : ⊢ (inv e) $\doteq$ e}

4.3 Opening a Pillar

For the student/developer view described above we need the operation of *opening a pillar* that allows the developer to access the internals of the symbols and axioms in the face as formalized in one of the pillars. We model this by creating a new theory for each pillar that combines the symbol aggregation and name abstraction of the face theory with the implementation details of that pillar.

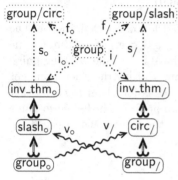

Fig. 8. Opening a pillar

Concretely, given a realm r and a pillar p we induce a theory r/p that is generated following the same procedure as the face (i.e. from Definition 4) but without omitting the definitions. For symbols in a different pillar we generate the definition by translating it over the view from that pillar into p. Effectively, we obtain the symbol definitions *as seen from p* which corresponds to the intuition of opening a pillar.

In the theory graph, we represent this as 1. a structure s from the pillar p that adds its symbols to r/p but with the renamings used in the face generation and, 2. a view v : r → r/p that formalizes the relation that r/p is an implementation for the face.

Figure 9 shows the theory group/circ representing the opening of pillar circ in the realm of groups above. It has the same symbols as the face but with all symbols that are non-primitive in circ having a definition. For the two symbols originating in the second pillar (/ and slash/inv_thm) their definition is obtained by translating over v_o.

theory group/circ = {
  G : type, $\circ$: G → G → G, e : G, inv : G → G,
  / : G → G → G = λa,b.a $\circ$ (inv b),
  circ/inv_thm : ⊢ (inv e) $\doteq$ e = trans e_{right} inv_{ax},
  slash/inv_thm : ⊢ (inv e) $\doteq$ e = e_{ax_2} e,
  $\vdots$
}

Fig. 9. Developer view example

The resulting theory graph is shown in Fig. 8 where the content of each theory is omitted. Note that the theories group, group/circ, group/slash as well as the structures s_o and $s_/$ and the views $i_o, i_/, f_o$ and $f_/$ are all induced.

5 Searching the Knowledge Space of the LATIN Logic Atlas

As a first application of the concepts described in this paper we build a system for searching the knowledge space (i.e. flattened theory graph) of the highly modular LATIN [Cod+11] library.

MathWebSearch. For searching, we use our MATHWEBSEARCH system [KMP12], which is a content-oriented search engine for mathematical expressions that indexes formula-URL pairs and provides a web interface querying the formula index via unification. The implementation of a math-literate web service that conducts such searches is very simple: instead of harvesting formulae from a formal digital library directly as in [Ian+13], we flatten the library first, and then harvest formulae. As discussed in Sect. 2, MMT flattening gives the induced constants information complete MMT URIs which we directly use for formula harvests. Then, we need to replace the human-oriented search front-end of MWS, i.e. the input of search queries and the presentation of search results. This can be used for:

Instance Search. E.g. to find all instance of associativity we can issue the query $\forall x, y, z : \boxed{S}.(x\boxed{\text{op}}y)\boxed{\text{op}}z = x\boxed{\text{op}}(y\boxed{\text{op}}z)$, where the $\boxed{-}$ are query variables that can be instantiated in the query. In the library from Fig. 2 we would find the associativity axiom SemiGrp/assoc, its directly inherited versions in Monoid, to Ring and in particular the version U?IntArith?c/g/assoc.

Applicable Theorem Search. Where universal variables in the index can be instantiated as well; this was introduced for a non-modular formal library in [Ian+13]. Here we could search for $3 + 4 = \boxed{R}$ and find the induced statement U?IntArith?c/comm with the substitution $R \mapsto 4+3$, which allows the user to instantiate the query and obtain the equation $3 + 4 = 4 + 3$ together with the justification U?IntArith?c/comm that can directly be used in a proof.

Induced Statements in the LATIN Library. The LATIN atlas is written in an extension of the TWELF encoding [RS09] of LF [HHP93], so it is natural to use an extension of LF notation with query variables for input. Therefore, we use the MMT notation language and interpretation service described in [IR12] to transform LF-style input into MMT objects and subsequently to MWS queries.

We implemented library flattening as described in Sect. 3 in MMT and applied it to the LATIN library. The flattening (once) of the LATIN library increases the number of declarations from 2310 to 58847 (a factor of 25.4) and the total size of the library from 123.9 MB to 1.8 GB (a factor of 14.8). As expected, the multiplication factor depends on the level of modularity of the library. For instance, the highly modular math sub-library containing mainly algebraic structures increases from 2.3 MB to 79 MB thus having a multiplication factor of 34.3, more than double the library average. The size of the MWS harvests also increases considerably, from 25.2 MB to 539.0 MB.

Explaining URIs of Induced Statements. The presentation of the MMT URIs requires some work as well: while the MMT system can directly dereference the MMT URI and thus be used to present the induced statement, humans want a justification that is more understandable than a MMT URI. Fortunately, this can be generated from the MMT URI by a simple template-based algorithm. Let us consider the search result U?IntArith?c/g/assoc from the instantiation search above, where we take U to be `http://cds.omdoc.org/cds/elal`. The first step is to localize the result in the theory U?IntArith with the sentence

$$\text{Induced statement } \forall x, y, z \ : \ \mathbb{Z}.(x + y) + z \ = \ x + (y + z)$$
$$\text{found in } \underline{\texttt{http://cds.omdoc.org/cds/elal?IntArith}} \ (\underline{\text{subst}}, \tag{1}$$
$$\underline{\text{justification}}).$$

Here the underlined fragments carry hyperlinks, the second pointing to the justification:

$$\underline{\text{IntArith}} \text{ is a } \underline{CGroup} \text{ if we interpret } \circ \text{ as } + \text{ and } G \text{ as } \mathbb{Z}. \tag{2}$$

which can be directly inferred from the information associated to the morphism c in the MMT URI. Then we skip over g, since its assignment is trivial and generate the sentence.

$$\underline{CGroups} \text{ are } \underline{SemiGrps} \underline{\text{ by construction}} \tag{3}$$

and finally we ground the explanation by the sentence

$$\text{In} \quad \underline{SemiGrps} \quad \text{we} \quad \text{have} \quad \text{the} \quad \text{axiom}$$
$$\underline{\text{assoc} : \forall x, y, z : G.(x \circ y) \circ z = x \circ (y \circ z)} \tag{4}$$

The sentences (1) to (4) can be generated from templates, since the MMT system gives access to the necessary information: source and target theory as well as the assignment ψ' for (2), the fact that the path from SemiGrp to CGroup[2] only consists of inclusion that triggers the template for (3) and the original formulation of the axiom assoc.

The resulting search interface is shown in Fig. 10 and is available at [FS]. Note that we make use of another peculiarity of the MMT system in this explanation: all constants in the theory graph carry notation declarations [KMR08], which can be used to generate human-readable presentations of arbitrary formal objects in the graph.

---

[2] In fact these theory identifiers are not adequate for explanations. We conjecture that verbalization of the primary symbol of the respective theory would be the right choice here – see [Koh14] for these concepts – but leave studying this to future work.

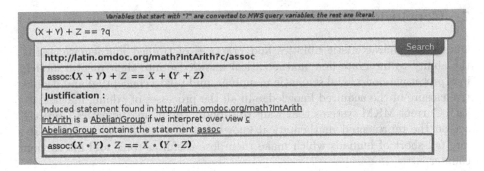

Fig. 10. The FLATSEARCH web interface for LATIN

6 Future Work: Realm-Supported Workflows in the Open Archive of Formalizations

A natural application area for induced realms as described in Sect. 4 is the integration of formal libraries, so that results in any of them can be used to prove new theorems in any other. This is the aim of the *Open Archive of Formalizations* [OAF] which integrates several, large, formal mathematical libraries. It uses the LATIN logic atlas [Cod+11] as a logical basis and imports the libraries as based on a dedicated LATIN meta-logic. Views between the base logics relate the libraries themselves at the foundational level but at a higher level, the important mathematical concepts remain unassociated.

Even though the underlying domains should form realms straddling the libraries, in practice, different libraries define core concepts (e.g. real numbers, functions, etc.) differently. Therefore, even after importing them to a common foundation, libraries remain effectively segregated. To achieve genuine interoperability we need to associate the equivalent concepts from each library with each other and establish a new library containing the *merged* concepts. This is a direct application of the ideas from Sect. 4. First, we can associate concepts by declaring realms. Then, we can generate the induced faces graph that provides the interface to the *merged*, integrated library. The operation of opening a pillar described in Sect. 4.3 allows access to the implementation details of each individual library where needed. Finally, the resulting interface can be refined via alignments.

The different induced realm graphs in Fig. 4 can be seen as special lenses that allow users with different roles to see the underlying archive according to their preferences and needs. This is similar to how an experienced human would present the materials if she were aware of the user role; therefore we can see the induced realms as a form of mathematical literacy.

Applying the above to the OAF is still future work but the challenges encountered during the project provided the main motivation for the work described in this paper.

7 Conclusion and Future Work

One of the characteristic abilities and practices of trained mathematicians is the ability to integrate new mathematical knowledge into their mental model, interpret it via non-trivial semantic mappings, and take a conceptual and deductive closure of the acquired knowledge in all the processes of "doing mathematics". Current MKM systems that want to support "doing mathematics" directly act on the represented mathematical knowledge they are fed with and therefore fall short of humans which make them less useful as tools and interaction counterpart.

The main hypothesis of the work presented here is the idea that running classical MKM algorithms on a suitably structured "mathematical knowledge space" (MKS) which extends the represented knowledge by a class of "induced knowledge items" will let them approximate mathematical literacy. We test this hypothesis on two classes of knowledge items in the context of theory graphs.

In a first case study we extend a theory graph with statements induced by views in the theory graph of the formal LATIN library. Indexing this in the math-specific, but otherwise illiterate MathWebSearch engine turns it into the FLATSEARCH engine that gives us results that approximate mathematical literacy. In the second case study, we build on a realm-structured knowledge collections and turn it into a MKS by inducing realm faces and into a personal MKS by opening pillars as needed. Here the induced knowledge items are theories, structures, and views in the theory graph.

In both cases much of the heavy lifting has been done by special URIs, that serve as systematic identifiers of induced elements. In the case of FLATSEARCH, these URIs are the MMT URIs already introduced in [RK13]. They are all we need to explain the results in terms of the original LATIN graph. In the realms case study we took great care to introduce new URIs for all induced knowledge items. It speaks for the strength and versatility of the MMT design that the realm-based URIs can be interpreted and justified in the MMT framework.

In the future, we plan to extend the "math literacy via induced knowledge structures" approach proposed in this paper with more facets and applications. We conjecture that the crucial step in such extensions will be the availability of some form of systematic naming scheme that uses the structural parts of the original knowledge to name induced knowledge items.

One extension that seems immediately profitable is to extend flattening and realms to flexiformal representations (representations of mathematical knowledge at flexible levels of formality; see [Koh13]) and apply it to traditional mathematical documents. [Lau07] revealed a theory graph of 51 theory nodes and 107 theory morphisms of which 12 were views, but 63 had non-trivial assignments in the first 35 pages of Bourbaki's Algebra. Applying FLATSEARCH to this graph would solve of the problems readers face with the Bourbaki books – which are otherwise well-liked for their structured approach: particular mathematical structures and objects can only be understood if one already knows all the material they depend on. One author even said that

Bourbaki was a dinosaur, the head too far away from the tail. Explaining: [...] You could say "Dieudonné what is the result about so and so?" and he would go to the shelf and take down the book and open it to the right page. After Dieudonné retired no one was able to do this. So Bourbaki lost awareness of his own body [Ric]

A flexiformalization of the Bourbaki books together with an extension of MMT that can deal with flattening of informal texts would go a long way to alleviate these problems.

Acknowledgments. This work has been supported by the German Research Council (DFG) under grant KO 2428/13-1.

References

[CFK14] Carette, J., Farmer, W.M., Kohlhase, M.: Realms: a structure for consolidating knowledge about mathematical theories. In: Watt, S.M., Davenport, J.H., Sexton, A.P., Sojka, P., Urban, J. (eds.) CICM 2014. LNCS, vol. 8543, pp. 252–266. Springer, Heidelberg (2014). http://kwarc.info/kohlhase/submit/cicm14-realms.pdf

[Cod+11] Codescu, M., Horozal, F., Kohlhase, M., Mossakowski, T., Rabe, F.: Project abstract: logic atlas and integrator (LATIN). In: Davenport, J.H., Farmer, W.M., Urban, J., Rabe, F. (eds.) MKM/Calculemus 2011. LNCS, vol. 6824, pp. 289–291. Springer, Heidelberg (2011). https://kwarc.info/people/frabe/Research/CHKMR_latinabs_11.pdf

[FS] FlatSearch Demo. http://cds.omdoc.org:8181/search.html (Accessed on 23 April 2015)

[HHP93] Harper, R., Honsell, F., Plotkin, G.: A framework for defining logics. J. Assoc. Comput. Mach. **40**(1), 143–184 (1993)

[Ian+13] Iancu, M., et al.: The Mizar mathematical library in OMDoc: translation and applications. J. Autom. Reasoning **50**(2), 191–202 (2013)

[IR12] Iancu, M., Rabe, F.: (Work-in-Progress) An MMT-based user- interface. In: Kaliszyk, C., Lüth, C. (eds.) Workshop on User Interfaces for Theorem Provers (2012)

[KK09] Kohlhase, A., Kohlhase, M.: Spreadsheet interaction with frames: exploring a mathematical practice. In: Carette, J., Dixon, L., Coen, C.S., Watt, S.M. (eds.) Calculemus/MKM 2009. LNCS (LNAI), vol. 5625, pp. 341–356. Springer, Heidelberg (2009). http://kwarc.info/kohlhase/papers/mkm09-framing.pdf

[KMP12] Kohlhase, M., Matican, B.A., Prodescu, C.-C.: MathWebSearch 0.5: scaling an open formula search engine. In: Campbell, J.A., Jeuring, J., Carette, J., Dos Reis, G., Sojka, P., Wenzel, M., Sorge, V. (eds.) CICM 2012. LNCS (LNAI), vol. 7362, pp. 342–357. Springer, Heidelberg (2012). http://kwarc.info/kohlhase/papers/aisc12-mws.pdf

[KMR08] Kohlhase, M., Müller, C., Rabe, F.: Notations for living mathematical documents. In: Autexier, S., Campbell, J., Rubio, J., Sorge, V., Suzuki, M., Wiedijk, F. (eds.) AISC/Calculemus/MKM 2008. LNCS (LNAI), vol. 5144, pp. 504–519. Springer, Heidelberg (2008). http://omdoc.org/pubs/mkm08-notations.pdf

[Koh13] Kohlhase, M.: The flexiformalist manifesto. In: Voronkov, A., et al. (eds.) 14th International Workshop on Symbolic and Numeric Algorithms for Scientific Computing (SYNASC 2012), Timisoara, Romania, pp. 30–36. IEEE Press (2013).http://kwarc.info/kohlhase/papers/synasc13.pdf

[Koh14] Kohlhase, M.: A data model and encoding for a semantic, multilingual terminology of mathematics. In: Watt, S.M., Davenport, J.H., Sexton, A.P., Sojka, P., Urban, J. (eds.) CICM 2014. LNCS, vol. 8543, pp. 169–183. Springer, Heidelberg (2014). http://kwarc.info/kohlhase/papers/cicm14-smglom.pdf

[KRSC11] Rabe, F., Kohlhase, M., Sacerdoti Coen, C.: A foundational view on integration problems. In: Davenport, J.H., Farmer, W.M., Urban, J., Rabe, F. (eds.) MKM 2011 and Calculemus 2011. LNCS (LNAI), vol. 6824, pp. 107–122. Springer, Heidelberg (2011). https://svn.eecs.jacobs-university.de/svn/eecs/archive/msc-2007/blaubner.pdf

[Lau07] Laubner, B.: Using Theory Graphs to Map Mathematics: A Case Study and a Prototype. MA thesis. Bremen: Jacobs University, August 2007. https://svn.eecs.jacobs-university.de/svn/eecs/archive/msc-2007/blaubner.pdf

[OAF] The OAF Project & System. http://oaf.mathhub.info (Accessed on 23 April 2015)

[OSV07] Odersky, M., Spoon, L., Venners, B.: Programming in Scala. Artima (2007)

[Rab08] Rabe, F.: The MMT system. https://svn.kwarc.info/repos/MMT/doc/html/index.html.2008

[Ric] Émilie Richter. Nicolas Bourbaki. http://planetmath.org/NicolasBourbaki.html

[RK13] Rabe, F., Kohlhase, M.: A scalable module system. Inf. Comput. **230**, 1–54 (2013). http://kwarc.info/frabe/Research/mmt.pdf

[RS09] Rabe, F., Schürmann, C.: A practical module system for LF. In: Cheney, J., Felty, A. (eds.) Proceedings of the Workshop on Logical Frameworks: Meta-Theory and Practice (LFMTP), pp. 40–48. ACM Press, New York (2009)

Strategies for Parallel Markup

Bruce R. Miller[✉]

Applied and Computational Mathematics Division,
National Institute of Standards and Technology, Gaithersburg, MD, USA
bruce.miller@nist.gov

Abstract. Cross-referenced parallel markup for mathematics allows the combination of both presentation and content representations while associating the components of each. Interesting applications are enabled by such arrangements: interaction with parts of the presentation to manipulate and query the corresponding content; enhanced search indexing. Although the idea of such markup is hardly new, effective techniques for creating and manipulating it are more difficult than it appears. Since the structures and tokens in the two formats often do not correspond one-to-one, decisions and heuristics must be developed to determine in which way each component refers to and is referred to by components of the other representation. Conversion between fine and coarse-grained parallel markup complicates XML identifier (ID) assignments. In this paper, we will describe the techniques developed for LaTeXML, a TeX/LaTeX to XML converter, to create cross-referenced parallel MathML. While not yet considering LaTeXML's content MathMLto be truly useful, the current effort is a step towards that continuing goal.

1 Introduction

Parallel markup for mathematics provides the capability of providing alternative representations of the mathematical expression, in particular, both the presentation form of the mathematics, i.e. its appearance, along with the content form, i.e. its meaning or semantics. Cross-linking between the two forms provides the connection between them such that one can determine the meaning associated with every visible fragment of the presentation and, conversely, the visible manifestation of each semantic sub-expression. Thus cross-linked parallel markup provides not only the benefits of the presentation and content forms, individually, but support many other applications such as: hybrid search where both the presentation and content can be taken into account simultaneously; interactive applications where the visual representation forms part of the user-interface, but supports computations based on the content representation.

Of course, the *idea* of parallel markup is hardly new. The m:semantics element has been part of the MathML specification [1] since the first version, in 1998! What seems to be missing are effective strategies for creating, manipulating and using this

---

© Springer International Publishing Switzerland 2015 (outside the US)
M. Kerber et al. (Eds.): CICM 2015, LNAI 9150, pp. 203–210, 2015.
DOI: 10.1007/978-3-319-20615-8_13

markup. Fine-grained parallelism is when the smallest sub-expressions are represented in multiple forms, whereas with coarse-grained parallelism the entire expression appears in several forms. Fine-grained parallelism is generally easier to create initially, and particularly when one deals with complex 'transfix' notations, or wants to preserve the appearance, but can infer the semantic intent of each sub-expression. Coarse-grained is often required by applications which may understand only a single format, or are unable to disentangle the fined-grained structure. HTML5 [3] only just barely accepts coarse-grained parallel markup, for example, by ignoring all but the first branch. Conversion from fine to coarse-grained is not inherently difficult, it can be carried out by a suitable walk of the expression tree for each format. But what isn't so clear is how to maintain the associations between the symbols and structures in the two trees. Indeed, there is typically no one-to-one correspondence between the elements of each format. Fine-grained parallelism, by itself, doesn't guarantee a clear association between all the symbols between the branches.

Our context here is LaTeXML, a converter from TeX/LaTeX to XML, and thence to web appropriate formats such as HTML, MathML and OpenMath. Input documents range from highly semantic markup such as sTeX [4], to intermediate such as used in the Digital Library of Mathematical Functions (DLMF) [6], to fairly undisciplined, purely presentational, markup as found on arXiv [8]. TeX induces high expectations for quality formatting forcing us to preserve the presentation of math. Meanwhile, the promise of global digital mathematics libraries and the potential reuse of a legacy of mathematics material encourages us to push as far as possible the extraction of content from such documents. At the very least, we should preserve whatever semantics is available in order to enable other technologies and research, such as LLaMaPuN [2], to resolve the remaining ambiguities. Getting ahead of ourselves, our first 'mini' strategy is just that: create fine-grained parallel markup at the first opportunity that presentation and semantics are available, in order to preserve them both.

In this paper, we describe the markup used in LaTeXML both for macros with known semantics, and for the result of parsing, and strategies for conversion to cross-linked, parallel markup combining Presentation MathML (pMML) and Content MathML (cMML). It should be noted that this does not mean that LaTeXML is producing useful quality cMML; the current work is a stepping stone towards that long-term goal.

Even in situations where only presentation markup is used, such as (currently) DLMF[1] where most symbols have always been hyper-linked to their definitions and the presentation-based search indexing is enhanced by the corresponding semantic content [7], the proposed strategies create a proper association between presentation and content resulting in a much cleaner implementation than the ad-hoc methods previously employed. Moreover, it is more complete, allowing the linkage to definitions to be extended to less textual operators such as binomials, floor, 3-j symbols, etc.

2   Motivation

Before diving into examples, a brief introduction to LaTeXML's internal mathematics markup, informally called XMath, is in order. This markup, inspired by OpenMath and both pMML and cMML, is intentionally hybrid in order to capture both the presentation and content properties of the mathematical objects throughout the step-wise processing from raw TeX markup, through parsing and, ultimately, semantic annotation. The main elements of concern are:

---

[1] http://dlmf.nist.gov/.

Listing 1.1. Internal representation of $a + F(a, b)$, after parsing to XMath (assuming F as a function)

```
<XMApp>
  <XMTok meaning=" plus"  role="ADDOP">+</XMTok>
  <XMTok role=" ID"  font=" italic">a</XMTok>
  <XMDual>
    <XMApp>
      <XMRef idref="m1.1" />
      <XMRef idref="m1.2" />
      <XMRef idref="m1.3" />
    </XMApp>
    <XMApp>
      <XMTok role="FUNCTION"  xml:id="m1.1"  font=" italic">F</XMTok>
      <XMWrap>
        <XMTok role="OPEN"  stretchy=" false">(</XMTok>
        <XMTok role=" ID"  xml:id="m1.2"  font=" italic">a</XMTok>
        <XMTok role="PUNCT">,</XMTok>
        <XMTok role=" ID"  xml:id="m1.3"  font=" italic">b</XMTok>
        <XMTok role="CLOSE"  stretchy=" false">)</XMTok>
      </XMWrap>
    </XMApp>
  </XMDual>
</XMApp>
```

XMApp generalized application (think m:apply or om:OMA);
XMTok generalized token (think m:mi, m:mo, m:mn, m:csymbol);
XMDual parallel markup container of the content and presentation branches;
XMRef shares nodes between branches of XMDual, via xml:id and idref attributes;
XMWrap container of unparsed sequences of tokens or subtrees (think m:mrow).

(Please see the online manual[2] for more details.)

By way of motivation, consider the simple example in Listing 1.1. The role attribute on tokens indicates the syntactic role that it plays in the grammar; in this case, we've asserted that F is a function, allowing the expression to be parsed. At the top-level, the sum requires no special parallel treatment since the presentation for infix operators is trivially derived from the content form (i.e. the application of '+' to its arguments). The application of F to its arguments benefits somewhat from parallel markup. This is a typical situation with the fine-grained XMDual: the content branch is the application of some function or operator (here F) to arguments (here a, b), but they are represented indirectly using XMRef to point to the corresponding sub-expressions within the presentation. While one could represent the delimiters and punctuation as attributes (as in MATHML's m:mfenced), that loses attributes of those attributes such as stretchiness, size or even color. A more compelling case is made by complex transfix notations or semantic macros, as we will shortly see.

However, this simple example already hints at a hidden complexity. Converting to either pMML and cMML is straightforward (given rules for mapping XMath elements to MATHML): simply walk the tree, depth-first, following each XMRef to the referenced node and choosing the first or second branch of XMDual for content or presentation, respectively. Even cross-linking is straightforward in the absence of XMDual, when the generated content or presentation nodes are 'sourced' from the same XMath node (F, a, and b, in the example): simply assign ID's to the source XMath node and the generated

[2] http://dlmf.nist.gov/LaTeXML/manual/.

Listing 1.2. MATHML representation of $a + F(a, b)$

```
<math display=" block" alttext=" a+F(a,b)" class=" ltx_Math" id=" m1">
  <semantics id=" m1a">
    <mrow xref=" m1.7.cmml" id=" m1.7">
      <mi xref=" m1.4.cmml" id=" m1.4">a</mi>
      <mo xref=" m1.5.cmml" id=" m1.5">+</mo>
      <mrow xref=" m1.6.cmml" id=" m1.6d">
        <mi xref=" m1.1.cmml" id=" m1.1">F</mi>
        <mo xref=" m1.6.cmml" id=" m1.6e">&ApplyFunction;</mo>
        <mrow xref=" m1.6.cmml" id=" m1.6c">
          <mo xref=" m1.6.cmml" id=" m1.6" stretchy=" false">(</mo>
          <mi xref=" m1.2.cmml" id=" m1.2">a</mi>
          <mo xref=" m1.6.cmml" id=" m1.6a">,</mo>
          <mi xref=" m1.3.cmml" id=" m1.3">b</mi>
          <mo xref=" m1.6.cmml" id=" m1.6b" stretchy=" false">)</mo>
        </mrow>
      </mrow>
    </mrow>
    <annotation-xml id=" m1b" encoding=" MathML-Content">
      <apply xref=" m1.7" id=" m1.7.cmml">
        <plus xref=" m1.5" id=" m1.5.cmml" />
        <ci xref=" m1.4" id=" m1.4.cmml">a</ci>
        <apply xref=" m1.6d" id=" m1.6.cmml">
          <ci xref=" m1.1" id=" m1.1.cmml">F</ci>
          <ci xref=" m1.2" id=" m1.2.cmml">a</ci>
          <ci xref=" m1.3" id=" m1.3.cmml">b</ci>
        </apply>
      </apply>
    </annotation-xml>
    <annotation id=" m1c" encoding=" application/x-tex">a+F(a,b)</annotation>
  </semantics>
</math>
```

nodes; record the association between them; afterwards, the presentation and content nodes that were sourced from the same ID are connected by getting an xref attribute referring each to the other. But with XMDual one has not only to determine when the generated nodes are related, one has to contend with extra tokens; in the example, the parentheses and comma appear only in the presentation. Presumably, those tokens should be associated with the *application* of F, as would the containing m:mrow. The desired result is shown in Listing 1.2.

A fuller illustration of the issues encountered in typical LATEX markup combines complex transfix notations and semantic macros, such as:

```
\left\langle\Psi\middle|\mathcal{H}\middle|\Phi\right\rangle
 + \defint{a}{b}{F(x)}{x}
```

This example, whose internal form is shown in Listing 1.3, involves quantum-mechanics notations, which LATEXML's parser is happily able to recognize. Additionally, we've introduced a semantic macro \defint to represent definite integration, which will be transformed to so-called 'Pragmatic' Content MATHML form, to enhance the illustration with a many-to-many correspondence. (The implementation of \defint is not difficult, but outside the scope of this article)

3 Main Strategies

We will see that a key part of the method is determining which nodes are visible to presentation, to content or to both branches. This is easily determined by an algorithm

Listing 1.3. Internal representation of $\langle\Psi|\mathcal{H}|\Phi\rangle + \int_a^b F(x)dx$ as XMath

```xml
<XMApp>
  <XMTok meaning=" plus" role="ADDOP">+</XMTok>
  <XMDual>
    <XMApp>
      <XMTok meaning="quantum−operator−product" />
      <XMRef idref="m2.5" />
      <XMRef idref="m2.6" />
      <XMRef idref="m2.7" />
    </XMApp>
    <XMWrap>
      <XMTok role="OPEN">⟨</XMTok>
      <XMTok role="ID" xml:id="m2.5">Ψ</XMTok>
      <XMTok role="CLOSE" stretchy="true">|</XMTok>
      <XMTok role="ID" xml:id="m2.6" font="caligraphic">H</XMTok>
      <XMTok role="OPEN" stretchy="true">|</XMTok>
      <XMTok role="ID" xml:id="m2.7">Φ</XMTok>
      <XMTok role="CLOSE">⟩</XMTok>
    </XMWrap>
  </XMDual>
  <XMDual>
    <XMApp>
      <XMTok meaning="hack−definite−integral" role="UNKNOWN" />
      <XMRef idref="m2.1" />
      <XMRef idref="m2.2" />
      <XMRef idref="m2.3" />
      <XMRef idref="m2.4" />
    </XMApp>
    <XMApp>
      <XMApp>
        <XMTok role="SUPERSCRIPTOP" scriptpos="post2" />
        <XMApp>
          <XMTok role="SUBSCRIPTOP" scriptpos="post2" />
          <XMTok mathstyle="display" meaning="integral" role="INTOP">∫</XMTok>
          <XMTok role="ID" xml:id="m2.1" font="italic">a</XMTok>
        </XMApp>
        <XMTok role="ID" xml:id="m2.2" font="italic">b</XMTok>
      </XMApp>
      <XMApp>
        <XMTok meaning="times" role="MULOP">×</XMTok>
        <XMDual xml:id="m2.3">
          <XMApp>
            <XMRef idref="m2.3.1" />
            <XMRef idref="m2.3.2" />
          </XMApp>
          <XMApp>
            <XMTok role="FUNCTION" xml:id="m2.3.1" font="italic">F</XMTok>
            <XMWrap>
              <XMTok role="OPEN" stretchy="false">(</XMTok>
              <XMTok role="UNKNOWN" xml:id="m2.3.2" font="italic">x</XMTok>
              <XMTok role="CLOSE" stretchy="false">)</XMTok>
            </XMWrap>
          </XMApp>
        </XMDual>
        <XMApp>
          <XMTok meaning="differential−d" role="DIFFOP" font="italic">d</XMTok>
          <XMTok role="UNKNOWN" xml:id="m2.4" font="italic">x</XMTok>
        </XMApp>
      </XMApp>
    </XMApp>
  </XMDual>
</XMApp>
```

like mark-and-sweep garbage collection. Simply traverse the tree from the root following the first branch of each XMDual to mark all nodes as visible to presentation, and repeat for the second branch to mark content visibility. The analogy is apt, as this also allows pruning of nodes that are not visible at all, as sometimes occurs with complex macro usage — another mini-strategy.

A few definitions will also be relevant. During the depth-first tree traversal transforming XMath into either pMML or cMML, the *current node* is the XMath node being transformed. The *current container* is the XMDual node (if any) containing the current node; in the context of this discussion, an XMath container is always the application of some function or operator. We'll call the latter the *current operator*.

We will ascribe to each generated target node (pMML or cMML) an XMath *source node* that can be considered 'responsible' for the target node. In the common, simplest case, a current node that is visible to both branches and generates a token node in the target can be used as the source node.

A special case occurs when the target MATHML element is a container; these generally do not correspond to symbols, and ought to be associated with the nearest application (think m:apply or m:mrow)[3]. In this case, the source should be the nearest ancestor XMDual of the current node, that is the *current container*.

Similar reasoning applies when a token (non-container) symbol is generated from an XMath token which is not visible to the opposite branch; typically the notational icing of some transfix. We presume that is the only visible manifestation of the *current operator*, and so that is taken as the source node. In the example, the angle brackets and vertical bars are thus ascribed to the quantum-operator-product operator.

In pseudo-code, the *source* node for a given *target* is thus:

```
if target is a container
  if current container exists
    current container
  else
    current
else if target is visible in both branches
  current
else if current container exists
  if current operator is hidden from presentation
    current operator
  else
    current container
else
  current
```

Once this ascription of source nodes is done, the cross-referencing between the generated targets is easily established: the xref of a pMML (cMML) node is the cMML (pMML, respectively) node that was ascribed to the same source XMath node; if multiple nodes were ascribed to that source node, the first target node, in document order, is the sensible choice. Applying this method to the example from Listing 1.3 yields Listing 1.4, where we can see, for example, that the angle brackets and vertical bars

---

[3] Exceptions are m:msqrt or m:menclose where they tend to represent *both* the application of an operation and yet are the only visible manifestation of the operator! However, we also note that a common use of cross-linking in HTML is to turn them into href links; but HTML does not allow nested links!.

Listing 1.4. MATHML representation of $\langle\Psi|\mathcal{H}|\Phi\rangle + \int_a^b F(x)dx$

```xml
<math display="block" alttext="..." class="ltx_Math" id="m2">
  <semantics id="m2a">
    <mrow xref="m2.13.cmml" id="m2.13">
      <mrow xref="m2.9.cmml" id="m2.9">
        <mo xref="m2.8.cmml" id="m2.8">&LeftAngleBracket;</mo>
        <mi mathvariant="normal" xref="m2.5.cmml" id="m2.5">&Psi;</mi>
        <mo xref="m2.8.cmml" id="m2.8a" stretchy="true" fence="true">|</mo>
        <mi xref="m2.6.cmml" id="m2.6" class="ltx_font_mathcaligraphic">&HilbertSpace;</mi>
        <mo xref="m2.8.cmml" id="m2.8b" stretchy="true" fence="true">|</mo>
        <mi mathvariant="normal" xref="m2.7.cmml" id="m2.7">&Phi;</mi>
        <mo xref="m2.8.cmml" id="m2.8c">&RightAngleBracket;</mo>
      </mrow>
      <mo xref="m2.10.cmml" id="m2.10">+</mo>
      <mrow xref="m2.12.cmml" id="m2.12c">
        <msubsup xref="m2.12.cmml" id="m2.12">
          <mo xref="m2.11.cmml" id="m2.11" symmetric="true" largeop="true">&int;</mo>
          <mi xref="m2.1.cmml" id="m2.1">a</mi>
          <mi xref="m2.2.cmml" id="m2.2">b</mi>
        </msubsup>
        <mrow xref="m2.12.cmml" id="m2.12b">
          <mrow xref="m2.3.cmml" id="m2.3c">
            <mi xref="m2.3.1.cmml" id="m2.3.1">F</mi>
            <mo xref="m2.3.cmml" id="m2.3d">&ApplyFunction;</mo>
            <mrow xref="m2.3.cmml" id="m2.3b">
              <mo xref="m2.3.cmml" id="m2.3" stretchy="false">(</mo>
              <mi xref="m2.3.2.cmml" id="m2.3.2">x</mi>
              <mo xref="m2.3.cmml" id="m2.3a" stretchy="false">)</mo>
            </mrow>
          </mrow>
          <mo xref="m2.11.cmml" id="m2.11a">&InvisibleTimes;</mo>
          <mrow xref="m2.12.cmml" id="m2.12a">
            <mo xref="m2.11.cmml" id="m2.11b">d</mo>
            <mi xref="m2.4.cmml" id="m2.4">x</mi>
          </mrow>
        </mrow>
      </mrow>
    </mrow>
    <annotation-xml id="m2b" encoding="MathML-Content">
      <apply xref="m2.13" id="m2.13.cmml">
        <plus xref="m2.10" id="m2.10.cmml"/>
        <apply xref="m2.9" id="m2.9.cmml">
          <csymbol xref="m2.8" id="m2.8.cmml" cd="latexml">quantum-operator-product</csymbol>
          <ci xref="m2.5" id="m2.5.cmml">normal-&Psi;</ci>
          <ci xref="m2.6" id="m2.6.cmml">&HilbertSpace;</ci>
          <ci xref="m2.7" id="m2.7.cmml">normal-&Phi;</ci>
        </apply>
        <apply xref="m2.12c" id="m2.12.cmml">
          <int xref="m2.11" id="m2.11.cmml"/>
          <bvar xref="m2.12c" id="m2.12a.cmml">
            <ci xref="m2.4" id="m2.4.cmml">x</ci>
          </bvar>
          <lowlimit xref="m2.12c" id="m2.12b.cmml">
            <ci xref="m2.1" id="m2.1.cmml">a</ci>
          </lowlimit>
          <uplimit xref="m2.12c" id="m2.12c.cmml">
            <ci xref="m2.2" id="m2.2.cmml">b</ci>
          </uplimit>
          <apply xref="m2.3c" id="m2.3.cmml">
            <ci xref="m2.3.1" id="m2.3.1.cmml">F</ci>
            <ci xref="m2.3.2" id="m2.3.2.cmml">x</ci>
          </apply>
        </apply>
      </apply>
    </annotation-xml>
    <annotation id="m2c" encoding="application/x-tex">...</annotation>
  </semantics>
</math>
```

are associated with the `quantum-operator-product` operator while the various m:bvar, m:lowlimit, etc., are properly associated with the integral, *not* the integral operator.

4   Outlook

We have described a set of strategies for generating parallel markup with cross-references consisting of encouraging fine-grained parallel structures, mark-and-sweep to detect nodes visible to presentation or content and to garbage-collect, and a method to determine related nodes within each branch of the parallel markup.

Parallel markup must also be adapted to larger structures such as eqnarray, and AMS alignments with intertext containing multiple formula and/or document-level text markup. While the fundamental issue is the same — separating presentation and content forms — this seems to demand a distributed markup that separates the presentation and content forms into distinct math containers. LaTeXML currently has an ad-hoc, but not entirely satisfactory solution for this, but we will experiment with adapting the methods described here. However, it remains to be seen whether cross-referencing across separate math containers can be made useful.

And, now that generating Content MathML is more fun, we must continue working towards generating *good* Content MathML. Ongoing work will attempt to establish appropriate OpenMath Content Dictionaries, probably in a FlexiFormal sense [5], improved math grammar, and exploring semantic analysis.

References

1. Ausbrooks, R., Buswell, S., Carlisle, D., Chavchanidze, G., Dalmas, S., Devitt, S., Diaz, A., Dooley, S., Hunter, R., Ion, P., Kohlhase, M., Lazrek, A., Libbrecht, P., Miller, B., Miner, R., Sargent, M., Smith, B., Soiffer, N., Sutor, R., Watt, S.: Mathematical Markup Language (MathML) version 3.0. W3C Recommendation, World Wide Web Consortium (W3C) (2010). http://www.w3.org/TR/MathML3
2. Ginev, D., Jucovschi, C., Anca, S., Grigore, M., David, C., Kohlhase, M.: An architecture for linguistic and semantic analysis on the arXMLiv corpus. In: Applications of Semantic Technologies (AST) Workshop at Informatik 2009 (2009)
3. Hickson, I., Berjon, R., Faulkner, S., Leithead, T., Navara, E.D., O'Connor, E., Pfeiffer, S.: HTML5. W3C Recommentation, World Wide Web Consortium (W3C) (2014). http://www.w3.org/TR/html5/
4. Kohlhase, M.: Using LaTeX as a semantic markup format. Math. Comput. Sci. **2**(2), 279–304 (2008)
5. Kohlhase, M.: The flexiformalist manifesto. In: Voronkov, A., Negru, V., Ida, T., Jebelean, T., Petcu, D., ane Daniela Zaharie, S.M.W. (eds.) 14th International Workshop on Symbolic and Numeric Algorithms for Scientific Computing (SYNASC 2012), pp. 30–36. IEEE Press, Timisoara (2013)
6. Miller, B.R., Youssef, A.: Technical aspects of the digital library of mathematical functions. Ann. Math. Artif. Intell. **38**(1–3), 121–136 (2003)
7. Miller, B.R., Youssef, A.: Augmenting presentation MathML for search. In: Autexier, S., Campbell, J., Rubio, J., Sorge, V., Suzuki, M., Wiedijk, F. (eds.) AISC/Calculemus/MKM 2008. LNCS (LNAI), vol. 5144, pp. 536–542. Springer, Heidelberg (2008)
8. Stamerjohanns, H., Kohlhase, M., Ginev, D., David, C., Miller, B.: Transforming large collections of scientific publications to XML. Math. Comput. Sci. **3**(3), 299–307 (2010)

Readable Formalization of Euler's Partition Theorem in Mizar

Karol Pąk$^{(\boxtimes)}$

Institute of Informatics, University of Bialystok, Bialystok, Poland
pakkarol@uwb.edu.pl

Abstract. We present a case study on formalization of a textbook theorem in a form that is as close to the original textbook presentation as possible. Euler's partition theorem, listed as #45 at Freek Wiedijk's list of "Top 100 mathematical theorems", is taken as the subject of the study. As a result new formal concepts including informal flexary (i.e. flexible arity) addition are created and existing ones are extended to go around existing limitations of the Mizar system, without modification of its core. Such developments bring more flexibility of informal language reasoning into the Mizar system and make it useful for wider audience.

Keywords: Operations on languages · Legibility of proofs · Euler's partition

1 Introduction

Famous mathematical theorems rarely occur with only one proof in informal mathematical practice. In the mathematical literature we can often find several formulations or even conceptually different proofs of the same theorem. However, the reader can easily compare proofs that have the same main idea. Such situations are not so popular in repositories of formal mathematical knowledge. Usually, one version of a theorem with one proof only is stored there. Additionally, comparing proofs created in different formal proof systems is not so trivial. Even, if we consider two declarative environments or two procedural ones, this problem does not seem much easier.

It comes as no surprise that the main idea of the formal proof is often different from all known informal proof variants of the theorem, even if the author tried to create a formal equivalent of a particular informal development. The experience of big proof formalization developments shows that proof script authors can often, given the set of definitions and theorems collected in the Mizar Mathematical Library (MML) [3], obtain a new, so far unknown, and sometimes simpler, proof of a particular statement [8,9]. Therefore, many authors compare informal proof variants to check the possible use of collected resources before starting their formalization effort.

The paper has been financed by the resources of the Polish National Science Center granted by decision n°DEC-2012/07/N/ST6/02147.

© Springer International Publishing Switzerland 2015
M. Kerber et al. (Eds.): CICM 2015, LNAI 9150, pp. 211–226, 2015.
DOI: 10.1007/978-3-319-20615-8_14

To illustrate such situations we can consider the development of Brouwer's fixed-point theorem [5]. This theorem has been used in the formalization of Jordan curve theorem in the Mizar [7]. But for this purpose, the 2-dimensional case was enough. Therefore, this statement has been provided by A. Korniłowicz, only for this case using basic arguments concerning the fundamental groups of the respective spaces [13]. Note that this approach for higher-dimensional cases requires incomparably more difficult facts about these groups. The same theorem was proved in a combinatorial way in the HOL Light by J. Harrison for n-dimensional case, based on Sperner's lemma [10]. The original approach to prove this lemma is based on intuitively clear facts about the standard n-dimensional simplex and its arbitrarily small subdivision. However, these facts are not so easy if we consider them formally. Therefore, he chose an alternative justification of this lemma, wherein the simplex structure is replaced by the cubic one. Note that simplex structure is explored in one of the approaches to prove Brouwer's invariance of the domain theorem that was selected to formalize in Mizar by K. Pąk [18]. Therefore, having a large collection of facts about simplices, Brouwer's fixed-point theorem has been redeveloped to the general case based on Sperner's lemma in the original approach.

In this paper, we present the results of an experiment where we formalize Euler's partition theorem in the original approach [1,6,22]. Obviously, we can obtain a very slick proof using definitions and theorems collected in the MML that looks more or less similar to the original proof. However, the point of this exercise was not to obtain "a formalization", but to see how a natural language proof can be expressed in the Mizar format. Therefore our aim was to recreate the main idea and steps of reasoning as closely as possible, sometimes work around the system's limitations, however without a modification of its core, to obtain the result that looks almost the same as the informal one. Furthermore, as a measure of "closeness", we consider also the sketch of the proof that is generated automatically from the Mizar proof scripts and is published in the journal *Formalized Mathematics*.

Structure of the Paper. In Sect. 2 we discuss several conceptually different proofs of Euler's partition theorem. We focus our attention on three approaches: the original one that was presented by L. Euler [6], the *Euler's bijective proof* that was presented by G.E. Andrews [1], and the approach basing on Sylvester's bijection created by J.J. Sylvester, and choose one. In Sect. 3 we analyze informal mathematical constructions that are used in the selected approach, and we propose an adaptation method of this construction to the formal language in a way that the obtained visual effect is as closely as possible to the informal one. In Sect. 4 we present a formalization of the proof in the selected approach written in the Mizar system. Finally, in Sect. 5 we conclude the paper and we discuss future work. Note additionally that each fragment of the Mizar proof scripts contained in this paper comes from Mizar theory files FLEXARY1.miz, EULRPART.miz available in the Mizar distribution.

2   Informal Proofs of Euler's Partition Theorem

Generally, a partition of a natural number n is a way of writing n as a sum of positive integers where the arrangement of the addends does not need to be determined. Denote by $\mathcal{O}_n$ the set of partitions of n into odd parts and similarly denote by $\mathcal{D}_n$ the set of partitions of n into distinct parts. Then Euler's partition theorem states that the cardinality of $\mathcal{O}_n$ is equal to the cardinality of $\mathcal{D}_n$ for all natural n. Euler presented a very slick proof in 1748 [6] by generating functions that can be sketched as follows:

$$
\begin{aligned}
1 + \sum_{n=1}^{\infty} |\mathcal{D}_n| x^n &= (1+x) \cdot (1+x^2) \cdot (1+x^3) \cdot (1+x^4) \cdot (1+x^5) \cdot \ldots \\
&= \frac{1-x^2}{1-x} \cdot \frac{1-x^4}{1-x^2} \cdot \frac{1-x^6}{1-x^3} \cdot \frac{1-x^8}{1-x^4} \cdot \frac{1-x^{10}}{1-x^5} \cdots \\
&= \frac{1}{1-x} \cdot \frac{1}{1-x^3} \cdot \frac{1}{1-x^5} \cdots \\
&= (1+x+x^2+x^3+\ldots) \cdot (1+x^3+(x^3)^2+(x^3)^3+\ldots) \\
&\quad \cdot (1+x^5+(x^5)^2+(x^5)^3+\ldots) \cdot \ldots \\
&= 1 + \sum_{n=1}^{\infty} |\mathcal{O}_n| x^n
\end{aligned}
\tag{1}
$$

However, in such an analytical proof, information that describes the relationship between relevant partitions of n is implicit and hard to grasp. Obviously, without this information the proof is complete, but there are many people who prefer to compare the cardinality of sets based on an explicit mapping that associates their elements. Such a bijective proof has also been described by Euler. It has been given by G.E. Andrews [1, pp. 149–150], and also by H.S. Wilf [22, p. 10] in the following form:

EULER'S BIJECTIVE PROOF: A partition into distinct parts can be written as

$$
n = d_1 + d_2 + \ldots + d_k.
\tag{3}
$$

Each integer d_i can be uniquely expressed as a power of 2 times an odd number. Thus, $n = 2^{a_1} O_1 + 2^{a_2} O_2 + 2^{a_3} O_3 + \ldots + 2^{a_k} O_k$ where each O_i is an odd number. If we now group together the odd numbers we get an expression like:

$$
\begin{aligned}
n &= (2^{\alpha_1} + 2^{\alpha_2} + \ldots) \cdot 1 + (2^{\beta_1} + 2^{\beta_2} + \ldots) \cdot 3 + (2^{\gamma_1} + 2^{\gamma_2} + \ldots) \cdot 5 + \ldots \\
&= \mu_1 \cdot 1 + \mu_3 \cdot 3 + \mu_5 \cdot 5 + \ldots
\end{aligned}
$$

In each series $(2^{\alpha_1} + 2^{\alpha_2} + \ldots)$, the α_i's are distinct (why?). Thus the sum is the binary expansion of some μ_j. We now see the partition of n into odd parts that corresponds, under this bijection, to the given partition (3) into distinct parts. It is the partition that contains μ_1 1's, μ_3 3's, etc. $\square$

Fig. 1. A bijective proof of Euler's partition theorem that is used in [22].

In this constructive proof only several simple facts are used (implicite): the existence and the uniqueness of conversion between a natural number and its binary equivalent; a positive number decomposition as the product of a power of two and an odd number.

It is also possible to prove this result by another partition transformation, where correspondence is visually apparent by some modification of the Ferrers diagram, called a *bent graph*. The construction of this graph is based on the observation that we can represent every odd number $2i + 1$ as one central point, and a column and a row that are built with i points. Then we may arrange the configurations of points in such a way that the central points are inserted into a diagonal line. In this way we obtain what we call the bent graph (see Fig. 2). Additionally, the symmetrical parts of this graph, located above and below the diagonal line, are called *regular graphs*. Sylvester's bijection that is defined on bent graphs has been presented in [20, pp. 287–288] and is formulated as follows: *Each of these graphs will be bounded by lines inclined to each other at an angle one-half of that contained between the original bounding lines, and each may be regarded as made up of bends fitting into one another.* The original proof also contains a sketch of a justification that the resulting partition consists of pairwise different numbers (see to the example presented in Fig. 2 as an illustration). Obviously, the justification presented in this form shows in a simple way the proof idea, but only the idea. Therefore, a formalization of such a reasoning could not reflect the original proof at a very high level of similarity.

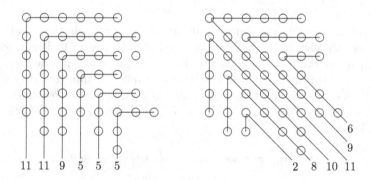

Fig. 2. An example that illustrates Sylvester's bijection, which maps a partition of 46 into odd numbers $(5, 5, 5, 9, 11, 11)$ to a partition of 46 into pairwise different numbers $(2, 6, 8, 9, 10, 11)$, used in [20].

Analysis of well-known justifications of Euler's partition theorem for the purposes of formal transcription shows that reflecting the original proof is possible in two mentioned approaches. However, the second justification, presented in Fig. 1, contains a more interesting informal construction, namely flexary (i.e. flexible arity, see [12]) addition with visible lower and upper bounds of summation, but also with only a lower bound available. Additionally, the formalization of this theorem in the HOL system imitates also the main idea of the second approach[1].

---

[1] For more details see https://code.google.com/p/hol-light/source/browse/trunk/100/euler.ml?r=2.

3 Formal Introduction of Informal Notations

To reflect the informal reasoning in the Mizar language, we discuss in this section definitions and notations used in the reasoning presented in Fig. 1. We divide this discusion into three subsections. In Subsect. 3.1 we choose concepts of a partition definition. Next, in Subsect. 3.2 we define operators that reflect in the Mizar system the informal flexary plus in both cases, finite and infinite. At the end, in Subsect. 3.3 we define a special kind of a matrix generalization to realize the rearrangement of the values of a finite sequence into a sequence of sequences that is used in the considered reasoning.

3.1 The Formal Definition of a Partition

The reasoning presented in Fig. 1 uses a partition of a natural number n (see (3)) defined as a finite sequence of positive integers that sum up to n. Note also that the order of the addends is not indicated. However, to count partitions of a number we need to opt for some type of arrangement, non increasing or non decreasing. The alternative solution is to represent a partition as a sequence of addends frequency, i.e. a sequence $(a_1, a_2, \ldots)$ that represent the partition $\{\underbrace{1, 1, \ldots, 1}_{a_1}, \underbrace{2, 2, \ldots, 2}_{a_2}, \ldots\}$. Obviously, we find this approach in the formulation "*the partition that contains μ_1 1's, μ_3 3's*" where the non-decreasing arrangement is suggested. Therefore, to simplify, we use only one method of arrangement in the definition of partition, which is formulated as a non-decreasing finite sequence of non-zero natural numbers that sum up to n, thus obtained by writing (in Mizar).

```
definition
    let n be Nat;
    mode a_partition of n -> non-zero non-decreasing natural-valued        (2)
        FinSequence means
            Sum it = n;
end;
```

Obviously, such a definition is adapted to represent simple modifications of partitions in an intuitive way, but requires a special attention in the formal approach. Note that we can neither simply include elements to a partition, nor modify its existing elements, without violating the arrangement. Such problems do not occur if we consider the *frequency representation* of partitions. In this case the realization of the mentioned operations on a partition is reduced to a simple modification that increases or reduces by one the value of one element in the frequency sequence. This approach has been used in the formalization of Euler's theorem in HOL, where the partition is defined as follows:

```
let partitions = store_name "partitions" new_definition
  'p partitions n <=> (!i. ~(p i = 0) ==> 1 <= i /\ i <= n) /\
                nsum(1..n) (\i. p(i) * i) = n';;
```

3.2 The Flexary Plus

Analyzing the first sentence of the proof, we find the Eq. (3) where we face two formalization problems. Obviously, the informal mathematical operators are present here, i.e. the flexary plus. This formula is usually formalized in the equivalent form $n = \sum_{i=1}^{k} d_i$. It can be written in the Mizar language as n = sum d, if we additionally assume that k is the length of d, or it can be written as n = sum (d|k), where k is an arbitrary natural number, d is a finite sequence of natural numbers, and d|k is the restriction of d to the set $\{1, 2, \ldots, k\}$. However, in our experiment we want to obtain a similar term represented as:

$$(d,2)+\ldots+(d,k). \tag{4}$$

Note also that the informal expression $d_2 + \ldots + d_k$ contains a hidden information that the finite sequence d has a second, a third, and up to a k-th element in the domain. The method of hiding this information in a term as (4) or getting around this problem is the second formalization problem. In the Mizar language we can resolve this, e.g. by creating a definition by cases or summarizing only these values that correspond to arguments in the domain of a finite sequence. In our experiment, we use a solution that gives the greater flexibility. The solution is based on the concept developed in the MML: a permissive definition of the function value, where it is assumed that the empty set is the value of a function outside its field, and also on the Peano number approach, where the empty set equals 0. However, this solution can be applied to summation of such D-valued sequence if D contains 0. We define the flexary plus as follows:

```
definition
  let k,n;
  let f,g be complex-valued Function;
  func (f,k) +...+ (g,n) -> complex number means
      h.(0+1) = f.(0+k) & ... & h.(n-'k+1) = f.(n-'k+k)      (5)
        implies
      it = Sum (h| (n-'k+1)) if f = g & k <= n
                            otherwise it = 0;
  end;
```

where h is a complex valued finite sequence, the operation $-'$ is the limited substraction of natural numbers, i.e. $a-'b$ is equal to $\max\{a - b, 0\}$, & ... & is the flexary logical conjunction (for more details see [11]). Note that the Mizar system's limitations prohibit the repetition of a locus in an operator expression when it is defined. Therefore, we cannot eliminate the function g, even if we want to consider only the case f = g.

The value of the defined above flexary plus is a complex number, but in the reasoning presented in Fig. 1 only the natural valued finite sequences are used, for which this value should be a natural number. To obtain such information about the value, the sufficient solution is to have the following registration in the environment of the Mizar article:

```
registration
   let n,k;
   let f be natural-valued FinSequence;                          (6)
   cluster (f,k) +...+ (f,n) -> natural;
end;
```

Based on the flexary plus definition, we can formulate and prove the first equality presented in Fig. 1 as $n = d.1 + (d,2)+...+(d,\text{len } d)$ for an arbitrary partition d of n, where `len d` is the length of d. Additionally, if we consider finite sequences a, O that represent the unique decomposition of d as a power of 2 times an odd number, we can formalize also the second equality as:

$$n = 2|\char`\^(a.1) * 0.1 + 2|\char`\^(a.2) * 0.2 +$$
$$(O\ (\#)\ 2|\char`\^a,3) +...+ (O\ (\#)\ 2|\char`\^a,\text{len } d), \qquad (7)$$

where (#) represents the product of functions and a, O have already been introduced to the reasoning in the preceding step which reads as follows:

```
consider O be odd-valued FinSequence,
      a be natural-valued FinSequence such that
A1: len O = len d = len a & d = O (#) 2|^a and      (8)
A2: d.1 = O.1*(2|^a.1) &...& d.len d = O.len d*(2|^a.len d);
```

Observe that the formula labeled by A2 is an equivalent formulation of the statement $d = O\ (\#)\ 2|\char`\^a$. This statement has been added only for improving the readability of dependencies occurring between d, O, and a. Note also that we could prove the equality (7) without the restriction on the length of d (that is equal to the length of O and a), e.g. for `len O` equal 0 we obtain simply that $O.1 = O.2 = 0$, since 1, 2 do not belong to the domain of O, but also $n = 0$, since 0-length sequence can be only the partition of 0.

Analyzing the next part of the reasoning presented in Fig. 1 we can observe that the flexary plus is used also in an unbounded form, without the upper bound of summation. Generally, this operation is used in the informal mathematical practice to speak conveniently about the sum of the terms of a sequence, where basing on several first terms we can precisely predict the others elements by analogy. Additionally, according to a popular informal convention, the information about the convergence of a sequence is often assumed *a priori*. However, this issue does not concern the reasoning presented in Fig. 1, where the unbounded flexary plus is used only in the context of finite sequences. For such kind of sequence, we can define this operator as the flexary plus with an upper bound, where the upper bound is greater than or equal to the maximum of the domain of the sequence. It is obtained by writing:

```
definition
   let n;
   let f be complex-valued Function;
   assume dom f /\ NAT is finite;
   func (f,n)+... -> complex number means                       (9)
         for k st for i st i in dom f holds i <= k holds
            it = (f,n) +...+ (f,k);
end;
```

Note that in the considered reasoning, only finite sequences are used, where the intersection of the domain and $\mathbb{N}$ is finite. Therefore, the introduction of such a definition to our experiment seems to be redundant. However, without this assumption, we cannot use this operator in the Mizar system if we need to substitute a term that is *a priori* a finite sequence on f, but we have not proved this statement yet. Obviously, such a possibility is very useful, if we want to formulate similarly a formal equivalents of an informal term. Moreover, we can reinforce the Mizar checker in such a way that the equality

$$(f,n)+... = (f,n) +...+ (f,len f); \qquad (10)$$

is automatically generated and added to every justification of a step, where the expression +... is used and a term substituted for f is a finite sequence. For this purpose we create the following redefinition:

```
definition
  let n;
  let f be complex-valued FinSequence;
  redefine func (f,n)+... -> complex number equals
    (f,n) +...+ (f,len f);
end;
```
(11)

We are aware that a definition that can generate automatically the summed sequence based only on the terms on endpoints (in a finite case) or two consecutive terms (in an infinite case) is a more interesting solution. However, such a solution requires a modification of the core Mizar system, as it has been done in the case of flexary logical operators for generalized conjunction and alternative (for more details see [11]). Such a solution goes beyond the point of this study.

3.3 Regrouping the Values of Sequence

In the considered reasoning, we come across another interesting informal procedure that consists of grouping the odd number. Obviously, partitioning of a set into non-empty subsets, according to some properties of its members is nothing new. However, to improve the readability of the defined partition, the proof authors add to the reasoning some exemplifications or even write elements of several members in such a way that the reader can easily find out other members by analogy. Note that such kind of exemplification is important for humans, but generally is unnecessary for the Mizar checker, except from the case where the existence of some kind of partition is proved.

Therefore, specially for our experiment, we define a specific kind of a finite sequence of finite sequences (a matrix generalization) over an odd valued finite sequence O, denoted in the Mizar language by odd_organization of O. This map is defined in such a way that the first sequence contains all arguments of O for which the value equal 1, the second sequence contains all arguments of O for which the value equal 3, etc., where the number of finite sequences is sufficient to cover the domain of O. Obviously, odd_organization of O is not uniquely determined by O, but we can use the global choice [14] (for more details see [7]).

Note that we can define `odd_organization` "directly", i.e. without subtypes and attributes. However, if another user of the MML will need to regroup the values in a different way, then probably he would have to provide some analogous properties. To avoid such situations in the MML, new types are defined as a restricted version of existing, more general types, if only the last ones exist. Therefore, we define the `odd_organization` in the following way. First we note that individual finite sequences in `odd_organization` have to be injective and determined, disjoint sets of values. Hence we introduce the following attribute:

```
definition
   let F be Function-yielding Function;
   attr F is double-one-to-one means
      for x1,x2,y1,y2 be object st                               (12)
         x1 in dom F & y1 in dom (F.x1) &
         x2 in dom F & y2 in dom (F.x2) & F_(x1,y1)=F_(x2,y2)
      holds x1 = x2 & y1 = y2;
end;
```

and a mode that reorganizes a finite set D into a finite sequence of finite sequences:

```
definition
   let D be finite set;
   mode DoubleReorganization of D -> double-one-to-one FinSequence of D*
      means Values it = D;
end;
```
$$(13)$$

Then we define a type where we have that in every individual finite sequence, elements are mapped to the same value, and such values in different finite sequences are different:

```
definition
   let f be finite Function;
   mode valued_reorganization of f -> DoubleReorganization of dom f means
   (for n ex x st
      x = f.it_(n,1) & ... & x = f.it_(n,len (it.n))) &       (14)
   for n1,n2,i1,i2 be Nat st
      i1 in dom (it.n1) & i2 in dom (it.n2) &
      f.it_(n1,i1) = f.it_(n2,i2)
   holds n1 = n2;
end;
```

and finally we define `odd_organization` as follows:

```
definition
   let f be odd-valued FinSequence;
   mode odd_organization of f -> valued_reorganization of f means   (15)
      2*n-1 = f.it_(n,1) & ... & 2*n-1 = f.it_(n,len (it.n));
end;
```

Based on this approach, we can prove in a more general form the properties of `odd_organization` that are needed to justify steps in the considered theorem.

Obviously this approach is much more difficult, but is consistent with the popularized direction of the development of the MML. According to this direction, the legible formulation and proving of a theorem is an important and challenging aim, when proof scripts are created for further development of the MML. However, no less important in this direction is extraction of definitions, creation of auxiliary theorems and notations in such a way that MML users will be able to adapt this knowledge for their own purposes.

4   The Theorem Formalization

In the reasoning presented in Fig. 1 the first and also the biggest part is the description of the transformation that maps a partition of a number into odd parts to a partition of the number into distinct parts. To adapt this fragment in a Mizar proof script we define this transformation as follows:

```
definition
    let n be Nat;
    let p be one-to-one a_partition of n;                          (16)
    func Euler_transformation p -> odd-valued a_partition of n means
```

where the value denoted by it can be determined by the condition:

```
for O be odd-valued FinSequence,a be natural-valued FinSequence,
    sort be odd_organization of O st
        len O = len p = len a & p = O (#) 2|^a                     (17)
for j holds card Coim(it,j*2-1) = ((2|^a)*.sort.j,1)+...
```

However, we decided on a more descriptive definition, where several formulas occur that describe some "exemplifications", only to improve the legibility of obtained condition. Note that the equivalence of such extended definition, presented below, with above ones is provided [15, (12)].

```
for j be Nat,O1 be odd-valued FinSequence,a1 be natural-valued FinSequence st
    len O1 = len d = len a1 & d = O1 (#) 2|^a1
for sort1 be DoubleReorganization of dom d st
    (1 = O1.sort1_(1,1) & ... & 1 = O1.sort1_(1,len (sort1.1))) &
    (3 = O1.sort1_(2,1) & ... & 3 = O1.sort1_(2,len (sort1.2))) &
    (3 = O1.sort1_(3,1) & ... & 5 = O1.sort1_(3,len (sort1.3))) &
for i holds
    2*i-1 = O1.sort1_(i,1) & ... & 2*i-1 = O1.sort1_(i,len (sort1.i))
holds
    card Coim(it,1) = (2|^a1).sort1_(1,1) +((2|^a1)*.sort1.1,2)+... &
    card Coim(it,3) = (2|^a1).sort1_(2,1) +((2|^a1)*.sort1.2,2)+... &
    card Coim(it,5) = (2|^a1).sort1_(3,1) +((2|^a1)*.sort1.3,2)+... &
    card Coim(it,j*2-1) = (2|^a1).sort1_(j,1) +((2|^a1)*.sort1.j,2)+...
```

$$(18)$$

Obviously to prove the correctness of this definition in the Mizar system, we have to justify that such a value exists and is unique. We selected the proof of the first conditions [15, (11)] to formally represent the informal description of Euler's transformation. For this aim, we constructed a reasoning, wherein all steps that are located on the first level of nesting (for more details see [7]) correspond to the selected fragments of the informal proof. Additionally, as a measure of correspondence we can analyze the generated automatically sketch of this reasoning presented in Fig. 3. Note that the full proof contains about 300 lines (for the full description see [15]), therefore we hide all nested reasonings and every list of statements that is used in a justification.

(11) Let us consider a one-to-one partition d of n. Then there exists an odd-valued partition e of n such that for every natural number j for every odd-valued finite sequence O_1 for every natural-valued finite sequence a_1 such that $\operatorname{len} O_1 = \operatorname{len} d = \operatorname{len} a_1$ and $d = O_1 \cdot 2^{a_1}$ for every double reorganization τ of $\operatorname{dom} d$ such that $1 = O_1(\tau_{1,1})$ and ... and $1 = O_1(\tau_{1,\operatorname{len}(\tau(1))})$ and $3 = O_1(\tau_{2,1})$ and ... and $3 = O_1(\tau_{2,\operatorname{len}(\tau(2))})$ and $5 = O_1(\tau_{3,1})$ and ... and $5 = O_1(\tau_{3,\operatorname{len}(\tau(3))})$ and for every i, $2 \cdot i - 1 = O_1(\tau_{i,1})$ and ... and $2 \cdot i - 1 = O_1(\tau_{i,\operatorname{len}(\tau(i))})$ holds $\overline{\operatorname{Coim}(e, 1)} = 2^{a_1}(\tau_{1,1}) + ((2^{a_1} \odot \tau)(1), 2) + \ldots$ and $\overline{\operatorname{Coim}(e, 3)} = 2^{a_1}(\tau_{2,1}) + ((2^{a_1} \odot \tau)(2), 2) + \ldots$ and $\overline{\operatorname{Coim}(e, 5)} = 2^{a_1}(\tau_{3,1}) + ((2^{a_1} \odot \tau)(3), 2) + \ldots$ and $\overline{\operatorname{Coim}(e, j \cdot 2 - 1)} = 2^{a_1}(\tau_{j,1}) + ((2^{a_1} \odot \tau)(j), 2) + \ldots$.
PROOF: $n = d(1) + ((d, 2) + \ldots + (d, \operatorname{len} d))$ by [16, (22)]. Consider O being an odd-valued finite sequence, a being a natural-valued finite sequence such that $\operatorname{len} O = \operatorname{len} d = \operatorname{len} a$ and $d = O \cdot 2^a$ and $d(1) = O(1) \cdot 2^{a(1)}$ and ... and $d(\operatorname{len} d) = O(\operatorname{len} d) \cdot 2^{a(\operatorname{len} d)}$. $n = 2^{a(1)} \cdot O(1) + 2^{a(2)} \cdot O(2) + ((O \cdot 2^a, 3) + \ldots + (O \cdot 2^a, \operatorname{len} d))$ by [16, (20)], [21, (25)]. Reconsider $\sigma = $ the odd organization of O as a double reorganization of $\operatorname{dom} 2^a$. Consider μ being a $(2 \cdot \operatorname{len} \sigma)$-element finite sequence of elements of $\mathbb{N}$ such that for every j, $\mu(2 \cdot j) = 0$ and $\mu(2 \cdot j - 1) = 2^a(\sigma_{j,1}) + ((2^a \odot \sigma)(j), 2) + \ldots$. Set $\alpha = a \cdot \sigma(1)$. Set $\beta = a \cdot \sigma(2)$. Set $\gamma = a \cdot \sigma(3)$. $n = (2^\alpha(1) + (2^\alpha, 2) + \ldots) \cdot 1 + (2^\beta(1) + (2^\beta, 2) + \ldots) \cdot 3 + (2^\gamma(1) + (2^\gamma, 2) + \ldots) \cdot 5 + ((\operatorname{id}_{\operatorname{dom} \mu} \cdot \mu), 7) + \ldots$ by [21, (29)], [16, (41)], [21, (25)], [4, (12)]. $n = \mu(1) \cdot 1 + \mu(3) \cdot 3 + \mu(5) \cdot 5 + ((\operatorname{id}_{\operatorname{dom} \mu} \cdot \mu), 7) + \ldots$ by [16, (42), (41), (25)]. Consider e being an odd-valued finite sequence such that e is non-decreasing and for every i, $\overline{\operatorname{Coim}(e, i)} = \mu(i)$. $n = \overline{\operatorname{Coim}(e, 1)} \cdot 1 + \overline{\operatorname{Coim}(e, 3)} \cdot 3 + \overline{\operatorname{Coim}(e, 5)} \cdot 5 + ((\operatorname{id}_{\operatorname{dom} \mu} \cdot \mu), 7) + \ldots$. $n = \sum C$ by [16, (20)], (9). For every j such that $1 \le j \le \operatorname{len} d$ holds $O(j) = O_1(j)$ and $a(j) = a_1(j)$ by [21, (25)], [19, (9)], [2, (4)]. For every j, $\overline{\operatorname{Coim}(e, j \cdot 2 - 1)} = 2^{a_1}(\tau_{j,1}) + ((2^{a_1} \odot \tau)(j), 2) + \ldots$ by [16, (42)], [21, (29)], [4, (72)], [16, (22)]. $\square$

Fig. 3. The sketch generated automatically of proof [15, (11)] that justifies the existence of `Euler_transformation`. The content of the sketch has not been changed with the exception of the order of bibliography items.

At the beginning, we introduce a partition d of n into distinct parts and represent that its elements sum up to n:

```
let d be one-to-one a_partition of n;
n = d.1 + (d,2)+...+(d,len d) proof...
```
(19)

Then we introduce two finite sequences that describe a decomposition of every element of partition p as the product of a power of two and an odd number. We formalize also the informal connection between these finite sequences and the number n that occur in the considered reasoning.

```
consider O be odd-valued FinSequence,a be natural-valued FinSequence such that
    len O = len d = len a & d = O (#) 2|^a and
    d.1 = O.1 * (2|^a.1) & ... & d.len d = O.len d * (2|^a.len d) by...
n = 2|^(a.1) * O.1 + 2|^(a.2) * O.2 + (O (#) 2|^a,3) +...+(O (#) 2|^a,len d)
    proof...
```

$$(20)$$

As it has been mentioned in Sect. 3.3 to select a method that groups together values of O we use the global choice: **the** odd_organization **of** O. Note that the same method is used to reorganize the values of 2|^a. It can be done since lengths of 2|^a and O are equal. However, we have to include this information in the type of this reorganization. In the Mizar system, such modification of the type can be realized as follows:

```
len (2|^a)=len O by...
then reconsider sort = the odd_organization of O as
    DoubleReorganization of dom (2|^a) by...
```

$$(21)$$

To formalize the unlabelled equality presented in Fig. 1 we have to introduce a finite sequence and three sets:

```
consider mu be (2*len sort)-element FinSequence of NAT such that
    for j holds mu.(2*j) = 0 &
        mu.(2*j-1) = (2|^a). sort_(j,1) + ((2|^a)*.sort.j,2)+... by...
set alpha = a*(sort.1), beta = a*(sort.2), gamma = a*(sort.3);
```

$$(22)$$

Then this equality can be formally formulated as follows:

```
n = ((2|^alpha).1+ (2|^alpha,2)+...) * 1 + ((2|^beta).1+ (2|^beta,2)+...) * 3 +
    ((2|^gamma).1+ (2|^gamma,2)+...) * 5 + ((id dom mu)(#)mu,7)+... proof...
n = mu.1 * 1 + mu.3 * 3 + mu.5 * 5 + ((id dom mu)(#)mu,7) +... proof...
```

$$(23)$$

Finally, to finish the informal contraction of value of p in Euler's transformation, we use the following 5 steps:

```
consider e be odd-valued FinSequence such that
    e is non-decreasing & for i holds card Coim(e,i) = mu.i by...
n = card Coim(e,1) * 1 + card Coim(e,3) * 3 +
    card Coim(e,5) * 5 + ((id dom mu)(#)mu,7) +... proof...
n = Sum e proof...
then reconsider e as odd-valued a_partition of n by...
take e;
```

$$(24)$$

However, these steps are not sufficient to finish a formal proof of the existence. For this purpose we need also to provide that the choice of e does not depend on a, O and sort. In the considered reasoning a part of this information is mentioned as "d_i can be uniquely expressed as a power of 2 times an odd number". We recreate this information proving that for every pair of finite sequences a1, O1 that satisfies d = O1 (#) 2|^a1 holds for j st 1 <= j & j <= len d holds O.j = O1.j & a.j = a1.j. Whereas the influence of sort selection for obtained value e is omitted. Therefore, to finish the proof, we provide this influence as an external auxiliary theorem [15, (5)] and only refering to this information from nesting reasoning that justifies the following proof step:

```
for j holds
    card Coim(e,j*2-1) = (2|^a1).sort1_(j,1) + ((2|^a1)*.sort1.j,2)+...
```
(25)

where sort1 is an arbitrary odd_organization of O1.

Obviously, in the situation described above, where only several steps at the first level of nesting do not have their informal counterparts in the text book proof, it has a negative consequence, i.e. the size of the full proof. There are three main reasons for that. Firstly, note that to obtain such a proof we encapsulate less important fragments of reasoning at the deeper levels of nesting as subreasonings. Additionally, if steps in two different nesting subreasonings refer to the same auxiliary fact, then to avoid duplication, generally, we try to locate this fact in a common top level of nesting in such a way that this fact is available from these subreasonings (for more details see [17]). However, this solution is inconsistent with our goal, where we try to remove auxiliary facts form the first level of nesting. The second reason is the consequence of faithfully reproducing the informal term without any restrictions that are *a priori* assumed in the informal context. Note that to resolve the problem of *a priori* assumptions we have to provide several facts for different cases, that are completely redundant if we resign from the mirroring. Let us focus on the term $2^{\alpha_1} + 2^{\alpha_2} + \ldots$ that occurs in Fig. 1 and can be formalized simply as sum (2|^alpha). But according to our purpose, we would like to obtain the term 2|^(alpha.1) + (2|^alpha,2)+... that unfortunately is equal to 1 (= 2|^0), if alpha is the empty finite sequence, where at the same time sum (2|^alpha)=0. Therefore, to resolve this problem we use (2|^alpha).1 instead of 2|^(alpha.1) in (23). The third reason is related to the previous one. Note that premises where we extract first few terms of summations are easily readable for a human, but difficult to use as premises. Often to use such premises we have to insert back these extracted terms and consider again the above-mentioned redundant cases.

This shows that the adaptation of the main idea from an informal proof to a formal one is not so trivial if we want to recreate it in a very precise way.

Since we defined Euler_transformation, we can prove that this transformation is a bijection. Note that the textbook proof contains only very sketchy justification of this property, that is formulated as follows: *In each series* $(2^{\alpha_1} + 2^{\alpha_2} + \ldots)$, *the α_i's are distinct (why?)*. We provide this fact in the following form [15, (13)]:

```
for O be odd-valued FinSequence, a be natural-valued FinSequence,
    s be odd_organization of O st len O = len a & O (#) 2|^a is one-to-one
holds a*.s.i is one-to-one
```
(26)

We remind that s.i is the finite sequence of all elements of the domain of O, for which the value under O is equals to 2*i-1. Then a*.s.i is the image of the elements of s.i under a. Based on this fact and the uniqueness of the binary number representation we provide in [15, (14)] that Euler_transformation is an injection:

```
for p1,p2 be one-to-one a_partition of n st
    Euler_transformation p1 = Euler_transformation p2          (27)
holds p1 = p2
```

Obviously, we provide also in [15, (15)] that `Euler_transformation` is a surjection, based on the existences of the binary number representation.

```
for e be odd-valued a_partition of n
    ex p be one-to-one a_partition of n st                     (28)
    e = Euler_transformation p
```

Based on the two above-described statements we can "easily" prove that the set of all natural valued finite sequences that are partitions of n into odd parts has the same size as the set of all natural-valued finite sequences that are partitions of n into distinct parts. This statement can be represented as follows:

```
card {p where p is Element of NAT*:p is odd-valued a_partition of n}
=card {p where p is Element of NAT*:p is one-to-one a_partition of n}   (29)
```

However, if we register in the Mizar environment, that there exists a set of all `a_partition of n` (for more details see [7]), we can represent (29) in more elegant form, used in [15, (16)], presented below:

```
card the set of all p where p is odd-valued a_partition of n
= card the set of all p where p is one-to-one a_partition of n          (30)
```

We can compare the obtained formulation of Euler's partitions theorem in the Mizar system with the formulation used in the HOL system:

```
let EULER_PARTITION_THEOREM = prove
('FINITE {p | p partitions n /\ !i. p(i) <= 1} /\
  FINITE {p | p partitions n /\ !i. ~(p(i) = 0) ==> ODD i} /\
  CARD {p | p partitions n /\ !i. p(i) <= 1} =
  CARD {p | p partitions n /\ !i. ~(p(i) = 0) ==> ODD i}'
```

5   Conclusion

In this paper we presented a formalization of a textbook theorem where we not only proved this theorem, but primarily tried to reflect the main idea of the informal proof with expressions that are available in a formal environment. We have created more expressive definitions and we extended existing ones to mirror informal mathematical language constructions in formal terms, working around the Mizar system's limitations, without modification of the core Mizar system. We have showed that an accurate formal reflection of informal terms obliges not only to introduce new concepts in our library, but also to conduct reasoning in a more difficult way. Our studies highlighted the differences between the way of conducting human legible reasoning and reasoning that is acceptable for a proof checker.

We have showed that appropriate use of flexary operators in formal reasonings can increase the legibility of obtained proof scripts, in effectively the same way as in mathematical textbook. However, based on the same example we have also showed a negative consequence of this method, namely the growth of the reasoning length. Obviously, going beyond the point of this study and modifying the core of the Mizar system, we can obtain a more natural definition of flexary operators, where we do not have to explicitly use the summed sequence.

Our effort allowed us to formulate similarly formal equivalents of the great majority of informal terms. Still a number of corner cases resisted our efforts. In particular handling sequences of length zero was problematic and we had to fall back to non-uniform treatment of the zero-length case for the sequence $2^{\alpha_1} + 2^{\alpha_2} + \ldots$

We believe that this study brings us closer to the situation that informal reasonings can be conducted in systems such as Mizar.

References

1. Andrews, G.E.: Number Theory, Dover edn. W. B. Saunders Company, Philadelphia (1971)
2. Bancerek, G.: Countable sets and Hessenberg's theorem. Formalized Math. **2**(1), 65–69 (1991)
3. Bancerek, G., Rudnicki, P.: Information retrieval in MML. In: Asperti, A., Buchberger, B., Davenport, J.H. (eds.) MKM 2003. LNCS, vol. 2594, pp. 119–131. Springer, Heidelberg (2003)
4. Byliński, C.: Functions and their basic properties. Formalized Math. **1**(1), 55–65 (1990)
5. Engelking, R.: General Topology. PWN - Polish Scientific Publishers, Warsaw (1977)
6. Euler, L.: Introduction to the Analysis of the Infinite Book I Translated by John D. Blanton. Springer, New York (1988)
7. Grabowski, A., Korniłowicz, A., Naumowicz, A.: Mizar in a nutshell. J. Formalized Reasoning **3**(2), 153–245 (2010)
8. Grabowski, A., Schwarzweller, C.: On duplication in mathematical repositories. In: Autexier, S., Calmet, J., Delahaye, D., Ion, P.D.F., Rideau, L., Rioboo, R., Sexton, A.P. (eds.) AISC 2010. LNCS, vol. 6167, pp. 300–314. Springer, Heidelberg (2010)
9. Grabowski, A., Schwarzweller, C.: Improving representation of knowledge within the mizar library. Stud. Logic Grammar Rhetoric **18**(31), 35–50 (2009)
10. Harrison, J.V.: A HOL theory of euclidean space. In: Hurd, J., Melham, T. (eds.) TPHOLs 2005. LNCS, vol. 3603, pp. 114–129. Springer, Heidelberg (2005)
11. Korniłowicz, A.: Tentative experiments with ellipsis in mizar. In: Campbell, J.A., Jeuring, J., Carette, J., Dos Reis, G., Sojka, P., Wenzel, M., Sorge, V. (eds.) CICM 2012. LNCS, vol. 7362, pp. 453–457. Springer, Heidelberg (2012)
12. Horozal, F., Rabe, F., Kohlhase, M.: Flexary operators for formalized mathematics. In: Watt, S.M., Davenport, J.H., Sexton, A.P., Sojka, P., Urban, J. (eds.) CICM 2014. LNCS, vol. 8543, pp. 312–327. Springer, Heidelberg (2014)
13. Korniłowicz, A., Shidama, Y.: Brouwer fixed point theorem for disks on the plane. Formalized Math. **13**(2), 333–336 (2005)
14. Leisenring, A.C.: Mathematical Logic and Hilbert's ε-Symbol. Gordon and Breach, New York (1969)
15. Pąk, K.: Euler's partition theorem. Formalized Math. **23**(2), 91–98 (2015). doi:10.2478/forma-2015-0009

16. Pąk, K.: Flexary operations. Formalized Math. **23**(2), 79–90 (2015). doi:10.2478/forma-2015-0008
17. Pąk, K.: Methods of lemma extraction in natural deduction proofs. J. Autom. Reasoning **50**(2), 217–228 (2013)
18. Pąk, K.: Topological manifolds. Formalized Math. **22**(2), 179–186 (2014)
19. Rudnicki, P., Trybulec, A.A.: Abian's fixed point theorem. Formalized Math. **6**(3), 335–338 (1997)
20. Sylvester, J.J., Franklin, F.: A constructive theory of partitions, arranged in three acts, an interact and an exodion. Amer. J. Math. **5**, 251–330 (1882)
21. Trybulec, W.A.: Non-contiguous substrings and one-to-one finite sequences. Formalized Math. **1**(3), 569–573 (1990)
22. Wilf, H.S.: Lectures on Integer Partitions (2000)

Automating Change of Representation for Proofs in Discrete Mathematics

Daniel Raggi[1]([✉]), Alan Bundy[1], Gudmund Grov[2], and Alison Pease[3]

[1] School of Informatics, University of Edinburgh, Edinburgh, Scotland
danielraggi@gmail.com
[2] School of Mathematical and Computer Sciences, Heriot-Watt University, Edinburgh, Scotland
[3] School of Computing, University of Dundee, Dundee, Scotland

Abstract. Representation determines how we can reason about a specific problem. Sometimes one representation helps us find a proof more easily than others. Most current automated reasoning tools focus on reasoning within one representation. There is, therefore, a need for the development of better tools to mechanise and automate formal and logically sound changes of representation.

In this paper we look at examples of representational transformations in discrete mathematics, and show how we have used Isabelle's Transfer tool to automate the use of these transformations in proofs. We give a brief overview of a general theory of transformations that we consider appropriate for thinking about the matter, and we explain how it relates to the Transfer package. We show our progress towards developing a general tactic that incorporates the automatic search for representation within the proving process.

Keywords: Change of representation · Transformation · Automated reasoning · Isabelle proof assistant

1 Introduction

Many mathematical proofs involve a change of representation from a domain in which it is difficult to reason about the entities in question to one in which some aspects essential to the proof become evident and the proof falls out naturally.

Many times the transformation makes it explicitly into the written proof, but sometimes it remains hidden as part of the esoteric process of coming up with the proof in the mathematician's mind. For a formal, mechanical proof, this can be problematic, not only because we need to account for the logical validity of the transformation, but because if we want a computational system to find a proof like a mathematician would, we need to be able to incorporate something

D. Raggi—This work has been supported by a scholarship from the Mexican Council of Science and Technology (CONACYT).

M. Kerber et al. (Eds.): CICM 2015, LNAI 9150, pp. 227–242, 2015.
DOI: 10.1007/978-3-319-20615-8_15

like the esoteric transformations going on inside the mathematician's mind into the mechanical search.

The importance of representational changes in mathematics is evidenced in historically notable works like Kurt Gödel's incompleteness theorems, where the proof involves matching (or encoding) meta-theoretical concepts like 'sentence' and 'proof' as natural numbers, or more recently Andrew Wiles' proof of Fermat's Last Theorem, which involves matching the Galois representations of elliptic curves with modular forms. This phenomenon is also seen in refinement based formal methods (e.g. VDM and B): one starts with a highly abstract representation that is easy to reason with, and then it is step-wise refined to a very concrete representation that can be implemented as a computer program. All of these transformations are justified by a general notion of morphism.

In this paper we give an overview of a general mathematical framework suitable for reasoning about representational changes in type-theoretic higher-order logics (these are transformations/morphisms between structures that land us in different theories). We see that the operation of Isabelle's *transfer* methods [6] fit into this notion of transformation. It is a way of mechanising inference between two domains, if the system is provided with a transformation by the user. We present a set of transformations we have identified as essential for reasoning in discrete mathematics, and show how we have used the *transfer* tool to implement mechanical proofs in Isabelle that use these transformations. We show our work towards automating the search for representation as a tactic for use within proofs in discrete mathematics in Isabelle.

2   Background

Isabelle/HOL is a theorem proving framework based on a simple type-theoretical higher-order logic [9]. It is one of the most widely used proof assistants for the mechanisation of proofs. Apart from ensuring the correctness of proofs written in its formal language, Isabelle has powerful automatic tactics like `simp` and `auto`, and through time it has been enriched with some internally-verified theorem provers like `metis` [7] and `smt` [11], along with a connection from the internal provers to some very powerful external provers like E, SPASS, Vampire, CVC3 and Z3 through the Sledgehammer tool [10].

The *Transfer* package was first released for Isabelle 2013-1 as a general mechanism for defining quotient types and transferring knowledge from the old 'representation' type into the new 'abstract' type [6]. However, their generalisation is not restricted to the definition of new quotient types, but allows the user to relate any two types by theorems of a specific shape called *transfer rules*. Some of these rules can be defined automatically when the user defines a new quotient type, but the user is free to add them manually, provided that they prove a preservation theorem. Central to this package, the *transfer* and *transfer'* tactics try to automatically match the goal sentence to a new one related by either equivalence or implication, inferring this relation from the transfer rules.

We have taken full advantage of the generality of the transfer package as a means of automating the translation between sentences across domains which are related by what we consider an appropriate and general notion of *structural transformation*. In Sect. 4 we give an overview of our notion of transformation and how the tactics of the transfer package are useful mechanisms for exploiting the knowledge of a structural transformation.

3   Overall Vision

The worlds of mathematical entities are interconnected. Numbers can be represented as sets, pairs of sets, lists of digits, bags of primes, etc. Some representations are only *foundational* and the reasoner often finds it more useful to discard the representation for practical use (e.g., natural number 3 is represented by $\{\emptyset, \{\emptyset\}, \{\emptyset, \{\emptyset\}\}\}$ in the typical ZF foundations, but this representation is rarely used in practice), and some are *emergent*; they only come about after a fair amount of accumulated knowledge about the objects themselves, but are more helpful as reasoning tools (e.g., natural numbers as bags of primes). Overall, we think that there is no obvious notion of 'better representation', and it's up to the reasoner to choose, depending on the task at hand. Thus, we envision a system where the representation of entities can be fluidly transformed.

We have looked at problems in discrete mathematics and the transformations commonly used for solving them. Below, we give one motivating example and show how we have mechanised the transformation in question inside Isabelle/HOL. Other motivating examples are briefly mentioned.

3.1   Numbers as Bags of Primes

Let us start with an example of the role of representation in number theory. Consider the following problem:

Problem 1. Let n be a positive integer. Assume that, for every prime p, if p divides n then p^2 also divides n. Prove that n is the product of a square and a cube.

A standard solution to this problem is to take a set of primes p_i such that $n = p_1^{a_1} p_2^{a_2} \cdots p_k^{a_k}$. Then we notice that the condition "if p divides n then p^2 also divides n" means that $a_i \neq 1$, for each a_i. Then, we need to find $x_1, x_2, \ldots, x_k$ and $y_1, y_2, \ldots, y_k$ where

$$(p_1^{x_1} p_2^{x_2} \cdots p_n^{x_k})^2 (p_1^{y_1} p_2^{y_2} \cdots p_n^{y_k})^3 = p_1^{a_1} p_2^{a_2} \cdots p_n^{a_k}$$

or simply

$$2(x_1, x_2, \ldots, x_k) + 3(y_1, y_2, \ldots, y_k) = (a_1, a_2, \ldots, a_k).$$

Thus, we only need to prove that for every $a_i \neq 1$ there is a pair x_i, y_i such that $2x_i + 3y_i = a_i$. The proof of this is routine.

The kind of reasoning used for this problem is considered standard by mathematicians. However, it is not so simple in current systems for automated theorem proving. The non-standard step is the 'translation' from an expression containing various applications of the exponential function into a simpler form in a linear arithmetic of lists, validated by the fundamental theorem of arithmetic.

The informal nature of the argument, in the usual mathematical presentation, leaves it open whether the reasoning is best thought as happening in an arithmetic of lists where the elements are the exponents of the primes, or perhaps a theory of bags (multisets) where the elements are prime numbers. The reader might find it very easy to fluidly understand how these representations match with each other and how they are really just different aspects of the same thing. Such ease supports our overall argument and vision: that to automate mathematical reasoning, we require a framework in which data structures are linked robustly by logically valid translations, where the translation from one to another is easily conjured up.

The *numbers-as-bags-of-primes* transformation that links each positive integer to the bag of its prime factors is valid because there are operations on each side (numbers and multisets) that correspond to one another. For example, 'divides' corresponds to 'sub(multi)set', 'least common multiple' corresponds to 'union', 'product' corresponds to 'multiset addition', etc. Furthermore, all the predicates used in the statement of Problem 1 have correspondences with well-known predicates regarding bags of primes. Thus, the problem can be translated as a whole. Other representations may not be very productive, e.g., try thinking about exponentiation in terms of lists of digits.

Table 1 shows more examples of number theory problems with their corresponding problem about multisets.

Table 1. Number theory problems and their multiset counterparts.

Problem in $\mathbb{N}$	Problem in multisets
Prove that there is a unique set $\{x, y, z\}$ with different x, y, z greater than 1, such that $xyz = 100$.	Prove that there is a unique way to partition $\{2, 2, 5, 5\}$ into three different non-empty parts.
Prove that in a set of 9 natural numbers, where none is divided by a prime larger than 6, there is a pair whose product is a perfect square.	Take 9 multisets whose only elements are 2, 3 and 5. Prove that two of the multisets have multiplicities with the same parity.

3.2 Numbers as Sets

Many numerical problems have *combinatorial proofs*. Theses are proofs where numbers are interpreted to be cardinalities of sets, and the whole problem can be converted to a problem about sets.

Enumerative combinatorics studies how sets relate to their cardinalities. As such, its theorems provide the link that allows us to translate numerical problems into finite set-theoretical problems.

Table 2 shows examples of arithmetic problems with their corresponding finite set theory problems. While the proofs of the numerical versions are not obvious at all (some of which are important results in basic combinatorics), the proofs of their finite set versions can be considered routine.

Table 2. Numerical problems and their set counterparts.

Problem in $\mathbb{N}$	Problem in sets
$\binom{n+1}{k+1} = \binom{n}{k} + \binom{n}{k+1}$	The set $\{x \subseteq \{0,1,\ldots,n\} : \lvert x \rvert = k+1\}$ can be partitioned into 2 parts: those that contain element n and those that don't
$\frac{n(n+1)}{2} = 1 + 2 + \cdots + n$	The set $\{x \subseteq \{0,1,\ldots,n\} : \lvert x \rvert = 2\}$ can be partitioned into n parts $X_1, X_2, \ldots, X_n$ where the largest element of each $x \in X_i$ is i
$2^{n+1} - 1 = \sum_{i=0}^{n} 2^i$	The power set of $\{0,1,\ldots,n\}$, excluding the empty set, can be partitioned into n parts $X_1, X_2, \ldots, X_n$ where the largest element of each $x \in X_i$ is i
$2^n = \sum_{i=0}^{n} \binom{n}{i}$	The power set of $\{1,\ldots,n\}$ can be partitioned into $n+1$ parts $X_0, X_1, \ldots, X_n$ where $\lvert x \rvert = i$ for every $x \in X_i$

3.3   Interconnectedness

We want to stress the importance of having fluidity of representations. For example, we talked about the ease with which we could think that the *numbers-as-bags-of-primes* transformation is actually a transformation of numbers to a theory of lists, where elements of the list are the exponents of the ordered prime factors. Inspired by this, we have mechanised many other simple transformations, but whose composition allows us to translate fluently from one representation to another. Our global vision of transformations useful in discrete mathematics, which we have mechanised[1], is represented in Fig. 1. It is worth mentioning that the diagram is not commutative and that it abstracts logical relations (information may be lost, so some paths can only be traversed in one direction).

In the next section we show how a notion of transformation that accounts for this kind of correspondence between structures can be applied in formal proofs using Isabelle's Transfer tool.

4   On Transformations and the Transfer Tool

In this section we give a brief overview of a very general theory of transformations. We do not claim originality of the essence of this theory. However, we believe that the presentation we give brings clarity to the problem. We explain how Isabelle's Transfer tool relates to it. Consider the following definitions:

---

[1] These can be found in http://homepages.inf.ed.ac.uk/s1052074/AutoTransfer/. They are updated regularly.

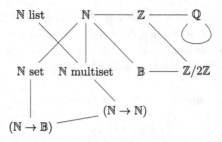

Fig. 1. Nodes stand for theories connected by transformations useful in discrete mathematics. Apart from the aforementioned transformations, it includes other simpler ones. Actually, some of these transformations, such as that connecting $\mathbb{N}$ list and $\mathbb{N}$ set, are polymorphic, but presented in the diagram as relating only to type $\mathbb{N}$. This is because the numbers-as-bags-of-primes transformation is not polymorphic.

Definition 1. *A **domain** is a class of entities and a set of types, where each entity of the domain corresponds to exactly one type.*

Definition 2. *A **transformation** from a domain $\mathscr{X}$ to a domain $\mathscr{Y}$ is a collection $\mathscr{R}$ where every $R \in \mathscr{R}$ is a relation $R : X \to Y \to \mathbb{B}$ between a type X of domain $\mathscr{X}$ and a type Y of domain $\mathscr{Y}$.*[2]

This relational notion of a transformation makes it possible to account for partial and multivalued mappings in a logic like Isabelle's HOL.

We consider a *structure* to be the class conformed by all the entities of the closure of a domain under a set of type constructors. In this work, we focus on structures containing type $\mathbb{B}$, generated with the *function type constructor* $\to$, because the basis of a higher order logic can be fully expressed under such a structure. Then, if our domain has entities of types A and B, its structure under $\to$ has all the entities of types $A \to B$, $A \to B \to A$, etc.

Preservation of structure is captured with the use of structural *relators*, which can be thought of as rules for extending relations (transformations) to the structures of their domains. In particular, given that our work is based on Isabelle/HOL and on the Transfer package, we focus on one relator.

Definition 3. *The **standard functional extension** of two relations $R_A : A \to A' \to \mathbb{B}$ and $R_B : B \to B' \to \mathbb{B}$ (written $R_A \Mapsto R_B$) is a relation that relates two functions $f : A \to B$ and $f' : A' \to B'$ whenever they satisfy the following property:*

$$\forall a : A.\ \forall a' : A'.\ [R_A\,a\,a' \longrightarrow R_B\,(f\,a)\,(f'\,a')]$$

We call the operator $\Mapsto$ the **standard function relator**. Intuitively, $(R_A \Mapsto R_B)\,f\,g$ means that f and g send arguments related by R_A to values related by R_B. This relator allows us to express how functions (and by extension relations) map to each other in a way that the structure of the domain is preserved.

---

[2] $\mathbb{B}$ stands for type of booleans.

For the numbers-as-bags-of-primes transformation, consider relation $\mathcal{F} : \mathbb{N} \rightarrow$ $\mathbb{N} \, \texttt{multiset} \rightarrow \mathbb{B}$, that relates every positive integer with the multiset of its prime factors.

Example 1. Let $* : \mathbb{N} \rightarrow \mathbb{N} \rightarrow \mathbb{N}$ be the usual multiplication and $\uplus : \mathbb{N} \, \texttt{multiset}$ $\rightarrow \mathbb{N} \, \texttt{multiset} \rightarrow \mathbb{N} \, \texttt{multiset}$ the 'addition' of multisets (in which the multiplicities are added per element). Then we have $(\mathcal{F} \Mapsto (\mathcal{F} \Mapsto \mathcal{F})) * \uplus$ (also written $(\mathcal{F} \Mapsto \mathcal{F} \Mapsto \mathcal{F}) * \uplus$).

Note that, by expanding the definition of $\Mapsto$ in $(\mathcal{F} \Mapsto (\mathcal{F} \Mapsto \mathcal{F})) * \uplus$ we get

$$\forall n_1. \forall N_1. \, \mathcal{F} \, n_1 \, N_1 \; \longrightarrow \; (\forall n_2. \forall N_2. \, \mathcal{F} \, n_2 \, N_2 \; \longrightarrow \; \mathcal{F} \, (n_1 * n_2) \, (N_1 \uplus N_2))$$

which is equivalent to

$$\forall n_1, n_2. \, \forall N_1, N_2. \, (\mathcal{F} \, n_1 \, N_1 \wedge \mathcal{F} \, n_2 \, N_2) \; \longrightarrow \; \mathcal{F} \, (n_1 * n_2) \, (N_1 \uplus N_2)$$

This demonstrates how nesting the operator $\Mapsto$ preserves its intuitive definition: 'related arguments map to related values'. In this particular case, this is true due to the law of exponents $p^a p^b = p^{a+b}$.

Furthermore, the matching of relations can also be expressed with the help of $\Mapsto$, using a boolean relation, as demonstrated by the example below with equivalence (boolean equality) $\texttt{eq} : \mathbb{B} \rightarrow \mathbb{B} \rightarrow \mathbb{B}$.

Example 2. Let $\texttt{div} : \mathbb{N} \rightarrow \mathbb{N} \rightarrow \mathbb{B}$ be the relation such that $\texttt{div} \, n \, m$ whenever n divides m (also written $n|m$), and $\subseteq : \mathbb{N} \, \texttt{multiset} \rightarrow \mathbb{N} \, \texttt{multiset} \rightarrow \mathbb{B}$ the relation such that $a \subseteq b$ whenever the multiplicity of each element of a is lesser or equal to its multiplicity in b. Then, we have $(\mathcal{F} \Mapsto \mathcal{F} \Mapsto \texttt{eq}) \, \texttt{div} \subseteq$, because n divides m if and only if every prime is contained at least as many times in the multiset-factorisation of m as it is in n.

Logical matches (preservation of truth values) can also be expressed across structures, e.g., $(\texttt{eq} \Mapsto \texttt{eq} \Mapsto \texttt{eq}) \, \texttt{imp} \, \texttt{imp}$ represents that implication $\texttt{imp}$ preserves truth if its arguments are replaced by equivalent ones. Other interesting logical matches can be expressed as well.

The general notion of transformation above tells us how theories will relate to one another. Isabelle's Transfer method is an algorithm for transforming a sentence using knowledge about one of these transformations. The simple standard function relator is at the basis of the method. We give a short introduction next.

4.1 Transforming Sentences with the Transfer Tool

When trying to prove a sentence β we want to find another sentence α such that $\alpha \longrightarrow \beta$, along with a proof for α. In particular, if β talks about a domain B and we know a structural transformation from a domain A to B, we might be able to find an α about A, such that $\alpha \longrightarrow \beta$.

Isabelle's *Transfer* tool provides a method for finding such α. The user has to provide theorems of the forms $R_1 \, a \, b$ or $(R_1 \Mapsto R_2) \, f \, g$ (and their proofs), i.e., instances of a structural transformation, and the tactics `transfer` and `transfer'` will try to automatically infer a sentence α such that $\alpha \longleftrightarrow \beta$ (in the case of `transfer`), or a weaker one such that $\alpha \longrightarrow \beta$ (in the case of `transfer'`).

Recall that the intuitive interpretation of $(R_1 \Mapsto R_2)\, f\, g$ is 'arguments related by R_1 are mapped to values related by R_2 by f and g. Thus, the first step of the transfer method is to search for a theorem of the structural transformation with the shape $(R_1 \Mapsto \mathsf{eq})\, p\, q$ in the case of `transfer` and $(R_1 \Mapsto \mathsf{imp})\, p\, q$ in the case of `transfer'`, where q is the property wrapping the sentence we want to prove. Finding it would imply that we can replace q by p provided that we can find that their arguments are related by R_1. Thus, the method searches recursively for rules in the structural transformation to prove this. The algorithm is analogous to type inference. It is based on the following derivation rules:

$$\frac{\mathscr{A}_{\mathscr{C}}^{*} \vdash (R_1 \Mapsto R_2)\, f\, g \quad \mathscr{A}_{\mathscr{C}}^{*} \vdash R_1\, x\, y}{\mathscr{A}_{\mathscr{C}}^{*} \vdash R_2\, (f\, x)\, (g\, y)}\,\text{elim}$$

$$\frac{\mathscr{A}_{\mathscr{C}}^{*}, R_1\, x\, y \vdash R_2\, (f\, x)\, (g\, y)}{\mathscr{A}_{\mathscr{C}}^{*} \vdash (R_1 \Mapsto R_2)\, (\lambda x.\, f\, x)\, (\lambda y.\, g\, y)}\,\text{intro}$$

where $\mathscr{A}_{\mathscr{C}}^{*}$ represents knowledge about the structural transformation. Practically, the user provides knowledge specific to this transformation (a set of theorems called *transfer rules*), and the algorithm includes in the search other general transfer rules such as $(\mathsf{eq} \Mapsto \mathsf{eq} \Mapsto \mathsf{eq})\, \mathsf{imp}\, \mathsf{imp}$. For more details of the actual implementation of the algorithm see [6].

5   Mechanising Transformations in Isabelle's HOL

In Sect. 3.1 we presented some problems in discrete mathematics which involve structural transformation. We have mechanised the transformations by proving the necessary transfer rules. The transfer tool allows us to use the transformations in proofs.

In this section we present a couple of examples from a larger catalogue of the transformations we have mechanised in Isabelle. The transformations we have formalised, as suggested in Fig. 1, are the following:

1. **Numbers-as-bags-of-primes**, where each natural number is related to the multiset of its prime factorisation.
2. **Numbers-as-sets**, where numbers are related to sets by the cardinality function.
3. **Sets-as-$\mathbb{B}$-functions**, where sets are seen as boolean-valued functions.
4. **Multisets-as-$\mathbb{N}$-functions**, where multisets are seen as natural-valued functions[3].

---

[3] This one is actually by construction using `typedef` and the Lifting package, which automatically declares transfer rules from definitions lifted by the user from an old type to the newly declared type.

5. **Sets-as-lists**, where sets are related to lists of their elements.
6. **Bits-from-integers**, where type bit is created as an abstract type from the integers.
7. **Bits-as-booleans**, where bits are matched to booleans.
8. **Q-automorphisms**, where rational numbers are stretched and contracted, parametric on a factor.
9. **Zero-or-some**, where natural 0 is related to bit 0 and positive natural numbers are related to bit 1.
10. **Multisets-as-lists**, where multisets are related to lists of their elements.
11. **Set-to-multiset**, where the functional representations of multisets and sets are related (this one, we get it for free from the zero-or-some transformation).
12. **Naturals-as-integers**, where naturals are matched to integers (this one was built by the developers of the transfer package, not us).
13. **Integers-as-rationals**, where integers are matched to rational numbers. Notice that composition of transformations leads to other natural transformations, like the simple relation between sets and multisets.[4]

Every transformation starts with a declaration and proof of *transfer rules*, which are sentences satisfied by the structural transformation.

5.1   Numbers as Bags of Primes

The relation at the centre of this transformation is $\mathcal{F} : \mathbb{N} \to \mathbb{N}\ \text{multiset} \to \mathbb{B}$, which relates every positive number to the multiset of its prime factors. It is defined as follows: $\mathcal{F}\,n\,M$ holds if and only if

$$(\forall x.\ \text{count}\ M\ x > 0 \longrightarrow \text{prime}\ x) \land n = \prod_{x \in M} x^{\text{count}\ M\ x}$$

The most basic transfer rules (instances of the structural transformation) are theorems such as $\mathcal{F}\ 6\ \{2,3\}$, whose proof are trivial calculations. Moreover, from the Unique Prime Factorisation theorem we know that $\mathcal{F}$ is bi-unique. Thus, we know that

$$(\mathcal{F} \Mapsto \mathcal{F} \Mapsto \text{eq})\ \text{eq}\ \text{eq}$$

i.e., that equality is preserved by the transformation.

From the fact that every positive number has a factorisation we have

$$((\mathcal{F} \Mapsto \text{revimp}) \Mapsto \text{revimp})\ \forall_{>0}\ \forall \qquad ((\mathcal{F} \Mapsto \text{revimp}) \Mapsto \text{revimp})\ \exists\ \exists_p$$
$$((\mathcal{F} \Mapsto \text{eq}) \Mapsto \text{eq})\ \forall_{>0}\ \forall_p \qquad\qquad ((\mathcal{F} \Mapsto \text{eq}) \Mapsto \text{eq})\ \exists_{>0}\ \exists_p$$

where **revimp** is reverse implication, $\forall_p$ is the bounded quantifier representing 'for every multiset where all its elements are primes' and $\forall_{>0}$ is the bounded quantifier representing 'for every positive number', and similarly for $\exists_p$ and $\exists_{>0}$.

---

[4] The mechanisation of these transformations have been submitted to the Archive of Formal Proofs, along with some examples of their use.

The mechanised proofs of these sentences follow relatively straightforward from the Unique Prime Factorisation theorem which is already part of Isabelle's library of number theory.

Furthermore, we proved the following correspondences of structure:

$$(\mathcal{F} \mapsto \mathcal{F} \mapsto \mathcal{F}) \ast \uplus \qquad\qquad (\mathcal{F} \mapsto \mathcal{F} \mapsto \mathcal{F}) \operatorname{lcm} \cup$$
$$(\mathcal{F} \mapsto \mathcal{F} \mapsto \mathcal{F}) \operatorname{gcd} \cap \qquad\qquad (\mathcal{F} \mapsto \mathcal{F} \mapsto \operatorname{eq}) \operatorname{div} \subseteq$$
$$(\mathcal{F} \mapsto \operatorname{eq} \mapsto \mathcal{F}) \operatorname{exp} \operatorname{smult} \qquad\qquad (\mathcal{F} \mapsto \operatorname{eq}) \operatorname{prime} \operatorname{sing}$$

Application in Proofs. We formalised the proof of Problem 1.

Let n be a positive integer. Assume that, for every prime p, if p divides n then p^2 also divides it. Prove that n is the product of a square and a cube.

Formally, we state this as

$$\forall n > 0. \ (\forall p > 0. \ \operatorname{prime} p \wedge p \operatorname{div} n \longrightarrow p^2 \operatorname{div} n)$$
$$\longrightarrow (\exists a > 0. \ \exists b > 0. \ a^2 \ast b^3 = n)$$

Notice that the quantifiers of n p, a and b are bounded (greater than 0). This is not necessary (e.g., p is prime, so it is redundant to say that it is positive), but it is convenient for the proof. If we want a proof for the unbounded version (which is also a theorem) we can divide in cases, when $n = 0$ and when $n > 0$. The case for $n = 0$ is trivial because then $a = 0$ and $b = 0$ are solutions. Thus, we prove prove directly the case for $n > 0$.

When we apply the transfer method to the sentence we get the following sentence about multisets:

$$\forall_p n. \ (\forall_p p. \ \operatorname{sing} p \wedge p \subseteq n \longrightarrow 2 \cdot p \subseteq n) \longrightarrow (\exists_p a. \ \exists_p b. \ 2 \cdot a + 3 \cdot b = n)$$

where $\forall_p$ is the universal quantifier bounded to prime numbers, and operator $\cdot$ represents the symmetric version of the multiplication previously referred to as smult (we present it as we do for reading ease).

The premise $(\forall_p p. \ \operatorname{sing} p \wedge p \subseteq n \longrightarrow 2 \cdot p \subseteq n)$ is easily proved to be equivalent to $\forall q. \ \operatorname{count} n q \neq 1$. Then it is sufficient to show

$$\forall_p n. \ (\forall q. \ \operatorname{count} n q \neq 1) \longrightarrow (\exists_p a. \ \exists_p b. \ 2 \cdot a + 3 \cdot b = n)$$

With a bit of human interaction, this can further be reduced to proving that, for every element of n, its multiplicity n_i (which the premise says is different from 1) can be written as $2a_i + 3b_i$, or formally:

$$\forall n_i : \mathbb{N}. \ n_i \neq 1 \longrightarrow \exists a_i. \ \exists b_i. \ 2a_i + 3b_i = n_i$$

This problem can actually be solved in a decidable part of number theory (Presburger arithmetic), for which there is a method implemented in Isabelle.

5.2 Numbers as Sets

At the centre of this transformation is the relation $\mathcal{C}$ where $\mathcal{C}\,A\,n$ holds if and only if $\mathtt{finite}\,A \wedge \mathtt{card}\,A = n$.

We first prove trivial cardinality properties like $\mathcal{C}\,\{1\cdots n\}\,n$, which allows us to consider standard representatives of numbers.

This relation is right-total but not left total, so we have the following two rules:

$$((\mathcal{C} \rightarrowtail \mathtt{imp}) \rightarrowtail \mathtt{imp})\, \forall\, \forall \qquad ((\mathcal{C} \rightarrowtail \mathtt{eq}) \rightarrowtail \mathtt{eq})\, \forall_{\mathtt{fin}}\, \forall$$

where $\forall_{\mathtt{fin}}$ is the universal quantifier restricted to finite sets. Furthermore, the relation is left-unique but not right-unique, so we have

$$(\mathcal{C} \rightarrowtail \mathcal{C} \rightarrowtail \mathtt{imp})\, \mathtt{eq}\, \mathtt{eq} \qquad (\mathcal{C} \rightarrowtail \mathcal{C} \rightarrowtail \mathtt{eq})\, \mathtt{eqp}\, \mathtt{eq}$$

where $\mathtt{eqp}$ is the relation of being equipotent, or bijectable.

Then, we have the following rules for the structural correspondence:

$$(\mathcal{C} \rightarrowtail \mathcal{C})\, \mathtt{Pow}\, (\lambda x.\, 2^x)$$
$$(\mathcal{C} \rightarrowtail \mathtt{eq} \rightarrowtail \mathcal{C})\, \mathtt{n\text{-}Pow}\, \left(\lambda n\, m.\, \binom{n}{m}\right)$$
$$(\mathcal{C} \rightarrowtail \mathcal{C} \rightarrowtail \mathtt{imp})\, \subseteq\, \leq$$
$$(\mathcal{C} \rightarrowtail \mathcal{C} \rightarrowtail \mathcal{C} \rightarrowtail \mathtt{imp})\, \mathtt{disjU}\, \mathtt{plus}$$

where $\mathtt{n\text{-}Pow}\,S\,n$ is the operator that takes the set of subsets of S that have cardinality n. Also, $\mathtt{disjU}\,a\,b\,c$ means $\mathtt{disjoint}\,a\,b \wedge a \cup b = c$ and $\mathtt{plus}$ is the predicative form of operator $+$.

We have mechanised combinatorial proofs, like the ones for the problems given in Table 2, of theorems using this transformation.

6   Automated Change of Representation

We have built a tactic that searches within the space of representations given a set of transformations. Then it tries to reason about each representation. Our goal is for it to embody our vision presented in Sect. 3. This is work in progress, but we address some simple requirements that we have already implemented and present our observations.

6.1 Transformations as Sets of Transfer Rules

As described in Sect. 4, we consider a transformation as a set of 'base' relations, and a structural extension of them. Then, *knowing* a transformation means knowing instances where the relations and their extensions (with respect to relators such as $\rightarrowtail$) hold. These instances of knowledge are what the Transfer package calls *transfer rules*. They are theorems that the user has to prove and, with enough of them provided, the transfer method will transform the goal to an equivalent, or stronger sentence in another domain.

In the traditional use of the Transfer method, there is a single attribute that encompasses all transfer rules. Given a goal, the Transfer method will try to derive an equivalent or stronger subgoal using all the rules with such attribute, with a simple inference mechanism (described in briefly in Sect. 4.1 and more detailed in [6]). We have packaged each of the transformations described in Fig. 1 as a set of transfer rules. Then, our tactic applies the transfer method one transformation at a time.

Transformation-Specific Language. Each transformation has a set of definitions that are linked by the transfer package. Some of them are defined only for use of the transformation, like `disjU` and `plus` (the predicative version of disjoint union of sets and addition of natural numbers, respectively), or bounded quantifiers. These are necessary for the transfer method to find matches, but theorems will not generally be stated in such terms. Our tactic normalises the language of the goal to suit the specific transformation that is going to be applied.

6.2 Reversing Transformations

We have implemented a tool to automatically *reverse* transformations. Let us explain this.

If we want to transform a sentence $p\,a$ to an equivalent one, the Transfer method will search for transfer rules $(R \mapsto \mathsf{eq})\,q\,p$ and $R\,b\,a$ for some R, q and b. If found, it will transform the sentence to the equivalent one $q\,b$. The fact that the sentences are equivalent means that if we had started with $q\,b$ as a goal, it would have been valid to transform it to $p\,a$. This means that, in theory, the same transfer rules can be used to do inference in one direction or the other, at least when the rules are regarding equivalence. The Transfer method does not do so: if one wants to use a transformation in both directions one has to define two distinct transformations, i.e., two distinct sets of transfer rules (in our example above one needs transfer rules $(R' \mapsto \mathsf{eq})\,p\,q$ and $R'\,a\,b$, where R' is the reverse of R). A transfer rule always has a 'reverse' version (although only equivalent ones retain full information), so we should be able to get these automatically. We have built a conversion tool that, given a set of transfer rules, will generate all their reverse rules (in a logically valid way, i.e., the reverse version is always equivalent to the original).

Our program uses the following rewrite rules:

$$R\,a\,b \Rightarrow (\mathsf{swap}\,R)\,b\,a$$
$$\mathsf{swap}\,(R_1 \mapsto R_2) \Rightarrow (\mathsf{swap}\,R_1 \mapsto \mathsf{swap}\,R_2)$$

where `swap` simply swaps the place of the arguments of a function. It is easy to see that these rules are valid. Moreover, $\mathsf{swap}\,R$ equals R when R is symmetric, which means that in some relations we can drop the `swap` function. Thus, our program drops `swap` from `eq` and turns `swap imp` and `swap revimp` into `revimp` and `imp`, respectively.

By reversing every transformation we can traverse every path in Fig. 1 in any direction (which does not mean that every sentence has a transformation to an equivalent one).

6.3 Search Between Representations

Our tactic searches the space of representations by applying each transformation, then reasoning within the theory where it arrived, and, if there are still open subgoals it will repeat the process iteratively.

Recall that transformations are relational. As such, the process is non-deterministic for each transformation, so there will be many branches per transformation. Apart from being non-deterministic, the transfer method will allow transformations of a sentence where some matches are left open, i.e., in the place of some constant we get a *schematic variable* that the user can instantiate manually, and prove their validity with the new instantiation. This can be handy, but our tactic favours branches with the lowest number of open subgoals, thus favouring complete matches; e.g., matches that will not leave any proof obligations open.

We have also noticed that the order in which the transformations are searched is crucial and have set an ad hoc order that favours the transformations we consider more interesting. Heuristics deserve further work, but that remains as a task for the future.

Discarding False Representations. Recall that our transformations do not necessarily yield equivalent sentences when applying the transfer algorithm (unless we restrict it to do so). Actually, the *numbers-as-sets* transformation can only be applied in useful ways if we allow the reduction of the goal to a strictly stronger subgoal (because, e.g., $A \subseteq B$ implies that $|A| \leq |B|$, but not the other way around, meaning that we can prove $|A| \leq |B|$ by showing first $A \subseteq B$, but we cannot prove $A \subseteq B$ by showing $|A| \leq |B|$). This can lead to false subgoals. Thus, our tactic calls the counterexample checker nitpick [1] and discards branches where a counterexample is found for one of its goals.

6.4 Overview

In a single step in the search, our tactic does the following:

1. Normalise to transformation-specific language.
2. Apply transformation.
3. If working with a transformation that generates a stronger subgoal, search for counterexamples and discard if they are found.
4. Apply `auto` tactic to transformed sentence.

The tactic can be applied recursively to search for a transformation to a domain more than one step away. When searching, the obvious stop condition is that the theorem has been proved, although there can be other good reasons to stop in a domain to allow the user to reason interactively.

Each of the 4 steps mentioned can have plenty of branches, so there is search involved. Branches with the least number subgoals are favoured, and the order in which the transformations are applied matters, but there are no clever heuristics involved.

Even though our observations about the trace of the search have led us to the current design and implementation of the tactic, the design is not yet complete and its implementation (although functional) is very much subject to change. There are still open questions regarding what search strategies, *stop* conditions, and reasoning tactics (between transformations) are the best, because these are subject to what evaluation criterion we should use. In Sect. 8 we discuss why this is problematic and how we are confronting it.

7   Related Work

Although representation is widely recognised as a crucial aspect of reasoning, to our knowledge there has been no attempt to incorporate the *automatic* search of representation into reasoning tools.

7.1   Institutions and HETS

The concept of Institution was introduced to as a general notion of logical system [5]. The Heterogeneous Tool Set (HETS) [8] was developed mainly to manage and integrate heterogeneous specifications. Based on the theory of Institutions, it links various logics, including Isabelle's HOL and FOL, and provides a way of translating between them. The uses of HETS have been to bring together various aspects of complex systems where different programming languages and reasoning tools are used for different parts of the system. We do not know of any uses of HETS where heterogeneity is taken advantage of as a means of finding proofs in one representation where other representations fail.

7.2   Little Theories and IMPS

"Little Theories" is the notion that reasoning is best done when it is modular [3]. IMPS is a an interactive proof system implemented based on the principles of Little Theories [4]. The modules, or 'little theories' of IMPS are small axiomatic theories connected by *theory interpretation*. Thus, it concerns different levels of abstraction of a theory, and not directly representation of the entities of the theory.

7.3   Uses of the Transfer Package

The use of the Transfer package has changed how new quotient types and subtypes are defined. This is what the *Lifting* package does [6]. As part of the lifting package, there is a way of automatically transferring definitions from an old type to a new type (e.g., multisets are defined as an abstract type from the type of N-valued functions).

The Lifting package has been the main application of the Transfer package, although the generality of their approach is acknowledged by the developers. Embodying this generality, they have built an Isabelle theory of transference from integers to natural numbers, very much in the spirit of the various transformations we have built ourselves.

8 Evaluation, Future Work and Conclusion

The main contributions presented in this paper are:

- We mechanised various useful transformations observed in proofs of discrete mathematics.
- We have proved example theorems using these transformations.
- We have identified some requirements for search over the space of representations, and implemented both a tool (for reversing transformations) and a tactic fulfilling the requirements.

Our tactic has yet to be evaluated properly. Below we examine some of the difficulties associated with this task.

What makes one proof better than another? There is no definite answer for this question. Simple measures, such as length, are important, but unsatisfactory as a whole. At the very least, we can agree that some proof is better than no proof. Thus, the simplest scenario for evaluation would be that in which our tactic that reasons within many representations finds proofs which cannot be found otherwise. Unfortunately, the current state of automatic theorem provers does not seem to be conducive to this. All the examples in which we have tested our techniques belong to either of the following classes:

1. They are so simple that they can be proved automatically[5] without the need of a transformation.
2. They are too complicated and require an intervention from the user to complete the proof, even after automatically applying a transformation.[6]

Thus, the proof-or-no-proof criterion is not applicable. Then, it is necessary to work on close analysis of interactive proofs with transformations and without them.

A venture for future research is the potential application of this framework for the transformation of geometric problems into algebraic representations, e.g., Gröbner bases [2], where there has been plenty of success in automated reasoning, or into SAT/SMT, which also have been an area of success in automation.[7]

---

[5] using Isabelle tactics like **auto**.

[6] The examples of this second (more interesting) class have been selected from either maths textbooks for undergraduate students, or from training material for contests such as the Mathematical Olympiads.

[7] We thank the anonymous referees of this paper for suggested these possibilities. They remain as future work.

Interestingly, we have an example (Pascal's theorem) that belongs to the class of problems where Isabelle's automatic tactics can find a proof, but where its proof using a transformation deserves attention. It is provable automatically (from the definition of the choose operator included in Isabelle's combinatorics library by its developers), but can be transformed using the numbers-as-sets transformations and proved only interactively there. Arguably, a combinatorial proof could be highly valued by mathematicians (or a scientist who analyses proofs), making this an example where the interactive proof deserves equal, or even more, attention than the automatic proof.

Furthermore, even in the case in which we had automatic proofs using the usual tactics (like Pascal's theorem, mentioned above), we have to consider that these tactics depend on background knowledge (in our case, this amounts to Isabelle's libraries, which have been vastly populated by users). This raises the question: are there ways in which we can measure success independently of the background theories? We think that this is partially achievable by building simpler theories, with some equal level of measurable simplicity, and testing tactics that incorporate representational change there. Even if impractical by itself, this might bring some scientific insight that might lead to better reasoning tactics and theorem provers in the future.

References

1. Blanchette, J.C., Nipkow, T.: Nitpick: a counterexample generator for higher-order logic based on a relational model finder. In: Kaufmann, M., Paulson, L.C. (eds.) ITP 2010. LNCS, vol. 6172, pp. 131–146. Springer, Heidelberg (2010)
2. Buchberger, B., Winkler, F.: Gröbner Bases and Applications, vol. 251. Cambridge University Press, Cambridge (1998)
3. Farmer, W.M., Guttman, J.D., Thayer, F.J.: Little theories. In: Kapur, D. (ed.) CADE 1992. LNCS, vol. 607, pp. 567–581. Springer, Heidelberg (1992)
4. Farmer, W.M., Guttman, J.D., Thayer, F.J.: IMPS: an interactive mathematical proof system. J. Autom. Reason. 9(11), 213–248 (1993)
5. Goguen, J.A., Burstall, R.M.: Institutions: abstract model theory for specification and programming. J. ACM (JACM) 39(1), 95–146 (1992)
6. Huffman, B., Kunčar, O.: Lifting and transfer: a modular design for quotients in Isabelle/HOL. In: Gonthier, G., Norrish, M. (eds.) CPP 2013. LNCS, vol. 8307, pp. 131–146. Springer, Heidelberg (2013)
7. Hurd, J.: System description: the Metis proof tactic. In: ESHOC, pp. 103–104 (2005)
8. Mossakowski, T., Maeder, C., Lüttich, K.: The heterogeneous tool set, HETS. In: Grumberg, O., Huth, M. (eds.) TACAS 2007. LNCS, vol. 4424, pp. 519–522. Springer, Heidelberg (2007)
9. Nipkow, T., Paulson, L.C., Wenzel, M.: Isabelle/HOL: A Proof Assistant for Higher-Order Logic. Springer, Heidelberg (2013)
10. Paulson, L.C., Blanchette, J.C.: Three years of experience with sledgehammer, a practical link between automatic and interactive theorem provers. Practical Aspects of Automated Reasoning (PAAR), 5th International Joint Conference on Automated Reasoning (IJCAR) (2010)
11. Weber, T.: SMT solvers: new oracles for the HOL theorem prover. Int. J. Softw. Tools Tech. Transf. 13(5), 419–429 (2011)

Performance Evaluation and Optimization of Math-Similarity Search

Qun Zhang[✉] and Abdou Youssef

Department of Computer Science,
The George Washington University, Washington, DC 20052, USA
qzhang@gwmail.gwu.edu

Abstract. Similarity search in math is to find mathematical expressions that are similar to a user's query. We conceptualized the similarity factors between mathematical expressions, and proposed an approach to math similarity search (MSS) by defining metrics based on those similarity factors [11]. Our preliminary implementation indicated the advantage of MSS compared to non-similarity based search. In order to more effectively and efficiently search similar math expressions, MSS is further optimized. This paper focuses on performance evaluation and optimization of MSS. Our results show that the proposed optimization process significantly improved the performance of MSS with respect to both relevance ranking and recall.

Keywords: Optimization · Performance evaluation · Math similarity search · Relevance ranking · Recall

1 Introduction

Math is a symbolic language layered in abstractions, rich in structure, and full of synonymy and polysemy among expressions. These characteristics make math search especially challenging, and call for similarity search. In a recent paper [11], we proposed an approach to math similarity search (MSS) for Strict Content MathML encoded math expressions. We identified conceptual factors of math similarity, and deduced a math similarity metric. For a query, MSS was designed to help us find math expressions with taxonomically similar functions, hierarchically similar structure, and/or semantically similar data types. It was found to be able to greatly improve the performance of a math search system.

Ideally, search performance evaluation requires a benchmark infrastructure. However, since math search is still young and evolving, initiatives to create such benchmarks have barely started, most notably the NTCIR [7] math task led by Aizawa and Kohlhase, et al. However, to the best of our knowledge, there is not yet a standard Strict Content MathML encoded math query benchmark infrastructure available to us. To create a full benchmark that can be agreed upon requires significant expertise. However, we do need some ground truth for our performance evaluation and system optimization. So for our research purpose, we leverage the DLMF [2] repository.

© Springer International Publishing Switzerland 2015
M. Kerber et al. (Eds.): CICM 2015, LNAI 9150, pp. 243–257, 2015.
DOI: 10.1007/978-3-319-20615-8_16

The rest of the paper starts with a brief summary of the related work in Sect. 2. It then presents a recap of the conceptual similarity factors and the similarity metric of MSS [11] in Sect. 3. Section 4 gives in detail the process of performance evaluation and optimization, and Sect. 5 presents the performance results. Finally, the conclusions are drawn in Sect. 6.

2   Background

An important aspect of search is relevance ranking. Therefore, comparison between ranking schemes is a significant piece of the search system performance evaluation. The various existing ranking comparison metrics can be classified into two categories: subjective metrics and statistical metrics.

The former is purely based on human judgment [5,10]. It is good for judging the ranking produced by a specific search tool.

Statistical comparison metrics are often based on correlation measurement, which is an extension of the distance measure between two permutations [1,3]. These works leverage some classical metrics such as Kendall's τ [6], and the standard correlation coefficient, also known as Spearman's ρ [9]. This type of ranking comparison metric can be automated. Thus, the statistical ranking comparison approach is selected for our research.

As mentioned earlier, search performance evaluation demands benchmarks. The NTCIR-11 Math-2 task [7] and its optional Wikipedia subtask [8] are the only significant ones for math search evaluation, as far as we know. The target dataset of NTCIR-11 Math-2 consists of about 100,000 scientific articles in ArXiv that are converted from LaTeX to XHTML. NTCIR-11 Math-2 provides some sample queries, but there are no expected results as reference for systematic evaluation. Rather, a pooling based relevance judgment is proposed to run manually by human reviewers' subjective assessment.

The NTCIR-11-Math-Wikipedia-Task uses the English Wikipedia as a test collection, which is more suitable for general users compared with the NTCIR-11 Math-2. It also provides a number of sample queries which include 2 content queries that are encoded in Content MathML. However, there are no expected results for participants to do self-evaluation, or any formal evaluation conducted by the organizers except a final judgment based on participants' oral presentations.

Both of the above research initiatives encode their dataset in XHTML, while our MSS is for Strict Content MathML encoded math expressions only. This makes it challenging for us to participate in these collective efforts at this moment. Additionally, none of the above tasks provides any reference results for us to use as ground truth to do performance evaluation; instead, the assessment is either done by their organizers in a pooling based manual process or not available at all. This prevents us from gaining performance evaluation benefits as of now. However, it is anticipated that these resources can be leveraged in the future once the Strict Content MathML encoded dataset and the full set of ground truth with both queries and expected results become available, so that more thorough and more objective performance evaluation can be done.

3   Math-Similarity Search (MSS)

The goal of MSS is, given a math expression that is encoded in Strict Content MathML, to identify a list of structurally and semantically similar math expressions from a library of Strict Content MathML encoded math expressions, and sort the list by similarity according to some similarity measure. In [11], we identified conceptual factors that capture aspects of math similarity, and came up with a similarity metric that takes all those factors into consideration.

3.1   Math Similarity Factors

We identified five major factors to math similarity measure, with focus on the structural and semantic aspects of math expression.

1. **Data Type.** This factor captures the superiority of functions and operators over arguments or operands (for our research purpose, there is no distinction between function and operator), and of structure over notation. For example, query $F = G\frac{m_1 m_2}{r^2}$, is more similar to $F = ke\frac{q_1 q_2}{r^2}$, than to a hypothetical expression $F = G \cdot m_1 + \frac{m_2}{r^2}$, containing the same set of operands and notations as the query.
2. **Taxonomic Distance of Functions.** Taxonomies of functions are available to us through the content dictionary (CD) attribute of Strict Content MathML. Functions in the same CD are assumed to be more similar than functions in different CDs.
3. **Match-Depth.** The higher in the parse tree an operator is, the more important it is. Hence, the more deeply nested a query is in a hit-expression, the less relevant the hit is. Therefore, match-depth is used as a similarity decaying multiplicative factor. One can utilize different models for this decay factor, such as the exponential model, the linear model, the quadratic model, or the logarithmic model.
4. **Query Coverage.** How much of the query is covered in the returned expression is important. In general, the greater the query coverage is, the greater the significance is to the similarity measure.
5. **Formula vs. Non-formula Expression.** Typically, formulas carry more weight than non-formula expressions, and equality more than inequalities. Therefore, formulas and equalities will be accorded more relevance weights. Note this factor is not a similarity factor, but it is included because of its effect on relevance ranking.

3.2   Math Similarity Metric

MSS takes Strict Content MathML parse trees as the primary model representing math expressions. The similarity between two math expressions is defined recursively based on the height of the corresponding parse trees. In the definitions below, various parameters will be utilized and later optimized. Note that the parameters δ, ζ, μ, and θ are all non-negative and < 1.

When the height is 0, the parse tree is a single node which can represent a constant, a variable, or a function. The similarity between two parse trees, T_1 and T_2 are defined as:

- If T_1 and T_2 are constants: $sim(T_1, T_2) = 1$, if $T_1 = T_2$; δ, otherwise.
- If T_1 and T_2 are variables: $sim(T_1, T_2) = 1$, if $T_1 = T_2$; ζ, otherwise.
- If T_1 and T_2 are functions:
  - $sim(T_1, T_2) = 1$, if T_1 and T_2 are the same function
  - $sim(T_1, T_2) = \mu$, if T_1 and T_2 are different functions of the same category in the taxonomy
  - $sim(T_1, T_2) = 0$, if T_1 and T_2 are different functions of different categories
- If T_1 and T_2 are not of the same data type:
  - $sim(T_1, T_2) = \theta$, if one is constant and the other is a variable
  - $sim(T_1, T_2) = 0$, if one is a function and the other is either constant or variable

Note that δ, ζ, μ, and θ are parameters that will be optimized experimentally.

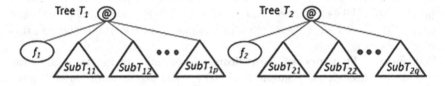

Fig. 1. Illustration of two trees T_1 and T_2 of height $h \geq 1$

When the heights of the two trees T_1 and T_2 are ≥ 1, each of them is composed of function apply operator as root, a leftmost child representing the function name, followed by a list of arguments which are sub-trees, as illustrated in Fig. 1. The similarity between T_1 and T_2 is defined as a weighted sum of the similarity between the two functions f_1 and f_2, and the similarity between the two lists of arguments:

$$sim(T_1, T_2) = \alpha \cdot sim(f_1, f_2) + \beta \cdot sim(\{SubT_{11}, SubT_{12}, \ldots, SubT_{1p}\},$$
$$\{SubT_{21}, SubT_{22}, \ldots, SubT_{2q}\}),$$

where $\alpha = \frac{\omega}{p+\omega}$, and $\beta = \frac{1}{p+\omega}$ are weighting factors, $\omega(>1)$ is a boost value (for boosting functions over arguments), and $0 \leq sim(\{SubT_{11}, \ldots, SubT_{1p}\}, \{SubT_{21}, \ldots, SubT_{2q}\}) \leq p$.

The similarity measurement between the two lists of arguments depends on whether the two functions, f_1 and f_2, are commutative. If both are non-commutative, the order of the arguments is observed. The similarity between the two lists of arguments is the sum of the similarities between the corresponding available pairs of argument sub-trees with one from each tree:

$$sim(\{SubT_{11}, SubT_{12}, \ldots, SubT_{1p}\}, \{SubT_{21}, SubT_{22}, \ldots, SubT_{2q}\}) =$$
$$\sum_{i=1}^{min(p,q)} sim(SubT_{1i}, SubT_{2i})$$

Otherwise, the similarity between the two lists of arguments is the similarity of the best match between the two lists:

$$sim(\{SubT_{11}, SubT_{12}, \ldots, SubT_{1p}\}, \{SubT_{21}, SubT_{22}, \ldots, SubT_{2q}\}) =$$
$$\max\{(\textstyle\sum_{i=1}^{p}(sim(SubT_{1i}, SubT_{2t(q,p,i)})))|t(q, p, i) \text{ is the } i\text{-th element of a}$$
$$p\text{-permutation of } q\}$$
$$\approx \textstyle\sum_{i=1}^{min(p,q)} \max\{sim(SubT_{1i}, SubT_{2\varphi(i)})| \text{ applying greedy approximation,}$$
$$\varphi(i) = 1, 2, \ldots, q \text{ and } \varphi(i) \notin \{\varphi(1), \varphi(2), \ldots, \varphi(i-1)\}\}$$

For partial match consideration, not only the similarity between T_1 and T_2 at their root level is measured, but also the similarity between entire tree T_1 and each single sub-tree of T_2, as well as the similarity between entire tree T_2 and each single sub-tree of T_1 are measured in order to find the best match among all of these possibilities. The final form of the similarity measure between T_1 and T_2, where T_1 is the parse tree of the query, is:

$$sim(T_1, T_2) = max\{$$
$$(\alpha \cdot sim(f_1, f_2) + \beta \cdot sim(\{SubT_{11}, \ldots, SubT_{1p}\}, \{SubT_{21}, \ldots, SubT_{2q}\})),$$
$$dp \cdot max\{sim(T_1, f_2), sim(T_1, SubT_{21}), sim(T_1, SubT_{22}), \ldots, sim(T_1, SubT_{2q})\},$$
$$cp \cdot max\{sim(f_1, T_2), sim(SubT_{11}, T_2), sim(SubT_{12}, T_2), \ldots, sim(SubT_{1p}, T_2)\}\}.$$

In the similarity measure, dp and cp are the match-depth penalty factors, dp corresponds to the depth of T_1 in T_2, and cp corresponds to the depth of T_2 in T_1. The parametrized values of dp and cp under the four match-depth decay models are of the form: (1) a^k for the exponential model; (2) $max(1-b \cdot k, \epsilon)$ for the linear model; (3) $max(1-c \cdot k^2, \epsilon)$ for the quadratic model; and (4) $max(1-d \cdot ln(k+1), \epsilon)$ for the logarithmic model; where k is the match-depth, ϵ is a small positive constant to ensure that the value of the decay factor stays positive, and a, b, c, and d are parameters to be determined experimentally for each of dp and cp. Also, the values of each of a, b, c, and d will differ depending on whether T_1 and/or T_2 are equality-formulas, inequality-formulas, or non-formulas.

4   Performance Evaluation and Optimization

Even though the preliminary implementation of MSS showed some promising advantages over the non-similarity based search, the many parameters of the MSS similarity metric must be optimized.

4.1   Evaluation Methodology

As the DLMF math digital library is among the few available and easily accessible for us, this research leverages the DLMF as the source for mathematical expressions repository, and we compare our similarity search to the DLMF search system. Since there is no Strict Content MathML encoding of the DLMF, we leveraged the first 4 chapters of the DLMF to create a test database of 600 equations, which we hand-encoded in Strict Content MathML. Then a cross-section

Table 1. The test queries

Query ID	Query Expression	Query ID	Query Expression
Q1	e^z	Q2	$\tan(z)$
Q3	$\int_a^b f(x)\,dx$	Q4	$f(z_0) = \dfrac{1}{2\pi i}\int_c \dfrac{f(z)}{z-z_0}\,dz$
Q5	$\ln(1+z) = z - \dfrac{z^2}{2} - \dfrac{z^3}{2} - \cdots$	Q6	$\dfrac{d}{dx}\int_a^x f(t)\,dt = f(x)$
Q7	$\int_c f(z)\,dz = 0$	Q8	$f(z) = c_0 + c_1 z + c_2 z^2 + \cdots$
Q9	$\lim \dfrac{\sin(z)}{z}$	Q10	$A\cosh(az) + B\sinh(az)$
Q11	$\sin^2 x + \cos^2 x = 1$	Q12	$\sin(x+y) = \sin x \cos y + \cos x \sin y$
Q13	$\int \sin(x)\,dx = -\cos(x)$	Q14	$\cosh(x) \le (\dfrac{\sinh(x)}{x})^3$
Q15	$\sinh(z) = \dfrac{e^z - e^{-z}}{2}$	Q16	$\delta(x-a)$
Q17	$a^2 + b^2$	Q18	$\dfrac{a(1-x^n)}{1-x} = a + ax + ax^2 + \cdots + ax^{n-1}$
Q19	$\sum a_j b_j \le (\sum a_j{}^p)^{1/p}(\sum b_j{}^q)^{1/q}$	Q20	$F(x) = \dfrac{1}{\sqrt{2\pi}}\int f(t)e^{ixt}\,dt$
Q21	$\Gamma(z)$	Q22	$\det[a_{ij}]$
Q23	$a \cdot b = \sum a_j b_j$	Q24	$\sqrt{a^2 + b^2}$
Q25	$e^{ix} + 1 = 0$	Q26	$e^x < \dfrac{1}{1-x}$
Q27	$\int \dfrac{dx}{1+e^x}$	Q28	$(f * g)(t)$
Q29	$\lim_{x\to\infty} x^a e^{-x}$	Q30	$\dfrac{d}{dx}\arctan(\sin(x^2))$
Q31	$\sin^2 x + \cos^2 x$	Q32	$\sin(x+y)$
Q33	$\int \sin(x)\,dx$	Q34	$a + ax + ax^2 + \cdots + ax^{n-1}$
Q35	$\sum a_j b_j \le$	Q36	$\int_c \dfrac{f(z)}{z-z_0}\,dz$
Q37	$\sin(u) + \sin(v)$	Q38	$\cos(u) + \cos(v)$
Q39	$\dfrac{d}{dz}\operatorname{arccot}(z)$	Q40	$\lim_{n\to\infty}\left(1 + \dfrac{1}{n}\right)^n$

of 40 queries with varying degrees of mathematical complexity and length were created and encoded in Strict Content MathML. They are listed in Table 1.

For each query in the test set, we identify the expected relevant expressions from the subset of DLMF source repository used as a test database, and further rank them manually in a relevancy based order by a group of human experts, which are then used as a ground truth. Each query is compared with the expressions in the test database, and a similarity value is computed with the MSS similarity metric described in Sect. 3. The full list of the expressions is then sorted in descending order of similarity. Depending on the number of relevant expressions in the ground truth for this particular query, the top "N" number of expressions is selected to form the MSS hit list for this query.

Up to this point, for any given query, there are three hit lists: one from the ground truth, another from the DLMF site returned by DLMF search, and a third from the similarity search MSS. We compare both the MSS hit list and the DLMF hit list with the ground truth list, based on the performance metric described in Sect. 4.2. The output of such comparison shows the alignment of a hit list with the ground truth.

4.2 Performance Metric

We evaluate the performance with respect to both recall and relevance ranking. The evaluation is done for each sample query, before the final average across all of the sample queries is computed.

The recall measure is fairly standard. In addition to the recall of all of the results in the ground truth hit list, we also measure the recall of the top 10 results relative to the ground truth.

For relevance ranking comparison, we measure the correlation between the ranked list that is returned from each search system, and the ground truth list for each sample query. The correlation is measured by the standard correlation coefficient Spearman's ρ, and Kendall's τ. Note that ρ focuses more on absolute order, i.e. where each hit is ranked, whereas τ focuses more on relative order of the hits, i.e. which comes before which. These two correlation metrics complement each other in serving as performance indicators.

Following the standard practice for correlation analysis, for both ρ and τ, we refer to their critical value tables to judge statistical significance of the correlation under evaluation. For any given sample query, depending on the size of its ground truth, i.e. the number of similar expressions expected to return, the corresponding critical value is looked up and compared with the computed correlation measure. The correlation appears statistically significant with the specific level of confidence only if the correlation measure reaches or exceeds the critical value of that confidence level. In our research, two different confidence levels are considered: 95 % and 99 %.

4.3 Optimization Concerns and Motivation

In [11] many parameters of the MSS metric need be optimized, such as taxonomic distance values (e.g. μ) between functions, function nodes type booster value ω, query coverage factor, etc. In the early implementation of the MSS metric, intuitive but still random values were used to initialize the parameters. After preliminary experimentation, some reasonable values were obtained, so that the MSS showed some performance advantage over non-similarity based math search. This was indicated in [11]. In this paper, we conduct a systematic experimentation based optimization to arrive at optimal values of all the parameters.

One of the similarity factors, the match-depth, is represented as a similarity-decaying multiplicative factor in our MSS metric. Several alternative models can be used for such decay, such as exponential decay, linear decay, quadratic decay, or logarithmic decay. Which of these decay models performs better for MSS metric, and how to configure these decay models along with other parameters of the MSS metric, become part of the concerns that motivate us to start the optimization process.

In the preliminary evaluation conducted in [11], the number of results in the ground truth for each sample query was normalized to 20. However, in reality the actual number of expressions in the test database that are similar to each query may not be the same - if less than 20, then the ground truth was padded with "weeds"; and if larger than 20, then normalizing the ground truth size to 20 would limit the view or window of expressions to examine. Either case can impact the overall system performance, e.g. recall. For this reason, we choose to remove the ground truth size normalization across all the queries, and leave the ground truth of all the sample queries as is.

4.4   The Optimization Process

As the number of parameters is quite large, and the metric to be optimized is non-linear, an exhaustive search of the solution space is prohibitive, and no fast technique that is guaranteed to converge to a global optimum of the parameters is available. Therefore, we resorted to a suboptimal, iterative optimization approach, which converged after a reasonable number of generations. The whole optimization process is outlined in Fig. 2.

Each generation has its own optimal trial value for each of the parameters, and its own best model for the match-depth factor. Within one generation, all the parameters are to be optimized. For each parameter, we step through different trial values with equal increments. For each selected trial value, we measure the MSS system performance for each sample query by using the performance metric to measure the recall and correlation between the MSS hit list and the ground truth list for that query. After that, an average of those recall and correlation measurements from each query execution is obtained, and referenced for the selection of best trial value for that specific parameter.

To decide when to stop the generational evolution, we need a "convergence" criterion, i.e. a measurement of the overall system performance across all of the queries for a given system configuration. This is similar to the best trial value selection for a specific parameter within a generation. Once such overall system performance of a generation does not show any improvement compared to the previous generation, the generational evolution terminates for the given match-depth model. However, the full optimization process does not stop until the above generational evolution is done for each the different match-depth models, and then the best model is identified at the end. Although the convergence values of the parameters may be only a local optimum, the resulting search performance, as shown in the next section, is a considerable improvement over the preliminary performance results [11] where the parameters were set to rough, intuitive estimates.

5   Optimization Results

In this section we show the findings of the optimization described in the previous section. Figures 3 and 4 show the recall of MSS, under the four different match-depth models, and the recall of DLMF, over the sample queries. Figure 3 shows

```
begin
for each different match-depth model
    for each generation of parameters do
        for each parameter
            for trial-value of current parameter
                for each query
                    run MSS search;
                    collect similarity measurement;
                    get hits;
                    conduct recall & correlation analysis;
                end-for
                evaluate average performance across all queries;
            end-for
            pick the best trial-value for current parameter
        end-for
        if (this generation did not improve over previous generation)
            stop generation evolution;
            flag "converged" for current model;
        end-if
    end-for
end-for
pick the best model set;
end
```

Fig. 2. Illustration of MSS optimization process

the overall recall, and Fig. 4 shows the recall relative to the top ten results. Both Figs. 3 and 4 indicate that (1) the logarithmic model is best among all the match-depth models, and (2) MSS with the logarithmic model outperforms the DLMF significantly in terms of recall: the overall recall and the top-10 recall of MSS are 0.8251 and 0.9075 (on average), which are quite high and 137 % and 109 % higher than those of DLMF search, respectively.

Similarly, Figs. 5 and 6 show the relevance ranking of MSS, under the four different match-depth models, and the relevance ranking of DLMF, over the same queries, where Fig. 5 shows the τ metric, and Fig. 6 shows the ρ metric. Both Figs. 5 and 6 have reference critical values marked out, where the green solid line indicates the critical values at 95 % confidence level, and the purple dotted line indicates the critical values at 99 % confidence level. The fact that the MSS's correlation measures are mostly above the critical values indicate that the improvement over DLMF search is statistically significant. Interestingly, both Figs. 5 and 6 indicate that the logarithmic match-depth model performs the best among all the match-depth model: its correlation measurements at the optimal setting are all higher than those of the other three match-depth models. Furthermore, MSS under the logarithmic match-depth model outperforms the DLMF in terms of relevance ranking: the average τ and ρ correlation measures (across all the 40 queries) are 0.4737 and 0.4928, respectively, which are 136 % and 207 % higher than those of DLMF search. In addition, as Figs. 3, 4, 5, and 6 show, MSS with the logarithmic match-depth model shows not only significantly greater relevance ranking and recall, but also more performance consistency than the DLMF across the sample queries.

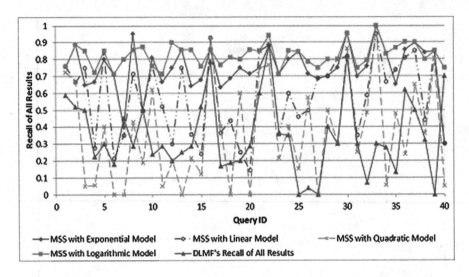

Fig. 3. Overall recall of MSS (with different Match-Depth models) vs. DLMF

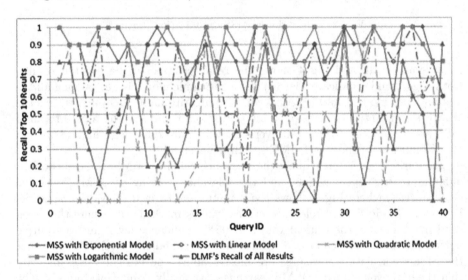

Fig. 4. Top 10 recall of MSS (with different Match-Depth models) vs. DLMF

The MSS with the logarithmic match-depth model shows its optimal performance after ten generations of optimization. The optimal values of the MSS parameters are listed in Table 2. Figure 7 shows the recall improvement throughout the optimization process, with respect to both the overall recall and the top-10 recall. Figure 8 shows the relevance ranking improvement throughout the optimization process, in terms of the τ and ρ correlation measures. Both of these figures also show the performance data of DLMF search, which appears constant

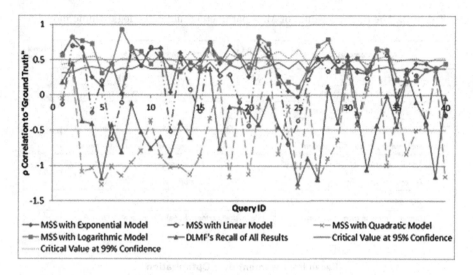

Fig. 5. ρ Correlation of MSS (with different Match-Depth models) vs. DLMF

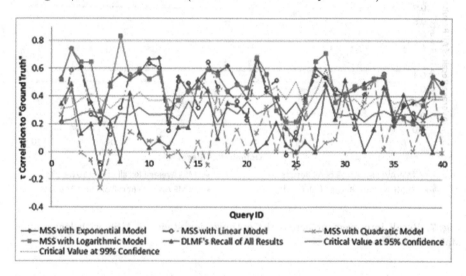

Fig. 6. τ Correlation of MSS (with different Match-Depth models) vs. DLMF

across the different generations of optimization of MSS. As shown in both Figs. 7 and 8, the performance improvement over the evolution of the initial five generations is more significant than that of the later five generations. Over the complete ten generations of the optimization process, all of the four performance metrics data improved substantially with more than 20 % improvement (over the non-optimized preliminary results of [11]), with the ρ correlation measure more than doubling.

Table 2. The optimal values of MSS parameters under the logarithmic model

MSS Parameter	Optimal Value	MSS Parameter	Optimal Value
μ (If Same CD)	0.5	ζ (Two Variables)	0.5
ϑ (Variable vs. Constant)	0.3	ω (If Function Weight)	1.3
δ (Two Constants)	0.7		
"d_0" in "$dp - max(1 - d_0 \cdot ln(k+1), \varepsilon))$" (Depth Penalty in Non-Formula Hit)	0.45	"d_3" in "$cp - max(1 - d_3 \cdot ln(k+1), \varepsilon))$" (Depth Penalty in Non-Formula Query)	0.5
"d_1" in "$dp - max(1 - d_1 \cdot ln(k+1), \varepsilon))$" (Depth Penalty in Equality Hit)	0.1	"d_4" in "$cp - max(1 - d_4 \cdot ln(k+1), \varepsilon))$" (Depth Penalty in Equality Query)	0.6
"d_2" in "$dp - max(1 - d_2 \cdot ln(k+1), \varepsilon))$" (Depth Penalty in Inequality Hit)	0.4	"d_5" in "$cp - max(1 - d_5 \cdot ln(k+1), \varepsilon))$" (Depth Penalty in Inequality Query)	0.7

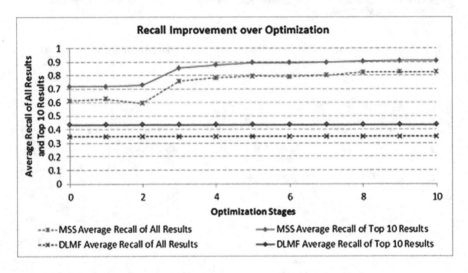

Fig. 7. Recall improvement across the optimization stages of MSS with logarithmic Match-Depth model

In order to avoid the problem of over-fitting of the parameters to the training data, and to obtain a more objective assessment of performance, we conducted cross-validation-based optimization and performance evaluation.

Generally in cross-validation, we train (optimize) a system on a dataset called the "training set", and perform testing on a different dataset called the "validation set" or "test set". To adopt the cross-validation technique in optimization and performance evaluation of MSS system, we partition all of the 40 sample queries randomly into two disjoint subsets of equal size, one of which is used as the training set for optimization and the other as test set for performance evaluation. Such cross validation is applied to MSS system with all the four different match-depth models. The performance results are shown in Table 3.

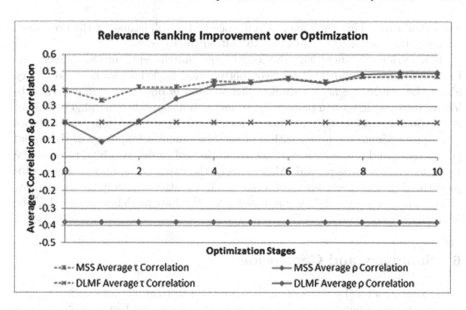

Fig. 8. Relevance ranking across the optimization stages of MSS with logarithmic Match-Depth model

Table 3. Performance result Without vs. With cross validation

Math Search System		Recall Metric		Relevance Ranking Metric	
		Avg. Overall Recall	Avg. Top 10 Recall	Avg. ρ Correlation	Avg. τ Correlation
MSS (Exponential Model)	W/O Cross-Validation	0.7467	0.8575	0.4091	0.4657
	W/ Cross-Validation	0.7785	0.87	0.4104	0.4490
MSS (Linear Model)	W/O Cross-Validation	0.5835	0.725	0.1824	0.4040
	W/ Cross-Validation	0.5904	0.735	0.1943	0.4168
MSS (Quadratic Model)	W/O Cross-Validation	0.3631	0.4225	-0.5625	0.1261
	W/ Cross-Validation	0.2796	0.27	-0.7955	0.0654
MSS (Logarithmic Model)	W/O Cross-Validation	0.8251	0.9075	0.4928	0.4737
	W/ Cross-Validation	0.8288	0.89	0.4807	0.4528
DLMF	on 40 Full Set	0.3476	0.435	-0.3812	0.2006
	on 20 Test Set	0.3729	0.47	-0.3561	0.2117

As shown in Table 3, with cross-validation the performance results of MSS largely agree with those obtained without cross-validation. Particularly, both with and without cross-validation, it turns out that the logarithmic match-depth decaying model is the best model of all the four alternatives, and under that model, all of our four performance metrics achieve the same level with as without cross-validation. Also, the same parameter values are reached with as without cross-validation. This lends more confidence to the robustness of the optimized values of the parameters, and more validation to our results.

It is curious to know why the logarithmic match-depth model delivers the best performance among all four competing models. While we have no theory about the reason, we note a similar phenomenon of objective stimulus vs. subjective response in human perception, of music and sound in general. The human perception of sound frequency, known as *pitch*, has been found in psychoacoustics to be proportional to the logarithm of the fundamental frequency of the sound [4]. As recall and relevance are subjective notions of (objective) hits, perhaps the same logarithmic laws prevail. Alternatively, the explanation may be much simpler: the degree of nested-ness of a match does matter but only to a sub-linear extent, and the only sub-linear model among the four models considered is the logarithmic one.

6   Summary and Conclusions

MSS takes structural aspects of math expressions into serious consideration, honors the greater significance of function over argument, and further differentiates functions by their taxonomic distance. With MSS, partial match of query can be detected and weighted based on the coverage. Additionally, equations (a.k.a. formulas) are differentiated from expressions (non-formulas) in MSS, and ranked higher. All of these make MSS discover more effectively the math expressions that are similar to queries, and rank them better.

In order to quantitatively evaluate the performance of MSS, relevance ranking and recall based performance metrics were used. Furthermore, a systematic optimization process was conducted which improved the performance of MSS substantially. Our experiments show that the MSS with a logarithmic match-depth decay model performs better than the MSS with other match-depth decay models. Finally, cross-validation was applied to the optimization and performance evaluation, and obtained the results that agreed with the conclusions drawn through the prior experiments. The performance data after optimization validate more sufficiently the advantage of math similarity search over non-similarity based search typified by the DLMF.

References

1. Bar-Ilan, J.: Comparing rankings of search results on the web. Inf. Process. Manag. **41**, 1511–1519 (2005). Elsevier
2. The Digital Library of Mathematical Functions (DLMF), the National Institute of Standards and Technology (NIST). http://dlmf.nist.gov/
3. Fagin, R., Kumar, R., Sivakumar, D.: Comparing top k lists. ACM-SIAM J. Discrete Math. **17**(1), 134–160 (2003)
4. Hartmann, W.: Signals, Sound, and Sensation. Springer, New York (1997)
5. Hawking, D., Craswell, N., Bailey, P., Griffiths, K.: Measuring search engine quality. Inf. Retr. **4**, 33–59 (2001). Springer Netherlands

6. Kendall, M.: Rank Correlation Methods. Hafner Publishing Co., New York (1955)
7. The 11th National Institute of Informatics Testbeds and Community for Information Access Research Workshop (2013–2014). http://ntcir-math.nii.ac.jp/
8. An optional free subtask of the NTCIR-11 Math-2 Task (2014). http://ntcir11-wmc.nii.ac.jp
9. Spearman, C.: The proof and measurement of association between two things. Am. J. Psychol. **15**, 72–101 (1904)
10. Vaughan, L.: New measurements for search engine evaluation proposed and tested. Inf. Process. Manag. **40**(4), 677–691 (2004). Elsevier
11. Zhang, Q., Youssef, A.: An approach to math-similarity search. In: Watt, S.M., Davenport, J.H., Sexton, A.P., Sojka, P., Urban, J. (eds.) CICM 2014. LNCS, vol. 8543, pp. 404–418. Springer, Heidelberg (2014)

Projects and Surveys

Mizar: State-of-the-Art and Beyond

Grzegorz Bancerek[1], Czesław Byliński[2], Adam Grabowski[3],
Artur Korniłowicz[3], Roman Matuszewski[4], Adam Naumowicz[3(✉)],
Karol Pąk[3], and Josef Urban[5]

[1] Association of Mizar Users, Białystok, Poland
bancerek@mizar.org
[2] Section of Computer Systems and Teleinformatic Networks,
University of Białystok, Białystok, Poland
bylinski@mizar.org
[3] Institute of Informatics, University of Białystok, Białystok, Poland
{adam,arturk,adamn,karol}@mizar.org
[4] Department of Applied Linguistics, Faculty of Philology, University of Białystok,
Białystok, Poland
romat@mizar.org
[5] Radboud University, Nijmegen, The Netherlands
josef.urban@gmail.com

Abstract. MIZAR is one of the pioneering systems for mathematics formalization, which still has an active user community. The project has been in constant development since 1973, when Andrzej Trybulec designed the fundamentals of a language capable of rigorously encoding mathematical knowledge in a computerized environment which guarantees its full logical correctness. Since then, the system with its feature-rich language devised to approximate mathematics writing has influenced other formalization projects and has given rise to a number of MIZAR modes implemented on top of other systems. However, the information about the system as a whole is not readily available to developers of other systems. Various papers describing MIZAR features have been rather incremental and focused only on particular newly implemented MIZAR aspects. The objective of the current paper is to give a survey of the most important MIZAR features that distinguish it from other popular proof checkers. We also go a step further and describe most important current trends and lines of development that go beyond the state-of-the-art system.

1 Introduction

The MIZAR [21,38] project is a long-term effort originally aimed at developing a computer environment to support mathematicians in preparing papers. Around 1973, Andrzej Trybulec, the leader of the project, has designed a language for writing formal mathematics. The implemented processor was intended to check written texts for logical consistency and correctness. For fifteen years numerous implementations of the system were developed in order to choose suitable underlying logic and expressive power of the language (PC – propositional calculus,

© Springer International Publishing Switzerland 2015
M. Kerber et al. (Eds.): CICM 2015, LNAI 9150, pp. 261–279, 2015.
DOI: 10.1007/978-3-319-20615-8_17

MSE – multi-sorted with equality, QC – quantifier calculus, etc.). An important issue was also to find the proper technology of automated cross-referencing between papers. The history of the first 30 years of MIZAR development has been described in [32]. All these experiments resulted in choosing the principles of the current MIZAR. The logical structure of the language is based on a variant of the classical first-order logic natural deduction system proposed by Jaśkowski [27]. The texts written in the language (called MIZAR articles) are organized into a data base – the MIZAR Mathematical Library, MML. The Tarski-Grothendieck set theory, which is basically ZFC set theory with the axiom of infinity replaced by Tarski's axiom of existence of arbitrarily large, strongly inaccessible cardinals forms the basis of doing mathematics in contemporary MIZAR.

Today there are a number of other projects committed to address problems related to computerized theorem proving developed at various research centers around the world. Most significantly: the HOL Light theorem prover[1], Isabelle[2], the Coq Proof Assistant[3], Metamath[4], ProofPower[5], Nqthm/ACL2[6], the PVS Specification and Verification System[7], and the Nuprl/PRL Project[8]. Each project is characterized by its own specifics implied by an assumed theoretical basis (e.g., type theory or set theory, classical logic versus intuitionistic logic or higher order logic) and main goals, towards which the project is geared (extracting programs from proofs, program verification, formalization of mathematics, automated theorem proving) [59]. Thanks to the discussions and research collaboration stemming from the QED [7] initiative, there has been a significant interplay between the projects. To name the most important cases of MIZAR's influence on other systems we can mention the Declare system developed by D. Syme [42], the MIZAR mode for HOL by J. Harrison [25], the Isar language for Isabelle by M. Wenzel [56], Mizar-light for HOL Light by F. Wiedijk [57], the declarative proof language (DPL) for Coq by P. Corbineau [16] and finally the miz3 proof interface for HOL Light [61] that combines both the procedural and declarative style of writing proofs. The MIZAR way of writing proofs was also the model for the notion of 'formal proof sketches' developed by F. Wiedijk [58].

However, the information about the fundamental aspects of the MIZAR system as a whole has not been readily available. To have a better understanding of MIZAR, the developers of other systems have had to collect the information scattered in scarce user reference materials or various MIZAR research papers describing particular newly implemented MIZAR aspects. With this paper we intend to give a concise survey of the most important MIZAR features that distinguish it from other popular proof checkers and show its current lines of development. Hopefully, this will become beneficial for further collaboration with

---

[1] http://www.cl.cam.ac.uk/~jrh13/hol-light/.
[2] http://www.cl.cam.ac.uk/research/hvg/Isabelle/.
[3] http://coq.inria.fr.
[4] http://us.metamath.org.
[5] http://www.lemma-one.com/ProofPower/index/.
[6] http://www.cs.utexas.edu/users/moore/acl2/.
[7] http://pvs.csl.sri.com.
[8] http://www.nuprl.org.

other similar projects. The work is organized as follows. In Sect. 2 we present the basics of the MIZAR language, in Sect. 3 we describe the main software components of the system. Section 4 presents the organization of MML. Next, we present the current MIZAR development in Sect. 5 and conclude with a vision of the project for the near future in Sect. 6.

2 Language

The MIZAR language encompasses both the grammar constructions that make use of a standardized list of reserved words and also the notation introduced by the authors of formalizations to encode concepts and notions.

In total, the current version of the language comprises 112 reserved words and 29 special symbols. However, the language is open in the sense that authors are allowed to extend it by introducing their own new symbols for the notions they define. Currently there are 8239 symbols used in the MIZAR Mathematical Library, including 722 predicate symbols, 1771 attribute symbols, 854 mode symbols, 4501 functor symbols, 36 left and right bracket symbols, 159 selector symbols, and 160 structure symbols[9]. In the MIZAR terminology, predicates are constructors of formulas, attributes are constructors of adjectives, modes are constructors of types, functors are constructors of terms. Selector and structure symbols are used to declare structural types and their fields. Symbol overloading is highly used in MIZAR to enable re-use of the symbols to denote different notions in much the same way it is done in pen-and-paper mathematics. For example the symbols * and + are used to denote 193 and 143 different operations in various fields of mathematics, respectively. The grammar of the MIZAR language[10] provides means to formulate mathematical statements in a way that resembles a formal exposition in the natural language. There are constructs to represent various kinds of formulas (quantifiers, standard logical connectives), reasoning methods (straightforward and diffused reasoning, proof by exhaustion) and all sorts of natural deduction proof steps (assumptions, conclusions, references, etc.).

However, there are numerous examples of constructs that mathematicians employ in their works to make the text more concise and less explicit in the number of trivial details. The so-called de Bruijn factor of the present formalizations, that measures the ratio between the length of the formal text and its informal counterpart, is still too high (it has been calculated as around 4 for typical examples and higher in the case of more complicated texts) [33]. Still, making the formal text as short as possible is not the ultimate goal, the readability of the formalization is equally important. Notable examples of informal constructs that are responsible for the discrepancy between informal and formal mathematics are the use of analogy, references to the reader's intuition, or various forms of ellipses. We will be seeking for new useful constructions extending

---

[9] Generated with MML Query, http://mmlquery.mizar.org.

[10] The description of the MIZAR grammar can be found at http://mizar.uwb.edu.pl/language/.

the formal language that would help to represent such linguistic structures. At this stage of the project, different sorts of ellipses, particularly those related to indexed variables, appear to be of high importance. Another open problem is how to introduce a convenient syntax for binding operators like integrals, sums, etc. The open, author-defined part of the MIZAR language poses a number of separate research questions. Although the authors are allowed to use different constructs supported by the language to express the same notions, the choice they make can be crucial. This concerns in particular the use of predicates, attributes, and modes that is most natural and closest to mathematical tradition. To some extent, they can be used interchangeably, and using each of them might present specific benefit. For example, the use of attributes offers a lot of simplification in reasoning. Modes allow to express predicative statements about objects that can be categorized in a natural way. On the other hand, predicates are most primitive and generic, and can be used to expand any notion of the three kinds. The current MIZAR language permits also the use of adjectives with visible arguments, e.g. `n-dimensional` (for an n-dimensional space), `X-defined` (for a function defined on some set X), `x-convergent` (for a sequence convergent to x) etc. In connection with the attribute clusters rounding-up automation [34], this enables a powerful Prolog-like computation mechanism.

3 Software

The heart of the MIZAR system is its proof checker, but it is rarely used today as a standalone program. Over the years, MIZAR has evolved into a complex proof assistant system [21] composed of many components that are used together to assist the user in various tasks related to formalizing mathematics. Apart from the core verification software, dedicated utilities support building the library and importing data from it for formalizations based on top of previous developments. A number of utilities distributed in the system's package help the users improve the quality of their formalizations by eliminating unnecessary or redundant parts. Several web-based services that allow to browse the available library as a semantically linked knowledge base, search its content with a dedicated query language, gather proof advice from automated theorem provers, or display the effects of the formalization in an automatically generated journal-style natural language representation. And although MIZAR texts can in principle be prepared with the aid of any ASCII text editor, several dedicated editors have been independently implemented, with MIZAR mode for Emacs being the most widely used and best integrated with the system's other elements [47]. Below we summarize the basic information about all these interlinked proof assistant components.

The MIZAR verifier consists of several modules responsible for checking various aspects of the correctness of MIZAR articles in a compiler-like manner. They are: SCANNER, PARSER, ANALYZER, REASONER, CHECKER, SCHEMATIZER.

SCANNER reads the source file and slices it into tokens. PARSER checks the syntactic correctness with respect to the grammar of the stream of tokens given by SCANNER and produces the abstract representation of the article in the form of stacked blocks and items, to be used by ANALYZER which identifies constructors and notations used on the basis of the type information imported from the environment. REASONER is responsible for checking if a proof tactic used by the author corresponds to the formula being proved. The checking is based on the internal representation of formulas in a simplified "canonical" form – their *semantic correlates*. CHECKER works as a classical disprover, additionally taking into account the type information associated with all terms, the properties of the employed constructors, equality calculus, etc. CHECKER also uses special built-in automation procedures for processing selected objects like e.g. complex numbers (direct computation) or boolean operations on sets. SCHEMATIZER processes *schemes* – statements that go beyond first-order logic, using free second-order variables to form infinite families of theorems, e.g. the scheme of mathematical induction.

Before CHECKER can start its work the text should be run through ACCOMMODATOR which imports knowledge either from the MML or a locally created data base. After the verification, the contents of an article can be extracted by EXPORTER and possibly incorporated into the MML.

3.1 XML Layer

MML is one of the largest corpora of nontrivial computer-understandable mathematics. This makes it into a unique resource for all sorts of semantic experiments and assistance tools that go beyond shallow natural-language treatment of mathematical texts. Such tools include database-like semantic search as done by MMLQuery [12], full automated theorem proving (ATP) over MML as done by the MPTP export [45,49] and the MizAR system based on it [28,52], or subsumption search based on ATP indexing data structures as used in MoMM [48] and in MathWebSearch [26]. Semantic parsing is also very useful for various functions provided by the Emacs authoring environment for MIZAR (MizarMode) [47], and e.g. its linking with MML Query [13].

Since parsing the advanced human-like MIZAR language is notoriously hard [15], a natural solution taken in 2004 was to make a complete and well-described separation between the parsing and semantic analysis stages (PARSER and ANALYZER) on the one hand, and proof checking and all kinds of other (possibly external) utilities and tools on the other hand. This resulted in a large reimplementation of MIZAR described in [46]. MIZAR started to use XML natively as its internal format produced by the early parsing and analysis processing stages. This format has been gradually extended, and now it contains a very complete semantically disambiguated form of a MIZAR article, as well as a description of the presentation-level syntax that allows various HTML-based presentations combining semantic information and tools with deeply hyperlinked MIZAR texts. The whole MIZAR internal library (items reusable in other articles) is now distributed in this format, and complete articles are translated to the format just by running the MIZAR verifier. An additional suite of open-source XSL-based translators from the native MIZAR XML format to richer or more targeted formats such as MPTP and HTML is developed and maintained by the XSL4Mizar project.[11]

3.2   MPTP and MizAℝ

There are several goals of the MPTP – MIZAR Problems for Theorem Proving – project [45,49], translating the MML to ATP formats. In short, the cooperation of modern ATP systems with large libraries of formalized mathematics is good both for the formalization efforts, providing strong proof assistance, cross-verification, automated theory refactorings, etc., and also for the ATP research, providing a large number of testing problems, allowing research of automated optimization on various mathematical domains and dealing with large knowledge bases, etc. Such cooperation is also the best candidate for merging the deductive (e.g., ATP) and inductive (e.g., machine learning) methods of Artificial Intelligence, because mathematics is (by definition) the most deductively developed science, and once we have a sufficient amount of such data, inductive methods can be applied too and combined with the deductive methods in novel ways [29,50,54,55].

The MPTP translation starts with the native MIZAR XML layer, which is first by XSL programs transformed to the extended TPTP (MPTP) Prolog-like format, which adds to TPTP MIZAR-like dependent types with attributes and subtyping, second-order constructs such as Fraenkel terms, and supposition-style proofs [44] based on Jaśkowski's natural deduction. The Prolog utilities then process this format, producing TPTP problems and proofs in various ways, typically either for large-scale ATP experiments over the whole translated MML, or in a fast interactive way used by the MizAℝ (Automated Reasoning for MIZAR) system.

MizAℝ is an online "cloud-based" remote-solving system which integrates several automated reasoning, artificial intelligence, and presentation tools with MIZAR and its authoring environment. The service provides ATP assistance to MIZAR authors in finding and explaining proofs, and offers generation of MIZAR problems as challenges to ATP systems. The system can be used on MIZAR goals directly from MizarMode just by typing by; after a goal that needs to be solved. This triggers the fast MPTP processing on the server and its parallellized solving using a combination of AI and ATP systems that give the author a 40 % chance of proving a top-level MIZAR theorem without any interactive assistance [28]. Another common way how to work with the system is via its web interface[12], where articles can be uploaded, remotely verified, hyperlinked, explored and interactively used for ATP experiments. Such remote-processing functionality is already close to the ideas of formal wikis for MIZAR [4,51], whose proper merging with MizAℝ is one of our next goals.

---

[11] http://github.com/JUrban/xsl4mizar.
[12] http://mizar.cs.ualberta.ca/~mptp/MizAR.html.

3.3 MML Query

MML Query is based on semantic on-line tool for searching, browsing and presentation of the evolving MML content [10]. The tool offers functionality that outranks commonly used grep-based utilities that often fail because of homonyms and overloading of symbols (and formats) heavily used in the MML. MML Query can also be used to build monographs – the uniform ordered semantic presentation of a specified piece of a theory which may be spread over the MML. The MML Query system also provides a text transformation processor MMLQT (MML Query Templates or MML Query Transformation) which is able to interpret the MML Query language to create ordered queries, version queries, and metadata queries, and to make searching with MML Query somewhat easier (non-expert searching, rough queries).

3.4 Formalized Mathematics Preview

Automatically generated natural language renderings of MIZAR articles can be previewed using a dedicated on-line service[13], which is also used for proof-reading papers in the *Formalized Mathematics* journal, see Sect. 4.3. The translation process [9] might be considered as a rewriting system, where the original MIZAR text is first reduced into an abstract form, which is later augmented with the information coming from the semantic analysis, and then the pattern translation of formulas and formats follows. The translation works on the basis of general and specific patterns. Several alternative patterns might be used for a given translation object. The current implementation of the translation system supports theorems, definitions, schemes, reservations and skeletal proof steps with references to selected most important facts. Finally, some metadata (a user provided summary in English and division into named sections) are incorporated to produce a form that resembles a standard journal paper.

Although the process of generating the journal papers is mechanized, the final result depends on the author or editor. The authors are encouraged to use the previewing facility and suggest any changes to the way their new notions are automatically translated using default patterns.

4 Mizar Mathematical Library

MML is a repository of articles covering various branches of mathematics and computer science. As of now, it contains over 1200 articles, over 10 thousand definitions, and approximately 50 thousand theorems. This collection has been written by over 250 authors. The acquired repository of formalized mathematics is considered one of the largest databases of this type [60]. All articles have been verified by the MIZAR checker and contain mathematical notions systematically defined on top of common axiomatics based on the Tarski-Grothendieck set theory. TG is a non-conservative extension of Zermelo-Fraenkel set theory and is

---

[13] http://fm.uwb.edu.pl/proof-read/.

distinguished from other axiomatic set theories by the inclusion of Tarski's axiom which states that for each set there is a Grothendieck universe it belongs to. Tarski's axiom implies the existence of strongly inaccessible cardinals, providing a richer ontology than that of conventional set theories such as ZFC.

4.1 Notable Formalizations

Through the years, when the MIZAR project evolved, the development of the repository of MIZAR texts (including the MIZAR language itself) was stimulated by large formalization projects. Among them, the most notable one was the formalization of *Compendium of Continuous Lattices* by Gierz et al. (mentioned in the second edition of the book issued under the title *Continuous Lattices and Domains*) in the years 1995–2003. This collective work of over a dozen of MIZAR authors resulted in 36 articles from the WAYBEL series reflecting faithfully the content of the book and 22 articles in the YELLOW series bridging the gap between the existing and desired state of the MML [11].

Another example of long-lasting cooperative work of a bigger group of authors was the formalization of the proof of the Jordan Curve Theorem continued from the very beginnings of MML until its successful finale – Artur Korniłowicz's "Jordan Curve Theorem" [30]. This may be considered as a part of a more general project: a formal encoding of general topology – also influential throughout the years. Among recently growing parts of mathematics represented in the MML we can list also functional analysis, lattice theory, and group theory.

The challenge which is stimulating not only for the MIZAR system, but also for other proof-assistants is the "Top 100 mathematical theorems" – the collection of important or interesting facts proposed at the edge of centuries by Paul and Jack Abad as "The Hundred Greatest Theorems". On the page maintained by Freek Wiedijk http://www.cs.ru.nl/F.Wiedijk/100/ one can find systems of computer formalization of mathematics ordered by the number of the items from that list which have been proven in these systems' libraries, covering 91 % of items altogether. Currently, among nine systems listed on the Wiedijk's page, the MIZAR system comes in second place with the total number of 62 items formalized.

4.2 MML Structure

The articles composing the library can be roughly divided into five parts. Although the parts are not formally separate, each of them requires slightly different management procedures:

- the axiomatics – currently containing three files with MML identifiers HIDDEN (introducing primitive notions: the root type object, membership relation in), TARSKI_0 and TARSKI_A – basically axioms of Tarski-Grothendieck set theory (actually Tarski's axiom A is the only exportable item in TARSKI_A);
- classical part, currently 323 items, not using the notion of a structure – pure set-theoretic part;

– structural part – all the other articles; this part deals with the notion of
a structure, e.g. algebraic structures such as groups, fields, vector spaces,
lattices, etc.;

– Encyclopedia of Mathematics in MIZAR (EMM) – currently 14 files with MML
identifiers starting with X; a collection of monographs;

– the formal model of random access Turing machines, started by Andrzej Trybulec and Yatsuka Nakamura in 1992.

The division into MML's classical and structural parts is an ongoing process as
some "classical" items are still being formalized. The process of such changes of
the library, called library revisions [23], is coordinated by the Library Committee
of the Association of MIZAR Users [5].

4.3   Formalized Mathematics

Although the MIZAR language is developed to be as close as possible to the language used in mathematical papers [22], it is still an artificial language, limited
in scope by a preset list of reserved words and allowed grammar constructions.
For a more complete popularization of formalized results, it is beneficial to make
the content of the repository accessible also in the form of conversational (collo-
quial or erudite) English, which would enable access to the base by persons not
familiar with the MIZAR language. An example of such accessibility is generating
articles for the journal *Formalized Mathematics* (ISSN 1426–2630, established in
1990) from the formalized articles contained in the MML [9]. Every submission
to MML is first reviewed in a standard journal manner by at least two (usually three) independent specialists; the reviews are on the double-blind basis.
MIZAR articles accepted to the MIZAR library are then automatically translated
into more human-readable LATEX format and published in *Formalized Mathematics*. The journal is published quarterly – with thirty as the approximate number
of MIZAR articles per volume.

5   Current Developments

All of the project's components undergo implementation and design changes
directed towards creating a better proof assistant environment. Most impor-
tantly, the verification system is being made stronger, the language more user-
friendly, the library better-organized and presentation methods more semantically
oriented.

5.1   Stronger Checker

The principal method of verification of informal mathematical papers is peer
review. The reviewers, who work in the same field, are capable of understanding
the mathematical text even if parts, deemed by the author obvious, are omitted,
or the author refers to an analogy to other examples of reasoning. The reviewers are also willing to accept minor errors (e.g., typographical) or imprecisions

stemming from the impossibility of attaining a coherent presentation of many notions and facts scattered throughout the vast literature. Nevertheless, from a computer system perspective, whose task is to semantically represent a given mathematical text, the above mentioned imprecisions are not acceptable.

In MIZAR automation is used to fill the gaps in the user provided declarative proofs, and its main role is to justify proof steps considered trivial by the human being. The current version of the checker provides several mechanisms that increase automation: *reductions* that reduce terms to their proper subterms [31]; *identifications* that identify notions defined within different theories; *properties of functors* that generate particular equalities representing chosen properties of terms (e.g. involutiveness, projectivity, commutativity, idempotence); *properties of predicates* that generate particular formulas representing chosen properties of relations (e.g. reflexivity, irreflexivity, symmetry, asymmetry, connectedness) [37]; and *definitional and functional expansions*. Moreover, there are attempts interfacing external dedicated computational systems (computer algebra systems, solvers, automated theorem provers) to strengthen the MIZAR notion of obviousness [2, 35, 36].

5.2 Improving Language Readability

The readability of MIZAR proof scripts is considered one of the most important factors of the formalization quality, but in practice enlargement of the database is rather orthogonal to the improvement of the formalization quality. Considering the current size of the library, manual editorial work on improving its legibility becomes infeasible. Therefore several aspects of proof legibility have been identified that can be approached in an automated fashion. Examples of such tasks include finding and removing inessential reasoning fragments or redundant premises in the justification of proof steps, analyzing the order of proof steps and reorganizing proof scripts in the MML according to a consistent style [40], or extracting fragments of reasoning in the form of lemmas or encapsulated nested proofs [24, 39].

5.3 Library Reorganization

Initially, the MML development was mainly geared towards volume parameters. Of prime importance was the mathematical result and the growth of the collection of the formalized theorems and proofs. First of all, attention was paid to the local quality of formalization, not to preserving the integrity of the knowledge in the whole base [41]. This approach permitted the accumulation of knowledge, but it did not guarantee taking full advantage of its amount. Currently, the MML development focuses on deciding the structure of the repository in such a way as to enable natural expansions while continuing easy access to the entire accumulated knowledge. The basic problems encountered while managing the development of this large repository are related to the preservation of the integrity of the information it contains [41]. For example:

- independently introducing by different authors different (sometimes incompatible) notations to denote the same (semantically equivalent) notions;
- independently developing the same theory by means of different notion apparatus;
- thematic dispersion of related knowledge in various sections of the repository.

Methods of finding out this type of situations in the MIZAR library are being worked on [39]. Integrity criteria and dedicated algorithms are implemented to assist error detection and the process of refactoring the database [23]. The need for database refactoring complies with the principles of database creation, where duplication and redundancy of information is avoided. At the initial stages of the MML creation, the focus was mainly on collecting as much formalized knowledge as possible to test various aspects of the system. Processing diverse data involved various modules of the system, which was crucial for determining directions of its further development by pointing out its stronger and weaker features. It also allowed to accumulate a considerable amount of formal knowledge. With the present size of the database, managing the library and also its applicability for users, especially new ones, requires developing and adopting a new approach.

In particular, methods to identify notations independently introduced by different authors that denote semantically equivalent notions are investigated. There are known cases of such definitions in the current library. For example, the notion of the exponentiation operation for numbers is denoted in separate formalizations as 'power(x, n)', 'x to_power n', 'x | ^ n'. In principle, the authors should be allowed to use the notation they prefer. However, this is a typical source of confusion, duplication, redundancy and an obstacle to efficient search for applicable facts in the library for other authors. Exploring more such cases requires statistical analysis as well as semantic matching of information. The considerations are based on the analysis of definitions in the simplest cases, but also on the analysis of the usage of selected notions in common contexts.

MIZAR developers have also detected cases where the same theories were independently developed by means of a different notion apparatus and have found ways for the best utilization of the independently developed results. For example, in the current library the group is a triple structure with its carrier, a binary operation and a pre-selected element serving as the group's unity. On the other hand, there is also the corresponding theory based on the ordered pair structure (the carrier and an operation) and the group's unity definable by means of an attached extra axiom expressed as an adjective ('unital'). A more complicated case is e.g. the development of lattice theory in terms of ordered sets on the one hand, and as an algebraic theory with two lattice operations on the other hand. For resolving such cases we can consider approaches based on selecting one from the concurrent developments and applying it to eliminate the rest, but also, as in the latter example, with finding ways to provide interoperability of different methods by identifying and encapsulating core components of all developments.

The development of the MML knowledge base has been incremental in its nature. Over two hundred authors who have contributed to its current content represent different backgrounds and skill levels (from students to university professors). As a result, the library suffers from thematic dispersion of related knowledge in various sections of the repository. For instance, numerous facts concerning simple set theory have been developed while proving some properties of digital circuits in a series of articles loosely connected with basic Boolean properties of sets. Responsible for that is partly the size of the library which makes it difficult for a researcher to grasp it as a whole, but also the authors' tendency or preference for specific approaches to developing mathematics. We investigate new methods based on knowledge exploration that can alleviate the problems. The research is directed towards more efficient, semantics based methods of searching for the information contained in the knowledge base. The main goal is to find ways how to unify (and generalize if needed) all relevant facts once a case of such dispersed knowledge is found in the library. This can be achieved by creating new specialized articles in selected fields, to enrich the repository and to test new language constructions and system properties, and also to define new directions of the development of the base.

5.4   More Semantic Representations

For the needs of the many forms of presenting the contents of the MIZAR library, used to popularize formalized knowledge within the mathematical community and on the Internet [22], translation rules that concern the improvement of the quality of article presentation, are being developed. Current research includes: the development of the system of transformation rules for the translation process using the XML/XSLT technology which will result in the design of a more flexible and easier modifiable software tool chain; forming a richer base of translation patterns including new categories for subjected phrases, patterns of mathematical formulae for new constructions, and variations of translation patterns dependent on the mathematical context; working out methods of presentation that take full advantage of the semantically linked information contained in MIZAR articles; an improvement of translation of proofs by extracting the references from proofs and a shallow translation of the proof (with the extraction applied for subproofs); the automatic generation of preliminaries for an article and each of its sections based on statistic analysis of the notation and terminology used and the theorems referred to and the subject classification automatically developed for the MML.

Currently, the MIZAR language and logic is mainly oriented at human users. The large number of human-friendly linguistic and logical features makes it unsuitable as a direct input for today's automated theorem systems, which work in relatively simple formalisms such as untyped first-order logic, and use simple Prolog-style languages such as the TPTP standard. Suitable layers and interfaces for correct bi-directional communication with such automated systems are being worked on, in particular we can mention the work with the TPTP format and its MIZAR-oriented MPTP extension [49]. Since 2005 MIZAR has been using an XML-based semantic internal layer, and this layer has been gradually enhanced to serve also a number of external applications. Objects on this layer are fully semantically disambiguated, i.e., there is no use of overloading, all term and formula constructors are linked to their definitions, and full types of terms are computed. This layer is already used for exporting the MIZAR formulas to ATP systems, but it is machine-oriented and so far cannot be used for importing the ATP proofs and for writing human-readable texts. Another issue is that this layer so far does not contain complete information about how proofs were done in MIZAR. This makes it difficult to replay the MIZAR proofs in other systems, and also to learn from such proofs. Many MIZAR mechanisms, such as the use of registrations, identities, requirements, and sometimes also definitions, are implicit, and they become explicit only during the process of verification.

The MIZAR developers have started research focused on producing a version of a "strict" semantic MIZAR layer [14], where no variables are reserved, all implicit mechanisms (registrations, identities, definitional expansions, etc.) used in the proofs can be made explicit before each proof (or even formula), and overloaded symbols are replaced by their unique synonyms. Such unique synonyms will be introduced either automatically, analogously to the current semantic names, or by suitable syntactic conventions in the MIZAR language.

This should allow at least an initial import of the ATP proofs and their verification in MIZAR, which is currently not possible, because such proofs may merge very different parts of the MML. Such different parts of the library are currently only mutually consistent on the semantic level, but it is a very nontrivial task (similar, e.g., to merging the notation of two different mathematical theories) to combine such parts also with respect to the overloaded notational mechanisms. Such a layer should in turn allow to construct a chain of MIZAR presentation improving utilities (similar to those that already exist for maintaining the MML) that will work on the verified proofs in this layer, and try to make the proofs more human-readable and mathematical by introducing common (possibly overloaded) notation, common names for variables (using reservations), common type mechanisms (such as registrations) for multiple proofs, etc.

Such an approach seems useful not just for importing the semantically encoded proofs produced by ATP systems, but also as a method for automatic merging of different parts of the MML. This is a common task that naturally arises when maintaining a large mathematical library like the MML, and which currently requires a lot of human effort, again because of clashing notational conventions. Such merged developments may first be exported into the "strict" semantic layer and verified there for correctness. After that, the presentation-improving utilities can attempt to automatically construct a common human-friendly notational layer for such merged articles.

Apart from importing and merging the semantically encoded mathematical parts produced by ATP systems, another application of such a layer is in splitting articles and producing a small independent article for each MIZAR theorem and definition. This is currently difficult to do automatically, due to mechanisms like reservations, etc. Having a small separate article for each MIZAR item again means that such small articles can be subjected to the number of existing MIZAR utilities, in particular those that detect the minimal set of (both notational and semantic) dependencies of an article [1,6]. Detecting such a minimal set is useful for various applications, ranging from training of premise-selection tools for large theory ATP systems, to experimental reverse mathematics assisted by ATP systems and automatically producing the strongest possible version of the theorems in the MML.

6 Future Mizar

Based on the successful long-term development of the MIZAR project, we are encouraged to believe that the project will eventually evolve into a widely-used computerized environment which could make the accumulated formalized mathematical knowledge accessible to a broad spectrum of users and in the future become a modern encyclopedia of mathematics.

For the realization of this long-term goal, it is imperative that first an effective information system for mathematics is formed, bridging the existing knowledge with computer capabilities of processing and searching for information. The fundamental element of this system is a language to represent mathematics in a computerized form. The specifics of the project is to define this language in such a way as to fulfill the above function. It is essential for this language to allow a uniform style in which mathematics will be done, at the same time not restricting the freedom of terminology usage and diverse methods of formalization. Furthermore, the formalization language should be close to the natural language, which would allow additional control of correctness of formalized texts, in particular the definitions of notions.

The key consideration will be defining criteria of readability of mathematical texts and proofs in a formalized form enabling the development of the base of mathematical knowledge, its accessibility and processing at various levels by a possibly wide group of users. To illustrate the readability of developments carried out with the use of the most popular state-of-the-art systems we can look e.g. at the statement of the Fundamental Theorem of Algebra and compare it to its Wikipedia entry: *Every non-constant single-variable polynomial with complex coefficients has at least one complex root.*

Coq:
```
forall f : CCX, nonConst _ f -> {z : CC | f ! z [=] Zero}.
```
HOL Light:
```
|- !a n. a(0) = Cx(&0) \/ ~(!k. k IN 1..n ==> a(k) = Cx(&0))
    ==> ?z. vsum(0..n) (\i. a(i) * z pow i) = Cx(&0)
```
Isabelle:
```
~constant(poly p) ==> z::complex. poly p z = 0
```
MIZAR:
```
for p being Polynomial of F_Complex st len p > 1 holds
    p is with_roots;
```

From the above samples it can be seen that, no matter which system we consider, there is still a significant difference in the readability of the formal and informal (natural language) representation. Improving the readability of formalized texts would allow better communication with the mathematical community and their greater engagement in the project. The participation of active mathematicians is particularly important for validating and standardizing the definitions of notions deposited in the base [43]. Involving more working mathematicians, who would be able to share their firsthand experience with using the language of mathematical publications on a daily basis, would result in the development and accessibility of a better language and system to formalize mathematics, and several forms of access to a wider audience of mathematical knowledge collected in the MIZAR Mathematical Library [18,19,33]. The accomplishment would be for diverse fields of science and education to benefit from such computer verified knowledge. The pre-processed database will also be used for research aimed at developing automated theorem proving systems (provers).

The ultimate, long-term goal, towards which the work on MIZAR is directed, is to construct a modern encyclopedia of mathematics. We believe that the MIZAR project is well positioned to start a new generation of encyclopedia. All major scientific encyclopedias are available in an electronic form and many, such as Wikipedia or Scholarpedia, solicit input from independent contributors, but the entered data is not verified. The information contained in the huge MIZAR Mathematical Library repository, verified, checked and cross-linked, can be used to build an encyclopedia, which is mathematical at first and later expanded to other sciences, an encyclopedia of entirely different merit, with exclusively formalized and verified data. As a source for citations of mathematical definitions and theorems, an MML based encyclopedia would be invaluable and unique for human users. On the other hand, the rich source of formal mathematical knowledge contained in the MML can be used to develop automated theorem proving methods and systems trained over the mathematics data, and to assist further development of mathematics over such large formal corpora [52,53]. Such automated methods can help with searching the large library, constructing new proofs automatically [20], finding alternative proofs [3], and help with re-structuring the proofs and theories.

References

1. Alama, J.: Mizar-items: exploring fine-grained dependencies in the Mizar mathematical library. In: Davenport, J.H., Farmer, W.M., Urban, J., Rabe, F. (eds.) MKM/Calculemus 2011. LNCS, vol. 6824, pp. 276–277. Springer, Heidelberg (2011). http://dx.doi.org/10.1007/978-3-642-22673-1_19

2. Alama, J.: Escape to Mizar from ATPs. In: Fontaine, P., Schmidt, R.A., Schulz, S. (eds.) Third Workshop on Practical Aspects of Automated Reasoning, PAAR-2012, Manchester, UK, 30 June–1 July 2012. EPiC Series, vol. 21, pp. 3–11. EasyChair (2012). http://www.easychair.org/publications/?page=1559779348

3. Alama, J.: Eliciting implicit assumptions of Mizar proofs by property omission. J. Autom. Reasoning 50(2), 123–133 (2013). http://dx.doi.org/10.1007/s10817-012-9264-3

4. Alama, J., Brink, K., Mamane, L., Urban, J.: Large formal wikis: issues and solutions. In: Davenport, J.H., Farmer, W.M., Urban, J., Rabe, F. (eds.) MKM/Calculemus 2011. LNCS, vol. 6824, pp. 133–148. Springer, Heidelberg (2011). http://dx.doi.org/10.1007/978-3-642-22673-1

5. Alama, J., Kohlhase, M., Mamane, L., Naumowicz, A., Rudnicki, P., Urban, J.: Licensing the Mizar mathematical library. In: Davenport, J.H., Farmer, W.M., Urban, J., Rabe, F. (eds.) MKM 2011/Calculemus 2011. LNCS, vol. 6824, pp. 149–163. Springer, Heidelberg (2011). http://dx.doi.org/10.1007/978-3-642-22673-1_11

6. Alama, J., Mamane, L., Urban, J.: Dependencies in formal mathematics: applications and extraction for Coq and Mizar. In: Campbell, J.A., Jeuring, J., Carette, J., Dos Reis, G., Sojka, P., Wenzel, M., Sorge, V. (eds.) CICM 2012. LNCS, vol. 7362, pp. 1–16. Springer, Heidelberg (2012). http://dx.doi.org/10.1007/978-3-642-31374-5_1

7. Anonymous: the QED manifesto. Bundy, A. (ed.) CADE 1994. LNCS, vol. 814. Springer, Heidelberg (1994)

8. Strotmann, A.: The categorial type of OpenMath objects. In: Asperti, A., Bancerek, G., Trybulec, A. (eds.) MKM 2004. LNCS, vol. 3119, pp. 378–392. Springer, Heidelberg (2004)

9. Bancerek, G.: Automatic translation in formalized mathematics. Mech. Math. Appl. 5(2), 19–31 (2006)

10. Bancerek, G.: Information retrieval and rendering with MML query. In: Borwein, J.M., Farmer, W.M. (eds.) MKM 2006. LNCS (LNAI), vol. 4108, pp. 266–279. Springer, Heidelberg (2006). http://dx.doi.org/10.1007/11812289_21

11. Bancerek, G., Rudnicki, P.: A compendium of continuous lattices in Mizar: formalizing recent mathematics. J. Autom. Reason. 29(3–4), 189–224 (2002)

12. Bancerek, G., Rudnicki, P.: Information retrieval in MML. In: Asperti, A., Buchberger, B., Davenport, J.H. (eds.) MKM 2003. LNCS, vol. 2594, pp. 119–132. Springer, Heidelberg (2003)

13. Bancerek, G., Urban, J.: Integrated semantic browsing of the Mizar mathematical library for authoring Mizar articles. In: Asperti, A., Bancerek, G., Trybulec, A. (eds.) MKM 2004. LNCS, vol. 3119, pp. 44–57. Springer, Heidelberg (2004)

14. Bylinski, C., Alama, J.: New developments in parsing Mizar. In: Campbell, J.A., Jeuring, J., Carette, J., Dos Reis, G., Sojka, P., Wenzel, M., Sorge, V. (eds.) CICM 2012. LNCS (LNAI), vol. 7362, pp. 427–431. Springer, Heidelberg (2012)

15. Cairns, P.: Informalising formal mathematics: searching the Mizar library with latent semantics. In: Asperti, A., Bancerek, G., Trybulec, A. (eds.) MKM 2004. LNCS, vol. 3119, pp. 58–72. Springer, Heidelberg (2004)

16. Corbineau, P.: A declarative language for the Coq proof assistant. In: Miculan, M., Scagnetto, I., Honsell, F. (eds.) TYPES 2007. LNCS, vol. 4941, pp. 69–84. Springer, Heidelberg (2008). http://dx.doi.org/10.1007/978-3-540-68103-8_5

17. Botana, F.: A symbolic companion for interactive geometric systems. In: Davenport, J.H., Farmer, W.M., Urban, J., Rabe, F. (eds.) MKM 2011 and Calculemus 2011. LNCS, vol. 6824, pp. 285–286. Springer, Heidelberg (2011)

18. Futa, Y., Okazaki, H., Shidama, Y.: Formalization of definitions and theorems related to an elliptic curve over a finite prime field by using Mizar. J. Autom. Reason. 50(2), 161–172 (2013). http://dx.doi.org/10.1007/s10817-012-9265-2

19. Gow, J., Cairns, P.: Closing the gap between formal and digital libraries of mathematics. In: Matuszewski, R., Zalewska, A. (eds.) From Insight to Proof: Festschrift in Honour of Andrzej Trybulec. Studies in Logic, Grammar and Rhetoric, University of Białystok, vol. 10(23), pp. 249–263 (2007). http://mizar.org/trybulec65/

20. Grabowski, A.: Efficient rough set theory merging. Fundamenta Informaticae 135(4), 371–385 (2014). http://dx.doi.org/10.3233/FI-2014-1129

21. Grabowski, A., Korniłowicz, A., Naumowicz, A.: Mizar in a nutshell. J. Formaliz. Reason. Spec. Issue: User Tutor. I 3(2), 153–245 (2010)

22. Grabowski, A., Schwarzweller, C.: Translating mathematical vernacular into knowledge repositories. In: Kohlhase, M. (ed.) MKM 2005. LNCS (LNAI), vol. 3863, pp. 49–64. Springer, Heidelberg (2006). http://dx.doi.org/10.1007/11618027_4

23. Grabowski, A., Schwarzweller, C.: Revisions as an essential tool to maintain mathematical repositories. In: Kauers, M., Kerber, M., Miner, R., Windsteiger, W. (eds.) MKM/CALCULEMUS 2007. LNCS (LNAI), vol. 4573, pp. 235–249. Springer, Heidelberg (2007). http://dx.doi.org/10.1007/978-3-540-73086-6_20

24. Grabowski, A., Schwarzweller, C.: Towards automatically categorizing mathematical knowledge. In: Ganzha, M., Maciaszek, L.A., Paprzycki, M. (eds.) Proceedings of Federated Conference on Computer Science and Information Systems - FedCSIS 2012, Wroclaw, Poland, 9–12 September 2012, pp. 63–68 (2012)

25. Harrison, J.: A Mizar mode for HOL. In: von Wright, J., Harrison, J., Grundy, J. (eds.) TPHOLs 1996. LNCS, vol. 1125. Springer, Heidelberg (1996). http://dl.acm.org/citation.cfm?id=646523.694700

26. Iancu, M., Kohlhase, M., Rabe, F., Urban, J.: The Mizar mathematical library in OMDoc: translation and applications. J. Autom. Reason. 50(2), 191–202 (2013). http://dx.doi.org/10.1007/s10817-012-9271-4

27. Jaśkowski, S.: On the Rules of Suppositions in Formal Logic. Studia Logica, Nakładem Seminarjum Filozoficznego Wydziału Matematyczno-Przyrodniczego Uniwersytetu Warszawskiego (1934). http://books.google.pl/books?id=6w0vRAAACAAJ

28. Kaliszyk, C., Urban, J.: MizAR 40 for Mizar 40 (2013). CoRR abs/1310.2805

29. Kaliszyk, C., Urban, J., Vyskočil, J.: Machine learner for automated reasoning 0.4 and 0.5 (2014). Accepted to PAAR 2014, CoRR abs/1402.2359

30. Korniłowicz, A.: Jordan curve theorem. Formaliz. Math. 13(4), 481–491 (2005)

31. Korniłowicz, A.: On rewriting rules in Mizar. J. Autom. Reason. 50(2), 203–210 (2013). http://dx.doi.org/10.1007/s10817-012-9261-6

32. Matuszewski, R., Rudnicki, P.: Mizar: the first 30 years. In: Mechanized Mathematicsand Its Applications, Special Issue on 30 Years of Mizar, vol. 4, no. 1, pp. 3–24 (2005)

33. Naumowicz, A.: An example of formalizing recent mathematical results in Mizar. J. Appl. Logic 4(4), 396–413 (2006). http://www.sciencedirect.com/science/article/pii/S1570868305000686

34. Naumowicz, A.: Enhanced processing of adjectives in Mizar. In: Grabowski, A., Naumowicz, A. (eds.) Computer Reconstruction of the Body of Mathematics. Studies in Logic, Grammar and Rhetoric, University of Białystok, vol. 18, no. 31, pp. 89–101 (2009)

35. Naumowicz, A.: Interfacing external CA systems for Grobner bases computation in Mizar proof checking. Int. J. Comput. Math. 87(1), 1–11 (2010). http://dx.doi.org/10.1080/00207160701864459

36. Naumowicz, A.: SAT-enhanced MIZAR proof checking. In: Watt, S.M., Davenport, J.H., Sexton, A.P., Sojka, P., Urban, J. (eds.) CICM 2014. LNCS, vol. 8543, pp. 449–452. Springer, Heidelberg (2014). http://dx.doi.org/10.1007/978-3-319-08434-3_37

37. Naumowicz, A., Byliński, C.: Improving MIZAR texts with *properties* and *requirements*. In: Asperti, A., Bancerek, G., Trybulec, A. (eds.) MKM 2004. LNCS, vol. 3119, pp. 290–301. Springer, Heidelberg (2004). http://dx.doi.org/10.1007/978-3-540-27818-4_21

38. Naumowicz, A., Korniłowicz, A.: A brief overview of MIZAR. In: Berghofer, S., Nipkow, T., Urban, C., Wenzel, M. (eds.) TPHOLs 2009. LNCS, vol. 5674, pp. 67–72. Springer, Heidelberg (2009). http://dx.doi.org/10.1007/978-3-642-03359-9_5

39. Pąk, K.: Methods of lemma extraction in natural deduction proofs. J. Autom. Reason. **50**(2), 217–228 (2013). http://dx.doi.org/10.1007/s10817-012-9267-0

40. Pąk, K.: Improving legibility of natural deduction proofs is not trivial. Logic. Methods Comput. Sci. **10**(3), 1–30 (2014). http://dx.doi.org/10.2168/LMCS-10(3:23)2014

41. Rudnicki, P., Trybulec, A.: On the integrity of a repository of formal mathematics. In: Asperti, A., Buchberger, B., Davenport, J.H. (eds.) MKM 2003. LNCS, vol. 2594, pp. 162–174. Springer, Heidelberg (2003)

42. Syme, D.: DECLARE: a prototype declarative proof system for higher order logic. Technical report, University of Cambridge (1997)

43. Trybulec, A., Korniłowicz, A., Naumowicz, A., Kuperberg, K.: Formal mathematics for mathematicians. J. Autom. Reason. **50**(2), 119–121 (2013). http://dx.doi.org/10.1007/s10817-012-9268-z

44. Urban, J., Sutcliffe, G., Trac, S., Puzis, Y.: Combining Mizar and TPTP semantic presentation and verification tools. Stud. Logic Gramm. Rhetor. **18**(31), 121–136 (2009)

45. Urban, J.: MPTP - motivation, implementation, first experiments. J. Autom. Reason. **33**(3–4), 319–339 (2004)

46. Urban, J.: XML-izing Mizar: making semantic processing and presentation of MML easy. In: Kohlhase, M. (ed.) MKM 2005. LNCS (LNAI), vol. 3863, pp. 346–360. Springer, Heidelberg (2006)

47. Urban, J.: MizarMode – an integrated proof assistance tool for the Mizar way of formalizing mathematics. J. Appl. Logic **4**(4), 414–427 (2006). http://dx.doi.org/10.1016/j.jal.2005.10.004

48. Urban, J.: MoMM – fast interreduction and retrieval in large libraries of formalized mathematics. Int. J. Artif. Intell. Tools **15**(1), 109–130 (2006). http://ktiml.mff.cuni.cz/urban/MoMM/momm.ps

49. Urban, J.: MPTP 0.2: design, implementation, and initial experiments. J. Autom. Reason. **37**(1–2), 21–43 (2006). http://dx.doi.org/10.1007/s10817-006-9032-3

50. Urban, J.: BliStr: The Blind Strategymaker (2014). Accepted to PAAR 2014, CoRR abs/1301.2683

51. Urban, J., Alama, J., Rudnicki, P., Geuvers, H.: A wiki for Mizar: motivation, considerations, and initial prototype. In: Autexier, S., Calmet, J., Delahaye, D., Ion, P.D.F., Rideau, L., Rioboo, R., Sexton, A.P. (eds.) AISC 2010. LNCS, vol. 6167, pp. 455–469. Springer, Heidelberg (2010)

52. Urban, J., Rudnicki, P., Sutcliffe, G.: ATP and presentation service for Mizar formalizations. J. Autom. Reason. **50**(2), 229–241 (2013). http://dx.doi.org/10.1007/s10817-012-9269-y

53. Urban, J., Sutcliffe, G.: ATP-based cross-verification of Mizar proofs: method, systems, and first experiments. Math. Comput. Sci. **2**(2), 231–251 (2008). http://dx.doi.org/10.1007/s11786-008-0053-7

54. Urban, J., Sutcliffe, G., Pudlák, P., Vyskočil, J.: MaLARea SG1 - machine learner for automated reasoning with semantic guidance. In: Armando, A., Baumgartner, P., Dowek, G. (eds.) IJCAR 2008. LNCS (LNAI), vol. 5195, pp. 441–456. Springer, Heidelberg (2008)

55. Urban, J., Vyskočil, J., Štěpánek, P.: MaLeCoP machine learning connection prover. In: Brünnler, K., Metcalfe, G. (eds.) TABLEAUX 2011. LNCS, vol. 6793, pp. 263–277. Springer, Heidelberg (2011)

56. Wenzel, M., Wiedijk, F.: A comparison of Mizar and Isar. J. Autom. Reason. **29**(3–4), 389–411 (2003). http://dx.doi.org/10.1023/A:1021935419355

57. Wiedijk, F.: Mizar light for HOL light. In: Boulton, R.J., Jackson, P.B. (eds.) TPHOLs 2001. LNCS, vol. 2152, pp. 378–394. Springer, Heidelberg (2001)
58. Wiedijk, F.: Formal proof sketches. In: Berardi, S., Coppo, M., Damiani, F. (eds.) TYPES 2003. LNCS, vol. 3085, pp. 378–393. Springer, Heidelberg (2004)
59. Gamboa, R.: ACL2. In: Wiedijk, F. (ed.) The Seventeen Provers of the World. LNCS (LNAI), vol. 3600, pp. 55–66. Springer, Heidelberg (2006)
60. Wiedijk, F.: Formal proof–getting started. Not. Am. Math. Soc. **55**(11), 1408–1414 (2008)
61. Wiedijk, F.: A synthesis of the procedural and declarative styles of interactive theorem proving. Logic. Methods Comput. Sci. **8**(1:30), 1–26 (2012)

Growing the Digital Repository of Mathematical Formulae with Generic LaTeX Sources

Howard S. Cohl[1]([⊠]), Moritz Schubotz[2], Marjorie A. McClain[1],
Bonita V. Saunders[1], Cherry Y. Zou[3], Azeem S. Mohammed[3],
and Alex A. Danoff[4]

[1] Applied and Computational Mathematics Division,
National Institute of Standards and Technology (NIST), Gaithersburg, MD, USA
{howard.cohl,marjorie.mcclain,bonita.saunders}@nist.gov
[2] Database Systems and Information Management Group,
Technische Universität, Berlin, Germany
schubotz@tu-berlin.de
[3] Poolesville High School, Poolesville, MD, USA
{chzou2009,azeemsm}@gmail.com
[4] Wootton High School, Rockville, MD, USA
aadanoff@gmail.com

Abstract. One initial goal for the DRMF is to seed our digital compendium with fundamental orthogonal polynomial formulae. We had used the data from the NIST Digital Library of Mathematical Functions (DLMF) as initial seed for our DRMF project. The DLMF input LaTeX source already contains some semantic information encoded using a highly customized set of semantic LaTeX macros. Those macros could be converted to content MathML using LaTeXML. During that conversion the semantics were translated to an implicit DLMF content dictionary. This year, we have developed a semantic enrichment process whose goal is to infer semantic information from generic LaTeX sources. The generated context-free semantic information is used to build DRMF formula home pages for each individual formula. We demonstrate this process using selected chapters from the book "Hypergeometric Orthogonal Polynomials and their q-Analogues" (2010) by Koekoek, Lesky and Swarttouw (KLS) as well as an actively maintained addendum to this book by Koornwinder (KLSadd). The generic input KLS and KLSadd LaTeX sources describe the printed representation of the formulae, but does not contain explicit semantic information. See http://drmf.wmflabs.org.

1 Introduction

Formula home pages are the principal conceptual objects for the DRMF project. These should contain the full context-free semantic information concerning individual orthogonal polynomial and special function (OPSF) formulae. The DRMF

© Springer International Publishing Switzerland 2015 (outside the US)
M. Kerber et al. (Eds.): CICM 2015, LNAI 9150, pp. 280–287, 2015.
DOI: 10.1007/978-3-319-20615-8_18

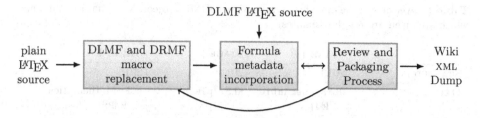

Fig. 1. Data flow of seeding projects. For most of the input LATEX source distributions, DLMF and DRMF macros are not incorporated. For the DLMF LATEX source, the DLMF macros are already incorporated.

is designed for a mathematically literate audience and should (1) facilitate interaction among a community of mathematicians and scientists interested in compendia formulae data for orthogonal polynomials and special functions; (2) be expandable, allowing the input of new formulae from the literature; (3) represent the context-free full semantic information concerning individual formulae; (4) have a user friendly, consistent, and hyperlinkable viewpoint and authoring perspective; (5) contain easily searchable mathematics; and (6) take advantage of modern MATHML tools for easy-to-read, scalably rendered content driven mathematics. In this paper we will discuss the DRMF seeding projects whose goal is to import data, for example, from traditional print media (cf. Fig. 1).

We are investigating various sources for seed material in the DRMF [3]. We have been given permission to use a variety of input resources to generate our online compendium of mathematical formulae. The current sources that we are incorporating into the DRMF are given as follows: (1) *NIST Digital Library of Mathematical Functions* (DLMF[1]) [1,6]; (2) Chaps. 1, 9, and 14 (a total of 228 pages with about 1800 formulae) from the Springer-Verlag book *"Hypergeometric Orthogonal Polynomials and their q-Analogues" (2010) by Koekoek, Lesky and Swarttouw* (KLS) [7]; (3) Tom Koornwinder's *Additions to the formula lists in "Hypergeometric orthogonal polynomials and their q-Analogues" by Koekoek, Lesky and Swarttouw* (KLSadd) [10]; (4) *Wolfram Computational Knowledge of Continued Fractions Project* (eCF); and the *Bateman Manuscript Project* (BMP) [4,5] (see Table 1). Note that the DLMF, KLS, KLSadd, and eCF datasets are currently being processed within our pipeline. For the BMP dataset, we have furnished high-quality print scans to Alan Sexton and are currently waiting on the math OCR generated LATEX output for this dataset which is currently being generated. In this paper we focus on DRMF seeding of generic LATEX sources, namely those which do not contain explicit semantic information.

2   Seeding with Generic LATEX Sources

DRMF seeding projects collect and stream OPSF mathematical formulae into formula pages. Formula pages are classified into those which list formulae in a

---

[1] We use the typewriter font in this document to refer to our seeding datasets.

Table 1. Overview of the first three stages of the DRMF project. Note that the numbers which are given are rough estimates.

	STAGE 1	STAGE 2	STAGE 3
STARTED IN	2013	2014	2015
DATASET	DLMF, semantic LaTeX	KLS, plain LaTeX	eCF: Mathematica BMP: book images
SEMANTIC ENRICHMENT	Identify constraints, substitutions, notes, names, proofs, ...	Add new semantic macros	Image recognition macro suggestion
TECHNOLOGIES	Manual review, rule-based approaches	Improved rules	Natural language processing and machine learning
NUMBER OF FORMULA HOME PAGES	500	1500	5000
HUMAN TIME PER FORMULA HOMEPAGE	10 min	5 min	1 min
TEST CORPORA CONTRIBUTION	Gold standard for constraint and proof detection	Gold standard for macro replacement	Evaluation metrics

broad category, and the individual formula home pages for each formula. Generated formula home pages are required to contain bibliographic information and usually contain a list of symbols, substitutions and constraints required by the formulae, proofs and formula names if available, as well as related notes. Every semantic formula entity (e.g., function, polynomial, sequence, operator, constant or set) has a unique name and a link to its definition or description.

For LaTeX sources which are extracted from the DLMF project, the semantic macros are already incorporated [11]. However, for generic sources such as the KLS dataset, the semantic macros need to be inserted in replacement for the LaTeX source which represents that mathematical object.

Here we give representative examples for the trigonometric sine function, gamma function, Jacobi polynomial and little q-Laguerre/Wall polynomials, which are rendered respectively as $\sin z$, $\Gamma(z)$, $P_n^{(\alpha,\beta)}(x)$, and $p_n(x;a|q)$. These functions and orthogonal polynomials have LaTeX presentations given respectively by \sin z, \Gamma(z), P_n^{(\alpha,\beta)}(x), and p_n(x;a|q). The semantic representations for these functions and orthogonal polynomials are given respectively by \sin@@{z}, \EulerGamma@{z}, \Jacobi{\alpha}{\beta}{n}@{x}, \littleqLaguerre{n}@{x}{a}{q}. The arguments before the @ or @@

symbols are parameters and the arguments after the @ or @@ symbol are in the domain of the functions and orthogonal polynomials. The different between the @ or @@ symbols indicates a specified difference in presentation, such as the inclusion of the parentheses or not in our trigonometric sine example. For the little q-Laguerre polynomials, one has three arguments within parentheses. These three arguments are separated by a semi-colon and a vertical bar. Our macro replacement algorithm indentifies these polynomials, and then extracts the information about what the contents of each argument is. Furthermore there are many ways in LATEX to represent open and close parenthese, our algorithm identifies these. Also, since the vertical bar in LATEX can be represented by '|' or '\mid', we search for both of these patterns. Our algorithm, for instance, also searches for and removes all LATEX white-space characters such as those given by \, \! or \hspace{}. There are many other details about making our search and replace work, which we will not mention here.

3   KLS Seeding Project

In this section we describe how we augment the input KLS LATEX source in order to generate formula pages (see Fig. 1). We are developing software processes input LATEX source to generate output LATEX source with semantic mathematical macros incorporated. The semantic LATEX macros that we are using (664 total with 147 currently being used for the DRMF project) are being developed by NIST for use in the DLMF and DRMF projects. Whenever possible, we use the standardized definitions from the NIST Digital Library of Mathematical Functions [6]. If the definitions are not available on the DLMF website, then we link to definition pages in the DRMF with included symbols lists. One main goal of this seeding project is to incorporate mathematical semantic information directly into the LATEX source. The advantage of incorporating this information directly into the LATEX source is that mathematicians are capable of editing LATEX whereas human editing of MATHML is not feasible. This enriched information can be further modified by mathematicians using their regular working environment.

For the 3 chapters of the KLS dataset plus the KLSadd dataset, a total number of 89 semantic macros were replaced a total of 3308 times. That's an average of 1.84 macros replaced per formula. Note that the KLSadd dataset is actively being maintained, and when a new version of it is published, in an automated fashion, incorporate this new information into the DRMF. This fraction will increase when more algebraic substitution formulae are included as formula metadata. The most common macro replacements are given as follows. The macro for the cosine function, Racah polynomial, Pochhammer symbol, q-hypergeometric function, Euler gamma function, and q-Pochhammer symbol were converted a total number of times equal to 117, 205, 237, 266, and 659. Our current conversions, which use a rule based approach, can be quite complicated due to the nature of the variety of combinations of LATEX input for various OPSF objects. In LATEX there are many ways of representing parentheses which are usually used for function arguments. Also, there are many ways to represent spacing delimiters which can mostly be ignored as far as representing the common semantic information for a mathematical function call. Our software canonicalizes these additional meaningless degrees of freedom and generates easy-to-read semantic LaTeX source and improves the rendering. Developing automatic software which performs macro replacements for OPSF functions in LaTeX is a challenging task. The current status of our rule-based approach is highly tailored to our specific KLS and KLSadd input LaTeX source.

Historically, the desired need for formal consistency has driven mathematicians to adopt consistent and unique notations [2]. This is extremely beneficial in the long run. We have interacted on a regular basis with the authors of the KLS and KLSadd datasets. They agree that our assumptions about consistent notations are correct and they consider using our semantic LaTeX macros in future volumes. Certainly the benefit of using these macros in communicating with different computer systems is clear.

Once semantic macros are incorporated, the next task is to identify formula metadata. Formula metadata can be identified within and must be associated with formulae. One must then identify semantic information for the formula within the surrounding text to produce formula annotations which describe this semantic information. There are annotations which can be summarized as constraints, substitutions, proofs and formula names if available, as well as related notes. The automated extraction of formula metadata is a challenging aspect of the seeding project and future computer implementations might use machine learning methods to achieve this goal. However, we have built automated algorithms to extract formula metadata. We have for instance identified substitutions by associating definitions for algebraic or OPSF functions which are utilized in surrounding formulae. The automation process continues by merging these substitution formulae as annotations in the original formulae which use them. Another extraction algorithm we have developed is the identification of related variables, understanding their dependencies and merging corresponding annotations with the pre-existing formula metadata. We have manually reviewed the printed mathematics to identify formula metadata. After we have exhausted our current rule-based approach for extracting the formula annotations, we will perform the manual insertion of the missing identified annotations into the LaTeX source. This will then be followed by careful checking and expert editorial review. This also evaluates the quality of our rule-based approach and creates a gold standard for future programs.

Once the formula metadata has been completely extracted from the text, then the remainder of the text should be removed and one is left with a list of LaTeX formulae with associated metadata. From this list (at the current stage of our project), we use this semantic LaTeX source to generate Wikitext. One of the features of the generated Wikitext is that we use a glossary that we have developed of our DLMF and DRMF macros to identify semantic macros within a formula and its associated metadata. Presentation and meaningful content MathML is generated from the DLMF and DRMF macros using a customized LaTeXML server (http://gw125.iu.xsede.org) hosted by the XSEDE project that includes all generated semantic macros. From this glossary, we generate symbols lists for each formula which uses recognized symbols. The generated Wikitext is converted to the MediaWiki XML-Dump format, which is then bulk imported to our wiki instance. Our DRMF Wiki has been optimized for MATHML-output. Because we are using Mathoid to render mathematical expressions [14], browsers without MATHML-support can display DRMF formulae within MediaWiki. However, some MATHML-related features (such as copying parts of the MATHML output) are not available on these browsers.

At the moment, There are 1282 KLS and KLSadd wikitext pages. The current number of KLS and KLSadd formula home pages is 1219 and the percentage of non-empty symbols lists in formula home pages is given by 98.6 percent. This number will increase as we continue to merge substitution formulae into associated metadata and as we continue to expand our macro replacement effort. We have detected 208 substitutions which originally appeared as formulae. We inserted these in an automated fashion into 515 formulae. The goal of our learning is to obtain a mostly unambiguous content representation of the mathematical OPSF formulae which we use.

4 Future Outlook

The next seeding projects which we will focus on are those which correspond to image and Mathematica inputs (see Table 1). We have been given permission from Caltech to use the BMP dataset within the DRMF. In the BMP dataset, the original source for data are printed pages of books. We are currently collaborating on the development of mathematical optical character recognition (OCR) software [15] for use in this project. We plan to utilize this math OCR software to generate LaTeX output which will be incorporated with the DLMF and DRMF semantic macros using our developed macro replacement software.

We are already developing for our next source, namely the incorporation of the Wolfram eCF dataset into the DRMF. We have been furnished the Mathematica source (also known as Wolfram language) for this dataset and we are currently developing software which translates in both directions from the Wolfram language to our semantic LaTeX source with DRMF and DLMF macros incorporated (cf. Table 1).

For the DLMF source, due to the hard efforts of the DLMF team for more than the past ten years, we already have semantic macros implemented, and all that remains is to extract the metadata from the text associated with formulae, removing the text after the content has been transferred, converting formulae information in tables to lists of distinct formulae, and generating formula home pages. We already have mostly achieved this for DLMF Chap. 25 on the Riemann Zeta function and are currently at work on Chaps. 5 (gamma function), 15 (hypergeometric function), 16 (generalized hypergeometric functions), 17 (q-hypergeometric and related functions) and 18 (orthogonal polynomials) which will ultimately be merged with the KLS and KLSadd datasets. Then we will continue to the remainder of the DLMF chapters.

Once semantic information has been inserted into the LaTeX source, there is a huge number of possibilities on how this information can be used. Given that our datasets are collections of OPSF formulae, we plan on taking advantage of the incorporated semantic information as an exploratory tool for symbolic and numerical experiments. For instance, one may use this semantic content to translate to computer algebra system (CAS) computer languages such as those used by Mathematica, Maple or Sage. One could then use the translated formulae while taking advantage of any of the features available in those software packages. We should also mention that the DRMF seeding projects generate real content MATHML. This has been a huge problem for Mathematics Information Retrieval research for many years [9,12]. One major contribution of the DRMF seeding projects is that they offer quite reasonable content MATHML.

From a methodological point of view, we are going to develop evaluation metrics that measure the degree of semantic formula enrichment. These should be able to evaluate new approaches such as mathematical language processing [13] and/or machine learning approaches based on the created gold standard. Additionally, we are considering the use of sTeX [8], in order to simplify the definition of new macros. Eventually, we can also develop a heuristic which suggests new semantic macros based on statistical analysis.

Acknowledgements. (The mention of specific products, trademarks, or brand names is for purposes of identification only. Such mention is not to be interpreted in any way as an endorsement or certification of such products or brands by the National Institute of Standards and Technology, nor does it imply that the products so identified are necessarily the best available for the purpose. All trademarks mentioned herein belong to their respective owners.) We are indebted to Wikimedia Labs, the XSEDE project, Springer-Verlag, the California Institute of Technology, and Wolfram Research Inc. for their contributions and continued support. We would also like to thank Roelof Koekoek, Tom Koornwinder, Roberto Costas-Santos, Eric Weisstein, Dan Lozier, Alan Sexton, Bruce Miller, Abdou Youssef, Charles Clark, Volker Markl, George Andrews, Mourad Ismail, and Dmitry Karp for their advice, invaluable assistance, and support.

References

1. NIST Digital Library of Mathematical Functions. http://dlmf.nist.gov, Release 1.0.9 of 2014–08-29. Online companion to [6]
2. Andrews, G.E., Askey, R., Roy, R.: Special Functions. Encyclopedia of Mathematics and its Applications, vol. 71. Cambridge University Press, Cambridge (1999)
3. Cohl, H.S., McClain, M.A., Saunders, B.V., Schubotz, M., Williams, J.C.: Digital repository of mathematical formulae. In: Watt, S.M., Davenport, J.H., Sexton, A.P., Sojka, P., Urban, J. (eds.) CICM 2014. LNCS, vol. 8543, pp. 419–422. Springer, Heidelberg (2014)
4. Erdélyi, A., Magnus, W., Oberhettinger, F., Tricomi, F.G.: Tables of Integral Transforms, vol. 1-2. McGraw-Hill Book Company Inc., New York-Toronto-London (1954)
5. Erdélyi, A., Magnus, W., Oberhettinger, F., Tricomi, F.G.: Higher Transcendental Functions, vol. 1-3. Robert E. Krieger Publishing Co., Inc., Melbourne (1981)
6. Olver, F.W.J., Lozier, D.W., Boisvert, R.F., Clark, C.W. (eds.): NIST Handbook of Mathematical Functions. Cambridge University Press, New York (2010). Print companion to [1]

7. Koekoek, R., Lesky, P.A., Swarttouw, R.F.: Hypergeometric Orthogonal Polynomials and their q-analogues. Springer Monographs in Mathematics. Springer, Berlin (2010). With a foreword by Tom H. Koornwinder
8. Kohlhase, M.: Using LaTeX as a semantic markup format. Math. Comput. Sci. **2**(2), 279–304 (2008)
9. Kohlhase, M., Sucan, I.: A search engine for mathematical formulae. In: Calmet, J., Ida, T., Wang, D. (eds.) AISC 2006. LNCS (LNAI), vol. 4120, pp. 241–253. Springer, Heidelberg (2006)
10. Koornwinder, T.H.: Additions to the formula lists in "Hypergeometric orthogonal polynomials and their q-analogues" by Koekoek, Lesky and Swarttouw. arXiv:1401.0815v2 (2015)
11. Miller, B.R., Youssef, A.: Technical aspects of the digital library of mathematical functions. Ann. Math. Artif. Intell. **38**(1–3), 121–136 (2003)
12. Nghiem, M.-Q., Kristianto, G.Y., Topić, G., Aizawa, A.: Which one is better: presentation-based or content-based math search? In: Watt, S.M., Davenport, J.H., Sexton, A.P., Sojka, P., Urban, J. (eds.) CICM 2014. LNCS, vol. 8543, pp. 200–212. Springer, Heidelberg (2014)
13. Pagel, R., Schubotz, M.: Mathematical language processing project. In: England, M., Davenport, J.H., Kohlhase, A., Kohlhase, M., Libbrecht, P., Neuper, W., Quaresma, P., Sexton, A.P., Sojka, P., Urban, J., Watt, S.M. (eds.) Joint Proceedings of the MathUI, OpenMath and ThEdu Workshops and Work in Progress track at CICM co-located with Conferences on Intelligent Computer Mathematics (CICM 2014). CEUR Workshop Proceedings, Coimbra, Portugal, 7–11 July, vol. 1186 (2014). http://CEUR-WS.org
14. Schubotz, M., Wicke, G.: Mathoid: robust, scalable, fast and accessible math rendering for wikipedia. In: Watt, S.M., Davenport, J.H., Sexton, A.P., Sojka, P., Urban, J. (eds.) CICM 2014. LNCS, vol. 8543, pp. 224–235. Springer, Heidelberg (2014)
15. Sexton, A.P.: Abramowitz and stegun – a resource for mathematical document analysis. In: Campbell, J.A., Jeuring, J., Carette, J., Dos Reis, G., Sojka, P., Wenzel, M., Sorge, V. (eds.) CICM 2012. LNCS, vol. 7362, pp. 159–168. Springer, Heidelberg (2012)

Formalizing Physics: Automation, Presentation and Foundation Issues

Cezary Kaliszyk[1], Josef Urban[2]([✉]), Umair Siddique[3], Sanaz Khan-Afshar[3], Cvetan Dunchev[4], and Sofiène Tahar[3]

[1] University of Innsbruck, Innsbruck, Austria
[2] Radboud University, Nijmegen, The Netherlands
Josef.Urban@gmail.com
[3] Concordia University, Montreal, Canada
[4] University of Bologna, Bologna, Italy

Abstract. In this paper, we report our first experiments in using learning-assisted automated reasoning for the formal analysis of physical systems. In particular, we investigate the performance of automated proofs as compared to interactive ones done in HOL for the verification of ray and electromagnetic optics. Apart from automation, we also provide brief initial exploration of more general issues in formalization of physics, such as its presentation and foundations.

1 Introduction: Formalization, Automation and Physics

Twenty years after the QED Manifesto [1], there is an encouraging progress in building computer-understandable and formally verified mathematical corpora. Large projects in mathematics include the completed formal proofs of the Kepler conjecture (Flyspeck) [8], the Odd Order theorem [7], the Four Color theorem [6], and verification of more than a half of the Compendium of Continuous Lattices textbook [3]. Verification of the seL4 kernel [15] and the CompCert compiler [17] show comparable progress in full-scale verification of complicated software. Such projects are often linked to advances in verification technology, and in particular to strong automation [9,11,16] that allows less verbose formal proofs and increases the general understanding intelligence of the formal proof assistants.

This ongoing progress brings closer the possibility of eventually expressing in a computer-understandable form all of today's scientific knowledge, and in particular the vast knowledge accumulated by exact sciences such as physics. Such a *Formalization of Physics* (FOP) project raises a number of interesting issues, ranging from philosophical and theoretical to very practical ones, on a scale that may eventually dwarf the current applications of formal verification. Just optical components are today a basis of a growing multi-billion business, technologies involving quantum-level phenomena become more and more important, the safety of space/air flight and other means of transport (particularly self-driving) may greatly benefit from formal treatment, and perhaps even more some of the big and dangerous "prides" of modern physics such as nuclear power

M. Kerber et al. (Eds.): CICM 2015, LNAI 9150, pp. 288–295, 2015.
DOI: 10.1007/978-3-319-20615-8_19

plants, tokamaks, and large hadron colliders. An interesting multidisciplinary problem is the formal analysis of engineering systems which requires formalized theories of Physics, Probability and Information Theory.

One of the first practical hurdles in FOP is the unfamiliarity with theorem proving in the Physics community. An attractive step that may reduce this gap is to wrap the internal complexities of tactical theorem proving systems in powerful high-level automation, user-friendly interfaces, collaborative reasoning platforms and proof advice systems. The main concrete contribution of this paper is to describe the first experiments in deploying and using such strong automation – the HOL^{y}Hammer system [12] – over the first formal physics developments. Section 2 briefly describes such projects in the area of *Formal Optics* (Formalization of Physics) and Sect. 3 describes first steps and experiments in using HOL^{y}Hammer for these developments. This initial experience leads us to discuss in Sect. 4 some wider and more concrete issues related to the present and future FOP project(s).

2 Formal Optics

Optical systems are becoming increasingly important by resolving many bottlenecks in todays communication, aerospace and biomedical systems. However, given the continuous nature of optics, the inability to efficiently analyze optical system models using traditional paper-and-pencil and computer simulation approaches sets limits especially in safety-critical applications.

In 2009, a project[1] was started at the Hardware Verification Group (HVG) of Concordia University in order to build a comprehensive framework for the formal analysis of optical systems. The project can be divided into three sub-projects:

- Formalization of Ray Optics in which light is considered as a ray, i.e., a simple geometrical line.
- Formalization of Electromagnetic Optics in which light is characterized as electromagnetic waves.
- Formalization of Quantum Optics in which light is characterized as a stream of photons.

Currently, fundamentals of ray optics, electromagnetic optics and quantum optics have been formalized [14] in HOL Light. This allowed the formal verification of some interesting and safety-critical optical systems such as optical resonators [19], laser resonator [13] and optical quantum flip gate [18]. In the sequel, we explore automation and presentation issues of these projects.

3 HOLyHammer and Formal Optics

HOL^{y}Hammer [12] is a recently developed online AI/ATP system for assisting formal (computer-understandable) verification done in HOL Light. The service

---

[1] http://hvg.ece.concordia.ca/projects/optics/.

allows its users to upload and automatically process an arbitrary formal development (project) based on HOL Light, and to attack arbitrary conjectures that use the concepts defined in some of the uploaded projects. The service uses several automated theorem provers (ATPs) combined with several premise selection methods trained on all the project proofs. The ITP (interactive theorem prover) and ATP proof data and theorems from different (possibly incompatible) projects and their versions are pooled together using a recursive content-based (MD5) naming of symbols and theorems, providing a large base of proofs to learn from. Authorized users can upload a new project against an arbitrary existing project (saved as standard and proof-recording checkpointed images), allowing fast processing of HOL Light projects that import large libraries such as the Multivariate Analysis. The system also provides version control and heuristic HTML-ization (cross-linking) of the uploaded projects. Users can ask parallel asynchronous queries to the service either from its web interface or directly from the HOL Light mode for Emacs. Below we describe the steps to deploy and test HOLʸHammer for Formal Optics.

3.1 Deployment

We have streamlined the HOLʸHammer installation and deployed it on a faster dedicated machine with 12 hyperthreading 2.6 GHz Xeons in Canada (U. of Alberta), which was serving so far the users of the similar online service for Mizar [9]. The HVG members were given access rights to upload their developments there, to update them, and their Emacs mode was configured to ask queries to this server. Such a dedicated/local HOLʸHammer installation is now quite easy and we hope that more users will use this option and we will eventually build a network of such online "hammer" installations that will further synchronize between them their proof data, projects, CPU-load, etc., in the spirit of large distributed formal wikis [2].

3.2 Experiments with Complete Automation

We have measured the strength of the HOLʸHammer automation on the Ray (Ray Optics) and EMF (Electromagnetic Optics) formalizations. These two projects are both based on HOL Light's Complex Multivariate Analysis, and they together contain 482 proved toplevel theorems and 125 definitions.[2] Table 1 shows the performance of 11 ATPs in proving the 482 theorems from their recorded HOL Light dependencies, and Table 2 shows the performance of various strategies that combine the three best ATPs with premise selection using learning from previous proofs[3]. The learning method used in all cases was distance-weighted k-nearest neighbor with IDF-weighted normalized term-based features [10]. The results are encouraging: the combined strength of the methods reaches nearly 50 % (239

---

[2] Many definitions are just abbreviations introducing proper physics terminology.

[3] The complete set of ATP inputs generated by HOLʸHammer and the corresponding ATP outputs are available at http://cl-informatik.uibk.ac.at/~cek/cicm15/data.tgz.

problems solved) in the first scenario when the premises are chosen by the user. 236 of these problems are already solved by one of the best three ATPs (Epar, Vampire 3.0, and Z3 4.0). The performance is 45 % (217 problems solved) in the fully automated mode when the relevant premises are chosen automatically by machine learning, and seven different combinations of premise selection and ATPs are needed for this. Note that there are 105 problems that Paradox found counter-satisfiable. This means that the incompleteness of the currently used HOL-to-FOL translation shows quite considerably on these problems, making more complete encodings an interesting problem to address in this context.

Table 1. ATP re-proving with 300 s time limit on the 482 Emf and Ray top-level problems

Prover	Theorem (%)	CounterSat (%)
Epar	219 (45.436)	0
Vampire 3.0	210 (43.568)	0
Z3 4.0	210 (43.568)	0
CVC4 1.3	201 (41.701)	0
Vampire 2.6	198 (41.079)	0
E 1.8	189 (39.212)	0
SPASS 3.5	154 (31.950)	0
Metis 2.3	152 (31.535)	0
iProver 1	116 (24.066)	0
Prover9 09.11a	114 (23.651)	0
Paradox 4.0	0 (0.000)	105 (21.784)
any	239 (49.585)	105 (21.784)

A brief review of the fully automatically solved problems shows that HOLʸHammer is particularly useful in automating proofs about complex vectors (used in the representation of planar waves) in Electromagnetic Optics, for example the following relation[4] between collinearity and orthogonality of complex vectors is proved by Epar using 17 other previous theorems:

```
∀x y:complex^N.
    collinear_cvectors x y ∧ ¬(x=cvector_zero) ∧ ¬(y=cvector_zero)
        ⟹ ¬(corthogonal x y)
```

An example of a fully automatically proved lemma in Ray Optics is a statement[5] about the stability of an optical resonator (represented by its ray transfer matrix) under certain conditions. In this case the AI/ATP found a relevant special lemma where most of the hard proving work was done, and which together with six auxiliary lemmas can be used to automatically prove the more general statement:

---

[4] http://mizar.cs.ualberta.ca/hh/ses/Emf202/cvectors.html#CORTHOGONAL_COLLINEAR_CVECTORS.

[5] http://mizar.cs.ualberta.ca/hh/ses/Ray203/resonator.html#STABILITY_LEMMA_GENERAL_SYM.

Table 2. ATP proving with k-NN premise selection and 300 s time limit on the 482 Emf and Ray top-level problems

Prover	Premises	Theorem (%)
Epar	1024	170 (35.565)
Epar	128	155 (32.158)
Vampire 3.0	128	121 (25.104)
Vampire 3.0	1024	119 (24.895)
E 1.8	128	104 (21.577)
Z3 4.0	128	103 (21.369)
Epar	32	102 (21.162)
Vampire 3.0	32	92 (19.087)
E 1.8	32	91 (18.880)
Z3 4.0	32	89 (18.465)
E 1.8	1024	68 (14.226)
Z3 4.0	1024	64 (13.389)
any		217 (45.021)

```
∀ (M:real^2^2) xi thetai.
 (det (M) = &1) ∧ ( −&1 < (M$1$1 + M$2$2) / &2) ∧ (M$1$1 + M$2$2)/&2 < &1
   ⟹ ∃(Y:real^2). ∀n.
      abs (((M pow 2) pow n ** vector [xi; thetai])$1) ≤ Y$1
   ∧ abs (((M pow 2) pow n ** vector [xi; thetai])$2) ≤ Y$2
```

3.3   Linking to Informal Physics Explanations

Formal mathematics as a science enjoys a remarkable property: it is in some sense fully "understood" by machines. Computers can correctly parse the formal definitions and statements, verify the proofs, and sometimes even find proofs independently of humans, regardless of any possible motivation and underlying intuition ivolved in proposing the definitions, theorems, proofs and theories. In this sense, formal mathematics is completely self-explanatory. While (physical) intuition may play varied part in formulation of various theories, such theories as formal mathematical objects are independent and decoupled from their (possible) underlying intuition. It is not unusual that for some abstract theory a new application is found, which has very little in common with the original intuition. Similarly, the popular term "abstract nonsense" refers to abstract arguments (e.g., in category theory) which are hard to link to any particular intuition. While some physicists (notably Feynman) criticized such decoupling from physical intuition as harmful, it is a fact that many mathematicians (to say nothing about computers) do mathematics without such links.

We believe that here is a real difference between (formal/abstract) mathematics and physics, and this difference really needs to be addressed by appropriate tools assisting formalization of physics. In physics, there is always first some underlying intuition about (part of) the real world, and this intuition is more or less perfectly captured by various abstract mathematical models. An important part of physics is the *informal* understanding of the (intended) correspondence between the physical phenomena and their formal models. This understanding however is not (yet) part of the actual formal code. In particular, those of us who are not experts in optics have found it significantly harder to understand some of the formal definitions modelling the physical systems and phenomena. While abstract concepts like sets, quasigroups, categories and topological spaces are acceptable to mathematicians as just such abstract concepts described by their formal definitions, taking an "optical resonator" to be just its formal definition does not seem to be right, because it forgets the "real" physical phenomenon that is linked to (and motivating) the particular choice of the formal model.

A solution that does not require much work from the formalizers (and which can even be done later by others) is to allow special comments in the formal text, that are during the HTML-ization turned into cross-links to informal explanations, in our case to Wikipedia. Such cross-links can be also harvested from the formalizations, thus providing an informal overview (and in some sense also high-level semantic anchors) of the physics topics dealt with in the formal code. About 20 such Wikipedia annotations have been inserted into the Ray Optics formalization,[6] making the resulting HTML presentation considerably easier to understand for some of us. Another very interesting informal resource that could provide such semantic anchors are the three volumes of Feynman's lectures that have been recently published online in a form that makes use of state-of-the-art informal presentation technologies such as MathJax.[7]

4   Some Issues and Considerations in Formal Physics

The tighter link between the formal mathematical theory and its underlying (physical) intuition is likely just one of several interesting differences between formalization of physics and formalization of mathematics. Clearly, the most obvious theoretical issue is whether it is possible to consistently formalize the whole of physics at all, and what should be the ultimate foundational framework for such formalization. For example, Beeson in [4] briefly derives (what he calls) a contradiction between quantum theory and general relativity that is apparently well-known to physicists, and which can perhaps be understood as quantum physics breaking some of the assumptions of general relativity about all possible worlds being regular solutions to Einstein's equations. There are probably several answers to this famous problem by current theoretical physics, the best-known involving various string and superstring theories for which we still lack enough experimental evidence.

This however just brings up the main issue with physics: it is about modelling the real world "well enough" which we do not fully know and probably

---

[6] See, e.g., http://mizar.cs.ualberta.ca/hh/ses/Ray203/resonator.html.

[7] http://www.feynmanlectures.caltech.edu/.

never will. As already the several approaches to the formalization of optics show, there are typically several models of the same phenomena. These models will often be "almost compatible" in terms of their predictions when used on their intended domain, e.g., the more complicated electromagnetic optics model will largely agree with the simpler ray optics model on an important class of optics problems. As one goes farther away from this class of problems, the predictions of these two models will disagree more and more. Some models designed for very different phenomena, such as the quantum-theoretical and relativistic, might quickly yield hard contradictions as soon as one tries to use both of them at once. A proper foundational framework should make such relations between the models as explicit as possible (e.g., by theorems exhibiting the asymptotic relations between the models and/or their incompatibilities and scope, perhaps enhancing by such explicit relations formalization frameworks such as Little Theories and Realms [5]), so that one can consistently and automatically combine the knowledge contained in them in the same way as the current large-theory AI/ATP methods do over large mathematical corpora.

An interesting related issue is to what extent such careful "theory engineering" could assist, emulate, or even replace "proper" mathematical solutions to inconsistencies in physics, such as the Dirac delta "function" (made consistent later by Schwartz's distributions), the physics way of treating the infinitesimals (made consistent by Robinson's ultraproduct models) or various approaches to counting with infinities (regularization, renormalization) in Feynman's diagrams.

There are also many practical issues and tasks that are already visible in our experiments. Physics is a heavy user of computation, and the pragmatic approach used sometimes by the HVG group is to just trust the results of computer algebra systems (e.g., using Mathematica to compute the numerical eigenvalues of the waveguide when there is no closed form solution [14]), temporarily adding them as axioms [14]. This is going to be a rich source of research problems for Calculemus-style projects, SMT solving, systems like MetiTarski, etc. In short, we suggest FOP as a rather exciting and very large and rewarding research topic whose automation, foundations and presentation issues will keep the formalization community busy in the next years, hopefully greatly expanding its current borders and methods.

Acknowledgements. Kaliszyk was supported by the Austrian Science Fund (FWF) grant P26201.

References

1. Boyer, R., et al.: The QED Manifesto. In: Bundy, Alan (ed.) CADE 1994. LNCS, vol. 814, pp. 238–251. Springer, Heidelberg (1994)
2. Alama, J., Brink, K., Mamane, L., Urban, J.: Large formal wikis: issues and solutions. In: Davenport, J.H., Farmer, W.M., Urban, J., Rabe, F. (eds.) MKM/Calculemus 2011. LNCS, vol. 6824, pp. 133–148. Springer, Heidelberg (2011)
3. Bancerek, G., Rudnicki, P.: A compendium of continuous lattices in MIZAR. J. Autom. Reasoning **29**(3–4), 189–224 (2002)

4. Beeson, M.: Constructivity, computability, and the continuum. In: Essays on the Foundations of Mathematics and Logic, Polimetrica, Milan, vol. 2 (2005)
5. Carette, J., Farmer, W.M., Kohlhase, M.: Realms: A structure for consolidating knowledge about mathematical theories. In: Watt, S.M., Davenport, J.H., Sexton, A.P., Sojka, P., Urban, J. (eds.) CICM 2014. LNCS, vol. 8543, pp. 252–266. Springer, Heidelberg (2014)
6. Gonthier, G.: The four colour theorem: engineering of a formal proof. In: Kapur, D. (ed.) ASCM 2007. LNCS (LNAI), vol. 5081, p. 333. Springer, Heidelberg (2008)
7. Gonthier, G.: Engineering mathematics: the odd order theorem proof. In: Giacobazzi, R., Cousot, R. (eds.) The 40th Annual ACM SIGPLAN-SIGACT Symposium on Principles of Programming Languages, POPL 2013, Rome, Italy, 23–25 January, pp. 1–2. ACM (2013)
8. Hales, T.: Dense Sphere Packings: A Blueprint for Formal Proofs. London Mathematical Society Lecture Note Series, vol. 400. Cambridge University Press, Cambridge (2012)
9. Kaliszyk, C., Urban, J.: MizAR 40 for Mizar 40. CoRR, abs/1310.2805 (2013)
10. Kaliszyk, C., Urban, J.: Stronger automation for Flyspeck by feature weighting and strategy evolution. In: Blanchette, J.C., Urban, J. (eds.) PxTP 2013. EPiC Series, vol. 14, pp. 87–95. EasyChair (2013)
11. Kaliszyk, C., Urban, J.: Learning-assisted automated reasoning with Flyspeck. J. Autom. Reasoning 53(2), 173–213 (2014)
12. Kaliszyk, C., Urban, J.: HOL(y)Hammer: Online ATP service for HOL Light. Math. Comput. Sci. 9(1), 5–22 (2015)
13. Khan-Afshar, S., Hasan, O., Tahar, S.: Formal analysis of electromagnetic optics. In: Proceedings of SPIE, vol. 9193, pp. 91930A–91930A-14 (2014)
14. Khan-Afshar, S., Siddique, U., Mahmoud, M.Y., Aravantinos, V., Seddiki, O., Hasan, O., Tahar, S.: Formal analysis of optical systems. Math. Comput. Sci. 8(1), 39–70 (2014)
15. Klein, G., Huuck, R., Schlich, B.: Operating system verification. J. Autom. Reasoning 42(2–4), 123–124 (2009)
16. Kühlwein, D., Blanchette, J.C., Kaliszyk, C., Urban, J.: MaSh: machine learning for sledgehammer. In: Blazy, S., Paulin-Mohring, C., Pichardie, D. (eds.) ITP 2013. LNCS, vol. 7998, pp. 35–50. Springer, Heidelberg (2013)
17. Leroy, X.: Formal verification of a realistic compiler. Commun. ACM 52(7), 107–115 (2009)
18. Mahmoud, M.Y., Aravantinos, V., Tahar, S.: Formal verification of optical quantum flip gate. In: Klein, G., Gamboa, R. (eds.) ITP 2014. LNCS, vol. 8558, pp. 358–373. Springer, Heidelberg (2014)
19. Siddique, U., Aravantinos, V., Tahar, S.: Formal stability analysis of optical resonators. In: Brat, G., Rungta, N., Venet, A. (eds.) NFM 2013. LNCS, vol. 7871, pp. 368–382. Springer, Heidelberg (2013)

A Survey on Retrieval of Mathematical Knowledge

Ferruccio Guidi and Claudio Sacerdoti Coen$^{(\boxtimes)}$

Department of Computer Science and Engineering – DISI,
University of Bologna, Bologna, Italy
{ferruccio.guidi,claudio.sacerdoticoen}@unibo.it

Abstract. We present a short survey of the literature on indexing and retrieval of mathematical knowledge, with pointers to 72 papers and tentative taxonomies of both retrieval problems and recurring techniques.

1 Purpose Driven Taxonomy of Retrieval Problems

Retrieval of mathematical knowledge is always presented as the low hanging fruit of Mathematical Knowledge Management, and it has been addressed in several papers by people coming either from the formal methods or from the information retrieval community. The problem being resistant to classical content search techniques [LRG13], it is usually addressed combining a small set of new ideas and techniques that are recurrent in the literature. Despite the amount of work, however, there is not a single solution that is the clearly winning on the others, nor convincing unbiased benchmarks to compare solutions. Some authors like [KK07] also suggest that the community should first better understand the actual needs of mathematicians from an unbiased perspective to improve the MKM technology as a whole. In this paper we collect a hopefully comprehensive bibliography, and we roughly classify the papers according to novel taxonomies both for the problems and the techniques employed. The only other surveys on the same topic are [AZ04], now outdated and focused mostly on (European) research projects that contributed to the topic in the 6th Framework Programme, [ZB12], which covers less literature in much greater detail without attempting a classification, [L13], which is focused on evaluation of mathematics retrieval, and [L10], which is written in Slovak.

We begin our discussion with a purpose driven taxonomy made of three different retrieval problems that deal with mathematical knowledge. Each problem is characterised by its own set of expectations and constraints, and adopting a solution to another problem may be infeasible or yield poor results. In the next sections we classify the papers according to an encoding based taxonomy (presentation vs. content vs. semantics) and to a taxonomy of techniques employed. Finally we point to the rich literature relative to the problem of ranking, and we touch the problem of evaluation of systems. We conclude with some notes on the availability of math retrieval systems.

© Springer International Publishing Switzerland 2015
M. Kerber et al. (Eds.): CICM 2015, LNAI 9150, pp. 296–315, 2015.
DOI: 10.1007/978-3-319-20615-8_20

1.1   Problem 1: Document Retrieval

Objective: A *human* is interested in recalling a *set of mathematical documents* (or fragments) that are related to a particular mathematical topic. Typically it is not the case that only one document provides the correct answer; on the contrary the user may be interested in a corpora of different documents that yield different, only partially overlapping information. In [Koh14] and other papers there are attempts at a classification of the information needs of users. However, at the moment only the system described in [ZKT08] tries to use the classification to improve the user experience.

Input: The human composes a query combining keywords (e.g. for topics [ACK08]), free text and mathematical formulae. Often the mathematical formulae are intended as examples of expressions related to the topic of interest. For example, a user interested in trigonometric identities can just enter one identity to retrieve them all. Or a formula showing a particular property of a special function can be used to disambiguate the special function among the ones with similar names.

The query can be composed using a very simple, Google-inspired, single line interface, or written using an ad-hoc query language (see [AY07b, AY08b, YA07] for some proposals), or by filling in some form. The first solution is the one preferred in the literature. In [Koh14] a comparison of the behaviour of mathematicians vs. other users highlighted that the professional mathematician is more interested in the precision of the output than the effort put into the input. Therefore mathematicians may use and appreciate more complex interfaces. On the contrary, other users are likely to prefer a simple, modern search interface.

Formulae can be entered in some textual syntax (e.g. LATEX, MathML), maybe with the help of on-the-fly formulae rendering [LSR14], or using graphical editors [MM06], or they can be acquired from hand-written snippets [AY08a]. The formula is likely to contain errors and ambiguities, for example if it is encoded at the presentational level (e.g. in Presentation MathML or LATEX), if it is acquired from hand-written text, or if the user only remembers it partially or in a wrong way. Errors and ambiguities are not a critical problem because formulae are just used to retrieve documents that contain *similar* formulae according to some similarity criterion. In [AY08a] the authors address the problem of combining and ranking results from different queries generated from ambiguous formulae due to errors in the recognition process. See also [ZB12] for a survey on the interaction between mathematical information retrieval and mathematical document recognition. Some authors [KT09] suggest that the visual presentation of the formula may sometimes be important in the definition of similarity, whereas in other situations it is the mathematical *content* of the formula that matters. Logically equivalent formulae whose content encoding is highly different are better considered less similar.

Once the search engine returns the result, the user may be given the opportunity to enhance the query by further filtering.

Output: The output is a *ranked list* of matching documents or document fragments (e.g. a chapter of a book, or a section of an article). When the query involves mathematical formulae, the ranking is determined by the similarity relation. The user must be given the possibility to quickly determine whether the matched document is interesting or not. Therefore the problem of how to present summaries of the selected documents in the result list is of fundamental importance [LSLM11, LSR14, MG08b, WG10, You05, You06, You07, You08]. Even highlighting correctly the bits of the summary that matches the query can make a significant difference in the user experience [LSLM11, LSR14, You05, You06]. The list of results must be the starting point for further investigations by the user. At least, all results must contain hyperlinks or other ways to retrieve the original document the summary points to. A study of user requirements in [ZKT08] suggests that results should be presented after clustering them according to their resource type (research paper, tutorial, slides, course, book, etc.). For example, a student may immediately decide to skip research papers, and a researcher may skip websites and tutorials.

Constraints: A balance must be obtained between *precision* (the fraction of retrieved documents that are relevant) and *recall* (the fraction of relevant documents that are retrieved). To maximise recall, precision is affected and many out of topic documents (false positives) are retrieved, penalising performance. Too many results are overwhelming and the user is likely to give attention only to the first ones in the list. Therefore, the search engine does not have to rank and produce summaries of documents with low scores. The ranking function is ultimately the one responsible for the perceived quality of the search engine.

Since the query is intended to be issued by a human, the performance of the search engine is not a critical requirement and up to a few seconds (or even minutes in some particular situations) may be acceptable. Nevertheless, modern textual search engines like Google are extremely fast, and the user is likely to expect the queries to be solved in less than a second.

1.2   Problem 2: Formula Retrieval

Objective: A program — more rarely a human — is interested in retrieving *all* formulae that are in some relation $\mathcal{R}$ with a query formula E. Sometimes the formula E can actually be a set of formulae. For example, E can be a goal to be proved automatically, and $T \mathcal{R} E$ when T is the conclusion of the statement of a theorem that can be instantiated to prove E. More precisely, T contains metavariables to be instantiated and $\mathcal{R}$ is one-sided unification up to some equational theory. The dual query is also used in the literature: E is a property (a statement containing metavariables), $\mathcal{R}$ is unification and the query finds all operations that satisfy the property E. By using several properties at once, the query can find all models of a given theory (e.g. all semirings in the library) [NK07], also up to renaming of constants and properties. An interesting application presented in [GK14], that uses techniques similar to [NK07], consists in matching concepts across libraries by first computing properties of an object in one library (i.e. patterns like commutativity of a binary operator) and then looking for objects in the other library that satisfy the same properties. To be more effective, properties are extracted from all libraries and concepts are matched according to a similarity measure to identify objects that satisfy a similar set of properties. A third example is obtained by choosing logical implication for $\mathcal{R}$. The query looks for all formulae that imply E.

Input: One or more formulae E that may or may not contain metavariables to be instantiated. Rarely, additional constraints can be expressed using keywords, classifications, free text, authors, etc. Formulae are not supposed to be ambiguous or contain errors. In [AGSC+06] ambiguity is resolved before performing the query using type checking and interaction with the user.

Some dedicated query languages are proposed to specify the structure of the formulae E [Ban06, BR03, BU04, GS03, KT10, Rab12]. They are implemented on top of relational databases or ad-hoc in-memory indexes.

Output: The query is meant to retrieve a set of formulae that satisfy a certain property. When the search is performed by a program, there is no need to present summaries of the document the formula occurs in. Even when a human issued the search, an hyperlink to the document may be sufficient.

In many situations the relation $\mathcal{R}$ can be extended to a ternary relation $T\mathcal{R}_\rho E$ meaning that T is related to E with score ρ, and the results can be ranked according to ρ. For example, if $\mathcal{R}$ reduces a proof of E to a proof of T, then the T's may receive a higher score if they are judged easier to prove.

Constraints: Maximisation of the recall is fundamental. The query should return all formulae that satisfy the query, even if they rank very low. Because searches are often basic operations of complex algorithms (e.g. automatic provers), speed is also critical. In several situations, the searches need to be performed in milliseconds.

To speed up the searches or when the relation $\mathcal{R}$ is undecidable, the search engine may use a second decidable relation $\mathcal{R}'$ such that $\mathcal{R} \subseteq \mathcal{R}'$. Using $\mathcal{R}'$, the query can return false positives, i.e. formulae T such that $T\mathcal{R}' E$ but not $T\mathcal{R} E$. For example, when E is a pattern and $\mathcal{R}$ is unification, $\mathcal{R}'$ may ignore the structure of the two formulae E and T and conclude $E\mathcal{R}'T$ when all symbols in E are also in T. Example: $f(xz, y + z)\,\mathcal{R}\,f(?, ?+?)$ and $f(y + z, xz)\,\mathcal{R}'\,f(?, ?+?)$, but not $f(y + z, xz)\,\mathcal{R}\,f(?, ?+?)$.

1.3 Problem 3: Document Synthesis

Objective: Composing a new mathematical document assembling fragments to be retrieved from a library. The most common occurrence is in educational software where a learning object must be assembled according to the expertise of the user, the topic of interest, etc. [BF03, LDM+08, LM06]. An unrelated example is the automatic generation of summaries and statistics for a mathematical library [BR03]. A final example is mining of formalised libraries, for example to build visual representations of the graph of dependencies over an axiom (to understand its implications) or a definition/statement (to understand the propagation of changes).

A variant is to solve a mathematical problem by composing mathematical (Web) services [CDT04]. Each service exposes metadata about the problem solved, the algorithm implemented, and its preconditions and postconditions.

Input: The query is not likely to involve mathematical formulae, and it is usually expressed using a query language over ontologies. A high level interface may hide the underlying query language. Sometimes the query is fixed once and for all, and needs to be run at regular periods.

Output: The expected output depends on the particular use case and it is usually made of a single result in place of a ranked list. The result may consist of a graph of objects and relations between them, or it may consist of the minimal information to build the expected document or solve the algorithmic problem.

Constraints: The constraints depend on the particular use case.

After an initial screening of the literature, we decided to analyse only papers about the first two problems, where formulae play a central role. Indeed, at a first glance most solutions to Document Synthesis employ standard query languages for ontologies, and only the ontologies themselves are math-specific (e.g. [CDT04, LDM+08]). Logic programming languages are also employed to represent what the user knows/ignores and the inference rules to assemble documents [BF03].

Moreover, we did not find in the literature convincing examples for the need of very expressive query languages to solve the Document Retrieval and the Formula Retrieval problems, where the kinds of queries are essentially fixed a priori. Moreover, evaluation of queries expressed in these languages are reported to be too slow to be used for Formula Retrieval. Sometimes additional techniques are employed for Document Synthesis, like semantical query reduction to relax the user provided query by allowing additional topics close to the one specified by the user [Lib13]. These techniques too seem to be very general and applicable to domains very different from that of mathematics.

Despite the strong interest of the community in the use of formulae in queries, studies on the behaviour of users [KK07, ZKT08] conclude that the added value may be low, and the finding is confirmed in [LM06, Mil13] where the logs of the DLMF and of ActiveMath search engines are analysed concluding that only few queries contain mathematical formulae, most are very simple ones, and such queries do not yield satisfactory results.

2   Encoding Based Taxonomy

Mathematical information can be encoded in a library at three different levels. The most shallow one is *presentation*. Presentation markup uses a finitary language to express the bi-dimensional layout of a formula, useful to present it to the reader. The standard XML language for presentation is *Presentation MathML*, and several tools can generate Presentation MathML from LaTeX (e.g. LaTeXML, Tralics), PDF files (e.g. Maxtract), handwritten text (e.g. InfTy Reader), digitised documents (e.g. InfTy Reader) or content markup (e.g. via XSLT stylesheets). We can therefore assume that the totality of the documents to be indexed are available in Presentation MathML.

The next level is *content*. At the content level, the structure of the formula is described, and symbols and operators appearing in it are linked to their entry in an ontology, called *content dictionary* in OpenMath terminology. The markup language is finitary, but the ontology is not since new mathematical entities can always be defined. The relation between content and presentation is one-to-many: the same presentation markup may represent different content expressions (*ambiguity*), and a content expression can be given different presentations according to the conventions of the community of readers, the language of the reader, but also for purely aesthetic reasons like constraints on the size of the formula. OpenMath and Content MathML are the two standard XML language for formulae at the content level. OMDoc is an attempt at standardising at the content level whole mathematical documents, comprising proofs. Content markup is currently mostly used for the exchange of formulae between systems, in particular CAS. There are no significant examples of large libraries of documents natively written using content markup. Nevertheless, there are tools like SnuggleTeX based on heuristics to semantically enrich (annotated) LaTeX documents or even MathML Presentation documents to content.

The last level is *semantics* and it is specific to libraries of formalised mathematical knowledge. The semantics level refines the content level by picking for every content level object one particular definition in a given logic. The definition chosen embeds the object with additional properties, e.g. computational properties. For example, addition over natural numbers can be defined in the Calculus of Inductive Constructions as a non-computable ternary predicate in logic programming style, or as a recursive function on the first argument — such that $0 + x$ and x become logically indistinguishable — or as a recursive function on the second argument — so that $0 + x$ and x are not indistinguishable, but only provably equal. Interactive theorem provers often provides an XML dump of their internal semantics representation.

Formula Retrieval is always formulated either on semantics markup or on content markup. Even when the semantic markup is available, it may be convenient to convert the library to content level by identifying alternative definitions of the same mathematical notion. In this way, it becomes possible to retrieve useful theorems on mathematically equivalent definitions, in the hope to reuse them after conversion to the definitions in use. One application of this technique is reuse of libraries across different systems based on the same logic.

The Document Retrieval problem is formulated in a way that is agnostic of the encoding. However, the user is likely to enter formulae in the query using a presentation language (mostly LaTeX, even if MathML starts to be used [LSLM11, LSR14, MG08b]). Some authors have provided evidence that precision is improved when exploiting parallel markup, even when the content part is automatically generated from the presentation part [NKTA14]. See [MY08] for motivations against content/parallel markup and in favour of a more lightweight encoding of content information in Presentation MathML. Other authors claim that precision can be lost by embracing content because sometimes the actual layout used in a presentation or the name used for variables are significant. Reference [GPBB14] in retrospect also described the choice of using Content MathML as a bad decision. Other authors dismiss indexing of Content MathML because of the non-availability of libraries or because of conversion from presentation to content being approximative and unreliable. Finally, [NKTA14] reports that automatic conversion of large formulae from Presentation to Content may be computationally unfeasible, and propose to limit the conversion to small ones.

Recently, the debate on presentation only vs. parallel markup seems to be solved in favour of the latter. For example, the system that scored better at the last NTCIR task reports better scores when applied to content markup generated from LaTeX w.r.t. presentation only markup [RSL14]. The authors second this observation already in [LSR13].

Moreover, several works in the literature that deal with presentation markup enrich it — in the document itself or in the indexes — with additional annotations to make explicit additional semantics that is latent in the library [Cai04] or in the text surrounding the formulae [GPBB14, KTHA14]. For example, in [KTHA14] artificial intelligence is applied to the whole document to recover from the text surrounding the formulae the name associated to the mathematical entities in the formula (e.g. "posterior probability", "derivative of f"). Reference [GPBB14] uses a cheaper approach by considering only one sentence around a formula, but it later observes that one sentence is often not sufficient and many relevant results are therefore missed. Another example is an analysis of co-occurrence of symbols in the corpus to identify related ones. It is shown that these techniques are important to augment recall or, sometimes, precision. In our view, like the heuristic based presentation-to-content translation, these are *attempts to infer and store partial semantics of mathematical expressions.* It may be questioned (see for example [MY08]) if the current content markups (OpenMath and Content MathML) are the right instruments to augment presentation markup with partial, approximate semantics, and if such additional semantics makes only sense in the indexes of search systems, or it may be serialised to an XML format for being reused by third parties.

Systems based on Content MathML, parallel markup or semantics appear in [Ban06, BR03, BU04, GK14, HS13, HKP14, KP13, L13, LSLM11, LSR13, LSR14, MG08a, MG08b, MM06, NCH12, NK07, Rab12, SLM13, YA09, ZY14].

3 Taxonomy of Techniques for Mathematical Retrieval

Implementations of solutions to mathematical search problems can be obtained combining one of the main techniques that will be presented in Sect. 3.2 with a choice of modular enhancement techniques from Sect. 3.1 used to improve precision, recall or both.

3.1 Modular Enhancement Techniques

The following techniques are general enough to be applied to solve both the Document and the Formula Retrieval problems, and by analysing the literature it seems that every system eventually applies all of them.

Segmentation. A preliminary step to indexing is segmentation of documents into chunks. Chunks are the unit of information to be returned to the user, with pointers to the parent document. Segmentation is trivial on formal mathematical documents, hard on web-pages, and intermediate on other resources like books or papers. See, for example, [ZKT08] for a discussion. Several systems implement segmentation; however, the last NTCIR competition has provided a data set of already segmented documents [AKO14] and that may hinder the study of segmentation techniques in the future.

Normalisation. To improve recall, both formulae in the query and the formulae in the library are put in normal form before indexing them. Having the same normal form is an equivalence relation $\equiv$, and the query retrieves formulae up to $\equiv$. For Formula Retrieval it is necessary that $\equiv \mathcal{R} \equiv \; \subseteq \mathcal{R}$. For Document Retrieval the $\equiv$ relation must be compatible with the similarity and ranking functions. When this is not the case, precision can be critically lowered.

Uses of normalisation include: repairing of broken XML/MathML generated by automatic conversion tools [MM07] (e.g. when the structure imposed by $<$ mrows $>$ is not compatible with the mathematical structure); removal of information that does not contribute to the semantics like comments, layout elements (spaces, phantoms and linebreaks), XML/MathML attributes (color, font, elements in other namespaces) [FLRS12, HHN08, MM07]; picking canonical representations of the same presentation/content when different MathML encodings are possible (e.g. msubsup vs. msup and msub, mfenced vs. use of two parentheses, applications of trigonometric functions with/without using parentheses, etc.) [AY07a, FLRS12, MM07]; replacing names of bound variables with unique numerical indexes (e.g. De Brujin indexes) to search up to α-conversion [MM07, NK07]; ignoring parentheses and ordering of arguments of associative/commutative operators [AY07a, MG08a, MG08b, NK07, SY07, SL11, YS06]; expressing derived notions exposing the derivation (e.g. replacing $x \geq y$ with $y \leq x$, $x \nleq y$ with $\neg(x \leq y)$, arcsin with $\sin^{-1}$, etc.) [AY07a, MM07]; capturing logical equivalence/type isomorphisms (e.g. writing formulae in prenex normal form, currification of functions) [Del00, GK14, NK07].

A normalised formula can be quite different from the original one, and that can be a symptom that the formula is not significant. Therefore in [RSL14] normalised formulae are weighted according to their similarity to the initial one, and weights are considered during the ranking phase with great results.

Approximation. Normalisation does not lose information, converting a document to an equivalent one. Many papers call "normalisation" an approximation phase where subformulae are replaced with constrained placeholders to allow the formula to be matched by similarity. For example: names of variables or constants can be replaced by a single name [GWHT14]; all numeric constants by a single identifier [GWHT14, MM07, SL11]; subformulae may be replaced by their type. For example, in [HKP14] type information is used to retrieve formulae by sorted unification, i.e. by constraints with type placeholders in patterns. Approximation improves recall. To limit the loss of precision, systems that approximate index both the original and the approximated formula (or even several instances at different levels of approximation). The effects of approximation are similar to those of query reduction, but approximation is more efficient because it works at indexing time.

Enrichment. Enrichment works on the library or on the query to augment the information stored/looked for in the index by inferring new knowledge from existing one. It can contribute to the solution of both the Document Retrieval and the Formula Retrieval problems. Typical examples of enrichment are: heuristically generating and storing content metadata from Presentation MathML [MY08]; automatic/interactive disambiguation of formulae in the queries to perform a precise query at the content or semantics level [AGSC+06, Ban06, BR03, BU04]; automatic inference of metadata from context analysis or usage analysis (latent semantics) [Cai04, KTHA14, WG10].

The most impressive application of enrichment is presented in [HQ14]. The aim is to search for geometrical constructions that are described using a procedural language (e.g. draw the segments connecting A with B, B with C, and A with C). Enrichment consists in replacing the procedural with a declarative description (e.g. ABC is a triangle). The same declarative description can be obtained by multiple procedural ones, and thus recall is greatly improved. The technique can also be seen as a form of normalisation (see Sect. 3.1) where the normal form is not unique (e.g. it may be the case that by analysing the hypothesis one could deduce that ABC is also an equilateral triangle even if that is not stated in the procedural description).

Query Reduction. Query reduction trades precision for recall by selectively dropping or weakening some of the constraints present in the query. Results obtained from reduced queries can be ranked after results from precise queries. In the literature it occurs in many forms in solutions to both the Formula Retrieval and the Document Retrieval problems: a constant can be weakened to other constants that co-occur frequently with the given one; constants that occur too frequently can be dropped from the queries; a formula may be required to match only the toplevel structure of the formula given as a query.

3.2   Main Techniques

The following techniques are mutually exclusive. Moreover, each technique performs better on only one of the two problems.

Reduction to Full-text Searches. The technology to perform full-text searches is very advanced and there are popular open software implementations with good performance like Apache Lucene/Solr and ElasticSearch. The benefits of reducing search for formulae to full-text searches are speed of execution of the queries and the combination of formula based and textual searches almost for free. The main drawback is that the precise structure of a formula is partially lost in the translations proposed in the literature, and that it is impossible or very hard to capture precisely the kind of relations $\mathcal{R}$ used for Formula Retrieval, unless $\mathcal{R}$ is approximated by a much coarser relation $\mathcal{R}'$. Therefore the technique has been successfully applied so far only to Document Retrieval [ACK08, GWHT14, GPBB14, HHN08, KTHA14, LM06, LSLM11, LSR14, Mil13, MY03, MM07, MG08a, MG08b, PZ14, MM06, SL11, You05, You07, You08].

All the proposals employ vectors to represent features, and compare features with weighted cosine distance. The usual approach consist in turning an expression into a (large) set of "sentences" that partially describe the formula. For example, in [KTHA14, TKNA13] a sentence is the set (ordered or not) of symbols found in either a path from the root of the formula to a leaf, or as children of the same node. Matching is then performed by a disjunctive query and results are ranked using TF-IDF and length normalisation. As the authors claim, the system "is *too* flexible: it is difficult to say where the relevant results stop and random matches begin; thus we predict higher recall but lower precision rates than exact match systems". Other authors extract sentences or n-grams that capture the formula more precisely. As a general remark, the clear impression we got from the literature is that the fewer features extracted, the lower the precision. All kinds of techniques can be used to extract the features, comprising regular expressions [ACK08] and finite state automata [NCH12].

Some systems cluster documents at indexing time (e.g. [ACK08]), and retrieve documents comparing the feature vector of the query with the centroid of the cluster. For example, documents about trigonometric functions are likely to be automatically clustered together. However most systems do not seem to cluster in advance, and prefer the flexibility of weights to capture similarity of features (e.g. similarity of occurrences of trigonometric functions).

According to the set of features extracted, the weighting function used, and the other modular techniques used in combination, the accuracy achieved by systems based on this technique range from extremely low to extremely high (see, for example, [AKO14]).

Structure-Based Indexing via Tries/Substitution Trees. Formula Retrieval can be solved with the data structures developed for automatic theorem proving to store libraries of lemmas and quickly retrieve formulae up to instantiation/generalisation. Pointers to all the statements are stored in the leaves of a tree that precisely encodes in its paths the statements. To match a formula, the tree is recursively traversed using the formula to drive the descent. The relations $\mathcal{R}$ that can be captured are only instantiation and generalisation of whole formulae. MathWebSearch [HKP14, KP13, KT10] are based on this approach. Retrieval of formulae is very fast, assuming that the index can be entirely stored in main memory.

This approach consistently maximises precision but presents poor recall. To accept larger relations, or to be applied to Document Retrieval, or to cope with too rigid queries, the technique needs to be integrated with other ideas. For example: to match subtrees of formulae in the library, every subtree of a lemma needs to be stored as well in the index; to solve unification problems up to an equational theory that admits normal forms, all formulae are normalised; to allow queries that use keywords or free text, a free-text search engine must be run in parallel and the results need to be combined in the ranking phase [HKP14, LDM+08].

Reduction to SQL or ad-hoc Queries. The third approach consists in approximating formulae via relations to be stored in a relational DB [AS04, GS03]. An alternative consists in storing the relations in ad-hoc indexes in memory, and it is employed when the indexes already exists for other purposes (typically in libraries of formalised knowledge) [Ban06, BR03, BU04]. The technique is applied to Formula Retrieval and the database can be reused for Document Synthesis without modifications. Approximated queries up to generalisation/instantiation can be made efficient [AS04] without requiring an index stored in main memory for Structure-Based Indexing (see page 10). Recall can be maximised by relaxing the representation of formulae as relations or by employing normalisation. Ad-hoc inverted indices for paths and to map each Content MathML node to its parent have also been used in [HS13]: the search engine is very fast, but the precision obtained is low.

Reduction to XML-based Searches. Some systems [AY07b, AY08b, YA07] that index MathML documents at the content level, base their searching capabilities on the existing XPath/XQuery technology. The system described in [SLM13], which is based on Stratosphere, is batch oriented, trades flexibility with performance, and it is essentially math-unaware (for example, it does not normalise the input in any way). Other systems [CDT04, LDM+08], that deal with ontologies indexed in the Ontology Web Language, rely on third-party OWL search engines implementing graph matching.

4   Ranking

Because users only inspect the first results returned by a query, precision when solving Document Retrieval is strongly determined by the ranking function. Ranking is also of paramount importance for Formula Retrieval: when the search retrieves the candidates for progressing in a proof, correctly ranking the results may dramatically cut the number of wrong proof attempts and backtracks. The ranking criterion for the two Problems is, however, very different: for the first problem similarity of formulae in the query and in the results should contribute significantly to the score; for the second problem the score should be determined by the intended use of the results. For example, a lemma L_1 that exactly matches the goal to prove and has no premise should always score better than a lemma L_2 that also exactly matches the goal, but that has hypotheses to be proved later.

Ranking according to the intended use for Formula Retrieval has received very little interest in the literature we examined. On the other hand, several papers explicitly address ranking for Document Retrieval. The consensus seem to be that a good ranking function needs to be sophisticated and that the usual metrics induced by reduction to textual searches are completely inadequate (see, for example, [You07]). All the proposed ranking techniques are strongly based on heuristics and, unfortunately, most of them are incomparable and hard to combine.

One class of metrics takes into consideration also the structure of the formulae involved and the enriched semantics, when available. For example, [ZY14] heavily exploits Content MathML to rank results by considering the taxonomic distance of constants (where close is approximated to being defined in the same content dictionary), the data type hierarchical level (matching a function is more significant than matching a numerical constant), matching depth (partially matching the formula at the top level is more significant than matching a deeply nested subformula), coverage (percentage of formula matched), kind of matched expression (formula vs. term). All this information needs to be computed and amalgamated employing some kind of heuristic algorithm. A second paper [SYM+14] confirmed that each one of the listed similarity feature factors significantly improves the ranking, but the last one, that still contributes, has lower relevance.

In [SL11] ranking is determined by the weights used during matching, and the authors claim that each document base and scientific field should have its own weighting function. Nevertheless, they "tried to create a complex and robust weighting function that would be appropriate to many fields".

In [You07] the author proposes a parameterised ranking function that works on mathematical documents (not only formulae), that seems applicable to enriched presentation and that weights a lot of additional information including keywords, the number of cross-references and their kind (e.g. definitional vs. propositional). Ranking employs a hybrid of scalarisation and vectorisation.

In [KT13] the authors propose to adopt tree edit distance to measure similarity of formulae. Most of the paper is about optimisations to improve efficiency of ranking because tree edit distance is hard to compute. The final proposal combines some clever memoisation and a procedure to quickly prune documents bounding their similarity scores with a lightweight computation. The paper also shows benchmarks comparing the processing time and success rate of most search and ranking algorithm in the literature, reimplemented by the authors and run on the same dataset. From the benchmarks the method proposed seems to be superior, but the implementations do not exploit relevant enriched information like cross-references, semantic proximity of definitions, etc. The benchmarks are therefore non conclusive.

In [NCH12] the authors employ a continuous learning ranking model after having extracted features from Content MathML mathematical formulae using a finite state automata. Benchmarks show their ranking to be superior than the ones used in classical ranking of textual documents. However, they do not compare with [KT13] or [You07] (that work on Presentation MathML).

Simpler approaches to ranking can be found in [YA09] (based on Subpath Set, reported to work well only on "simplified" Content MathML) and [KT09] that works on Presentation MathML and measures similarity as a function of the size of subtrees in common. The ranking metric used in [ACK08] is a TF-IDF modified with weights to assign more importance to some operators, but the details given to determine the weights are insufficient.

Ranking algorithms can be too complex to be incorporated in the search phase, for example when using Lucene technology. Moreover, they are typically slower than the search phase. Therefore several authors suggest to re-rank only the first results of the query, that employs a simpler ranking measure to determine the interesting candidates to be ranked more accurately [KT13, You07].

An algorithm to automatically categorise documents is presented in [ZKT08], where it is argued that clustering documents according to their category greatly improves the usefulness of the tool for the user.

5   Evaluation of Math Information Retrieval

Several papers present benchmarks on the systems proposed, and rarely compare them with reimplementations of the algorithms found in the literature (e.g. [KT13]). The significance of most of these benchmarks is unclear, because conflicting results are found in the literature, most techniques are not presented in sufficient details in the papers to be exactly reproduced, and systems are very sensitive to the kind of queries examined. The only alternative is to compare different tools on unbiased, standard benchmarks that are currently lacking.

The main issue is not to come up with large corpora of documents: at least for Document Retrieval on enriched Presentation MathML documents, a large corpus can be easily obtained converting documents from ArXiV, DLMF, PlanetMath, Wikipedia, etc. For Formula Retrieval, the existing libraries of interactive theorem provers, like Mizar and Coq, can be directly used after conversion. The problem is to determine large sets of *real world, interesting queries*, and to evaluate the results. Automatic evaluation is particularly hard in the domain of mathematics, whereas manual evaluation is limited to a tiny number of queries and runs. Formulating good sets of queries is also complex, because users with different mathematical background and motivations are likely to issue different queries. Moreover, what makes a query hard can just be the use of non standard mathematical notations, errors in the encoding of formulae, or formulation at the wrong level of abstraction. Reference [L13] discusses the problem at length and reviews the state of the art of evaluation of Math Information Retrieval before 2013, including the experience of the MIR workshop at CICM 2012 were two systems were compared on about ten hard queries proposed by the judges, and the conclusion was that the systems were too sensitive to the formulation of the query.

The situation is improving since 2013 with the creation of a math oriented task in the NTCIR initiative [AKO13, AKO14] that is attracting a small, but increasing number of participants [GWHT14, GPBB14, HS13, HKP14, KP13, KTHA14, LRG13, LAP+14, LSR13, PZ14, RSL14, SLM13, SYM+14, TKNA13]. The initiative is too young to come to definite conclusions and the current choice of tasks and queries is not granted yet to have significant coverage and to be unbiased. For example, in [KTHA14] the authors report that despite several improvements to the tool (quantified via NTCIR-11 runs), their tool scored lower than in NTCIR-10. They justify the phenomenon by noticing that "in NTCIR-11, query variables get much bigger emphasis, most topics feature complete and very particular formulae, and sub-formulae matching is not nearly as useful as before". Indeed, as reported in [AKO14] "the design decision . . . to exclusively concentrate on formula/keyword queries and use paragraphs as retrieval units . . . has also focused research away from questions like result presentation and user interaction. . . . few of the systems has invested into further semantics extraction from the data set. . . . We feel that this direction should be addressed more in future challenges". An effect of the bias towards search up to unification w.r.t. search up to similarity is observable in [PZ14] too: the system proposed works very well even if it works on Presentation MathML only and the set of features extracted is very simple (bag-of-symbol-pairs model, where a pair is made of two symbols in a father-son relation). The reason why it works well is that the authors also index approximated pairs where the child is a wildcard, and in the future works they are thinking at improving even more the handling of wildcards to score better. In comparison, most other systems based on feature vectors just replace wildcards in the query with $< \mathtt{m} : \mathtt{ci} >$ identifiers. The emphasis on formula queries is also to be evaluated considering the already cited works that conclude that users do not see (yet?) much value in them [KK07, ZKT08].

Finally, the NTCIR task does not cover distinctly the Document Retrieval and the Formula Retrieval problems, but only Document Retrieval with an emphasis on exact pattern matching of formulae that should be more distinctive of Formula Retrieval.

6   Availability of Math Retrieval Systems

Most of the systems described in the literature are research prototypes, and the majority of them are no longer working or no longer accessible. At the time this paper was written, the only ones for Document Retrieval with a running Web interface or code that can be downloaded are: (1) Design Science's MathDex (formerly MathFind) (2) NIST DMLF (3) MathWebSearch (4) MiAS (Math Indexer and Searcher), also used to search the EuDML (5) the system described in [PZ14]. In addition to those, the following commercial systems are also accessible: (a) Springer LaTeXSearch (b) Wolfram Alpha. Systems (a), (1), (2) and (5) are based on Presentation MathML; systems (3) and (4) can use either Presentation MathML or Content MathML/parallel markup, but work better on the latter; finally system (b) actually generates on the fly most of the result of the query, for example by plotting functions, computing their Taylor expansion, etc. It does not really qualify then as a search engine.

Most interactive theorem provers also have their own implementation of a search engine to solve Formula Retrieval. Most of the time, the implementation is embedded in the system and does not work on the whole library at once, with the exception of MML Query for Mizar.

7   Conclusions

Mathematical knowledge retrieval, the low hanging fruit of Mathematical Knowledge Management, is still far from being grasped. Despite the significant amount of work dedicated to the topic in the last 12 years, only a few systems are still available, and their precision and recall scores compared to other knowledge retrieval fields are low. Moreover, usability and user requirement studies suggest that queries containing formulae — the main focus of the majority of papers — are perceived by users as not very useful (yet?).

The main contributions of this paper have been providing an hopefully comprehensive bibliography on the subject, and presenting taxonomies for both mathematical retrieval problems and techniques. We believe that our purpose driven taxonomy can be useful in classifying papers, in clarifying the scope of application of techniques and in the much needed development of unbiased benchmarks for mathematical retrieval.

References

ACK08. Adeel, M., Cheung, H.S., Khiyal, S.H.: Math GO! prototype of a content based mathematical formula search engine. J. Theor. Appl. Inf. Technol. 4(10), 1002–1012 (2008)

AGSC+06. Asperti, A., Guidi, F., Coen, C.S., Tassi, E., Zacchiroli, S.: A content based mathematical search engine: Whelp. In: Filliâtre, J.-C., Paulin-Mohring, C., Werner, B. (eds.) TYPES 2004. LNCS, vol. 3839, pp. 17–32. Springer, Heidelberg (2006)

AKO13. Aizawa, A., Kohlhase, M., Ounis, I.: NTCIR-10 math pilot task overview. In: Proceedings of the 10th NTCIR Conference, Tokyo, Japan, pp. 654–661 (2013)

AKO14. Aizawa, A., Kohlhase, M., Ounis, I.: NTCIR-11 math 2 task overview. In: Proceedings of 10th NTCIR Conference, Tokyo, Japan, pp. 88–98 (2014)

AS04. Asperti, A., Selmi, M.: Efficient retrieval of mathematical statements. In: Asperti, A., Bancerek, G., Trybulec, A. (eds.) MKM 2004. LNCS, vol. 3119, pp. 17–31. Springer, Heidelberg (2004)

AY07a. Altamimi, M.E., Youssef, A.: A more canonical form of content MathML to facilitate math search. In: Proceedings of Extreme Markup Languages (2007)

AY07b. Altamimi, M.E., Youssef, A.S.: Wildcards in math search, implementation issues. In: Proceedings of the ISCA 20th International Conference on Computer Applications in Industry and Engineering, CAINE 2007, San Francisco, California, USA, 7–9 November, pp. 90–96 (2007)

AY08a. Ahmadi, S.A., Youssef, A.: Lexical error compensation in handwritten-based mathematical information retrieval. In: Proceedings of Towards Digital Mathematics Library, DML 2008, Birmingham, UK, 27 July, pp. 43–54. Masaryk University, Brno (2008)

AY08b. Altamimi, M.E., Youssef, A.S.: A math query language with an expanded set of wildcards. Math. Comput. Sci. **2**(2), 305–331 (2008)

AZ04. Asperti, A., Zacchiroli, S.: Searching mathematics on the web: state of the art and future developments. In: Karlsruhe, F (ed.) Proceedings of New Developments in Electronic Publishing of Mathematics, pp. 9–18 (2004)

Ban06. Bancerek, G.: Information retrieval and rendering with MML query. In: Borwein, J.M., Farmer, W.M. (eds.) MKM 2006. LNCS (LNAI), vol. 4108, pp. 266–279. Springer, Heidelberg (2006)

BF03. Baumgartner, P., Furbach, U.: Automated deduction techniques for the management of personalized documents. Ann. Math. Artif. Intell. **38**(1–3), 211–228 (2003)

BR03. Bancerek, G., Rudnicki, P.: Information retrieval in MML. In: Asperti, A., Buchberger, B., Davenport, J.H. (eds.) Mathematical Knowledge Management, pp. 119–132. Springer, Bertinoro (2003)

BU04. Bancerek, G., Urban, J.: Integrated semantic browsing of the mizar mathematical library for authoring mizar articles. In: Asperti, A., Bancerek, G., Trybulec, A. (eds.) MKM 2004. LNCS, vol. 3119, pp. 44–57. Springer, Heidelberg (2004)

Cai04. Cairns, P.: Informalising formal mathematics: searching the mizar library with latent semantics. In: Asperti, A., Bancerek, G., Trybulec, A. (eds.) MKM 2004. LNCS, vol. 3119, pp. 58–72. Springer, Heidelberg (2004)

CDT04. Caprotti, O., Dewar, M., Turi, D.: Mathematical service matching using description logic and OWL. In: Asperti, A., Bancerek, G., Trybulec, A. (eds.) MKM 2004. LNCS, vol. 3119, pp. 73–87. Springer, Heidelberg (2004)

Del00. Delahaye, D.: Information retrieval in a *Coq* proof library using type isomorphisms. In: Coquand, T., Nordström, B., Dybjer, P., Smith, J. (eds.) TYPES 1999. LNCS, vol. 1956, pp. 131–147. Springer, Heidelberg (2000)

FLRS12. Formánek, D., Líška, M., Růžička, M., Sojka, P.: Normalization of digital mathematics library content. In: Davenport, J., Jeuring, J., Lange, C., Libbrecht, P. (eds.) Joint Proceedings of the 24th OpenMath Workshop, the 7th Workshop on Mathematical User Interfaces (MathUI), and the Work in Progress Section of the Conference on Intelligent Computer Mathematics. CEUR Workshop Proceedings, vol. 921, pp. 91–103. Neuveden, Aachen (2012)

GK14. Gauthier, T., Kaliszyk, C.: Matching concepts across HOL libraries. In: Watt, S.M., Davenport, J.H., Sexton, A.P., Sojka, P., Urban, J. (eds.) CICM 2014. LNCS, vol. 8543, pp. 267–281. Springer, Heidelberg (2014)

GPBB14. Pinto, J.M.G., Barthel, S., Balke, W.-T.: QUALIBETA at the NTCIR-11 math 2 task: an attempt to query math collections. In: Proceedings of the 10th NTCIR Conference, Tokyo, Japan, pp. 103–107 (2014)

GS03. Guidi, F., Schena, I.: A query language for a metadata framework about mathematical resources. In: Asperti, A., Buchberger, B., Davenport, J.H. (eds.) Mathematical Knowledge Management, pp. 105–118. Springer, Bertinoro (2003)

GWHT14. Gao, L., Wang, Y., Hao, L., Tang, Z.: ICST math retrieval system for NTCIR-11 math-2 Task. In: Proceedings of the 10th NTCIR Conference, Tokyo, Japan, pp. 99–102 (2014)

HHN08. Hashimoto, H., Hijikata, Y., Nishida, S.: Incorporating breadth first search for indexing MathML objects. In: IEEE International Conference on Systems, Man and Cybernetics, SMC 2008, pp. 3519–3523, October 2008

HKP14. Hambasan, R., Kohlhase, M., Prodescu, C.: MathWebSearch at NTCIR-11. In: Proceedings of the 10th NTCIR Conference, Tokyo, Japan, pp. 114–119 (2014)

HQ14. Haralambous, Y., Quaresma, P.: Querying geometric figures using a controlled language, ontological graphs and dependency lattices. In: Watt, S.M., Davenport, J.H., Sexton, A.P., Sojka, P., Urban, J. (eds.) CICM 2014. LNCS, vol. 8543, pp. 298–311. Springer, Heidelberg (2014)

HS13. Hagino, H., Saito, H.: Partial-match retrieval with structure-reflected indices at the NTCIR-10 math task. In: Proceedings of the 10th NTCIR Conference, Tokyo, Japan, pp. 692–695 (2013)

KK07. Kohlhase, A., Kohlhase, M.: Reexamining the MKM value proposition: from math web search to math web ReSearch. In: Kauers, M., Kerber, M., Miner, R., Windsteiger, W. (eds.) MKM/CALCULEMUS 2007. LNCS (LNAI), vol. 4573, pp. 313–326. Springer, Heidelberg (2007)

Koh14. Kohlhase, A.: Search interfaces for mathematicians. In: Watt, S.M., Davenport, J.H., Sexton, A.P., Sojka, P., Urban, J. (eds.) CICM 2014. LNCS, vol. 8543, pp. 153–168. Springer, Heidelberg (2014)

KP13. Kohlhase, M., Prodescu, C.: MathWebSearch at NTCIR-10. In: Proceedings of the 10th NTCIR Conference, Tokyo, Japan, pp. 675–679 (2013)

KT09. Kamali, S., Tompa, F.W.: Improving mathematics retrieval. In: Proceedings of Towards Digital Mathematics Library, DML 2009, Grand Bend, Ontario, Canada, 8–9 July, pp. 37–48. Masaryk University, Brno (2009)

KT10. Kamali, S., Tompa, F.W.: A new mathematics retrieval system. In: Proceedings of the 19th ACM International Conference on Information and Knowledge Management, CIKM 2010, pp. 1413–1416. ACM, New York (2010)

KT13. Kamali, S., Tompa, F.W.: Structural similarity search for mathematics retrieval. In: Carette, J., Aspinall, D., Lange, C., Sojka, P., Windsteiger, W. (eds.) CICM 2013. LNCS, vol. 7961, pp. 246–262. Springer, Heidelberg (2013)

KTHA14. Kristianto, G.Y., Topić, G., Ho, F., Aizawa, A.: The MCAT math retrieval system for NTCIR-11 math track. In: Proceedings of the 11th NTCIR Conference, Tokyo, Japan, pp. 120–126 (2014)

L10. Líška, M.: Searching Mathematical Texts (2010)

L13. Líška, M.: Evaluation of Mathematics Retrieval (2013)

LAP+14. Lipani, A., Andersson, L., Piroi, F., Lupu, M., Hanbury, A.: TUW-IMP at the NTCIR-11 Math-2. In: Proceedings of the 11th NTCIR Conference, Tokyo, Japan, pp. 143–146 (2014)

LDM+08. Libbrecht, P., Desmoulins, C., Mercat, C., Laborde, C., Dietrich, M., Hendriks, M.: Cross-curriculum search for intergeo. In: Autexier, S., Campbell, J., Rubio, J., Sorge, V., Suzuki, M., Wiedijk, F. (eds.) AISC 2008, Calculemus 2008, and MKM 2008. LNCS (LNAI), vol. 5144, pp. 520–535. Springer, Heidelberg (2008)

Lib13. Libbrecht, P.: Escaping the trap of too precise topic queries. In: Carette, J., Aspinall, D., Lange, C., Sojka, P., Windsteiger, W. (eds.) CICM 2013. LNCS, vol. 7961, pp. 296–309. Springer, Heidelberg (2013)

LM06. Libbrecht, P., Melis, E.: Methods to access and retrieve mathematical content in ACTIVEMATH. In: Iglesias, A., Takayama, N. (eds.) ICMS 2006. LNCS, vol. 4151, pp. 331–342. Springer, Heidelberg (2006)

LRG13. Ray R. Larson., Chloe J. Reynolds., Fredric C. Gey.: The Abject Failure of Keyword IR for Mathematics Search: Berkeley at NTCIR-10 Math. In: Proceedings of the 10th NTCIR Conference, Tokyo, Japan, pp. 662–666 (2013)

LSLM11. Líška, M., Sojka, P., Líška, M., Mravec, P.: Web interface and collection for mathematical retrieval: WebMIaS and MREC. In: Proceedings of Towards Digital Mathematics Library, DML 2011, Bertinoro, Italy, 20–21 July, pp. 77–84. Masaryk University, Brno (2011)

LSR13. Líška, M., Sojka, P., Růžička, M.: Similarity search for mathematics: Masaryk university team at the NTCIR-10 math task. In: Proceedings of the 10th NTCIR Conference, Tokyo, Japan, pp. 686–691 (2013)

LSR14. Líška, M., Sojka, P., Růžička, M.: Math indexer and searcher web interface. In: Watt, S.M., Davenport, J.H., Sexton, A.P., Sojka, P., Urban, J. (eds.) CICM 2014. LNCS, vol. 8543, pp. 444–448. Springer, Heidelberg (2014)

MG08a. Misutka, J., Galambos, L.: Mathematical extension of full text search engine indexer. In: 3rd International Conference on Information and Communication Technologies: From Theory to Applications, ICTTA 2008, pp. 1–6, April 2008

MG08b. Mišutka, J., Galamboš, L.: Extending full text search engine for mathematical content. In: Proceedings of Towards Digital Mathematics Library, DML 2008, Birmingham, UK, 27 July, pp. 55–67. Masaryk University, Brno (2008)

Mil13. Miller, B.R.: Three years of DLMF: web, math and search. In: Carette, J., Aspinall, D., Lange, C., Sojka, P., Windsteiger, W. (eds.) CICM 2013. LNCS, vol. 7961, pp. 288–295. Springer, Heidelberg (2013)

MM06. Munavalli, R., Miner, R.: Mathfind: a math-aware search engine. In: Proceedings of the 29th Annual International ACM SIGIR Conference on Research and Development in Information Retrieval, pp. 735–735. ACM (2006)

MM07. Miner, R., Munavalli, R.: An approach to mathematical search through query formulation and data normalization. In: Kauers, M., Kerber, M., Miner, R., Windsteiger, W. (eds.) MKM/CALCULEMUS 2007. LNCS (LNAI), vol. 4573, pp. 342–355. Springer, Heidelberg (2007)

MY03. Miller, B.R., Youssef, A.: Technical aspects of the digital library of mathematical functions. Ann. Math. Artif. Intell. 38(1–3), 121–136 (2003)

MY08. Miller, B.R., Youssef, A.M.: Augmenting presentation MathML for search. In: Autexier, S., Campbell, J., Rubio, J., Sorge, V., Suzuki, M., Wiedijk, F. (eds.) AISC 2008, Calculemus 2008, and MKM 2008. LNCS (LNAI), vol. 5144, pp. 536–542. Springer, Heidelberg (2008)

NCH12. Nguyen, T.T., Chang, K., Hui, S.C.: A math-aware search engine for math question answering system. In: Proceedings of the 21st ACM International Conference on Information and Knowledge Management, pp. 724–733. ACM (2012)

NK07. Normann, I., Kohlhase, M.: Extended formula normalization for ϵ-retrieval and sharing of mathematical knowledge. In: Kauers, M., Kerber, M., Miner, R., Windsteiger, W. (eds.) MKM/CALCULEMUS 2007. LNCS (LNAI), vol. 4573, pp. 356–370. Springer, Heidelberg (2007)

NKTA14. Nghiem, M.-Q., Kristianto, G.Y., Topić, G., Aizawa, A.: Which one is better: presentation-based or content-based math search? In: Watt, S.M., Davenport, J.H., Sexton, A.P., Sojka, P., Urban, J. (eds.) CICM 2014. LNCS, vol. 8543, pp. 200–212. Springer, Heidelberg (2014)

PZ14. Pattaniyil, N., Zanibbi, R.: Combining TF-IDF text retrieval with an inverted index over symbol pairs in math expressions: the tangent math search engine at NTCIR 2014. In: Proceedings of the 11th NTCIR Conference, Tokyo, Japan, pp. 135–142 (2014)

Rab12. Rabe, F.: A query language for formal mathematical libraries. In: Campbell, J.A., Jeuring, J., Carette, J., Dos Reis, G., Sojka, P., Wenzel, M., Sorge, V. (eds.) CICM 2012. LNCS, vol. 7362, pp. 143–158. Springer, Heidelberg (2012)

RSL14. Růžička, M., Sojka, P., Líška, M.: Math indexer and searcher under the hood: history and development of a winning strategy. In: Proceedings of the 11th NTCIR Conference, Tokyo, Japan, pp. 127–134 (2014)

SL11. Sojka, P., Líška, M.: The art of mathematics retrieval. In: Proceedings of the 11th ACM Symposium on Document Engineering, pp. 57–60. ACM (2011)

SLM13. Schubotz, M., Leich, M., Markl, V.: Querying large collections of mathematical publications: NTCIR10 math task. In: Proceedings of 10th NTCIR Conference, Tokyo, Japan, pp. 667–674 (2013)

SY07. Shatnawi, M., Youssef, A.: Equivalence detection using parse-tree normalization for math search. In: Proceedings of the Second IEEE International Conference on Digital Information Management (ICDIM), Lyon, France, 11–13 December, pp. 643–648 (2007)

SYM+14. Schubotz, M., Youssef, A., Markl, V., Cohl, H.S., Li, J.J.: Evaluation of similarity-measure factors for formulae based on the NTCIR-11 math task. In: Proceedings of 10th NTCIR Conference, Tokyo, Japan, pp. 108–113 (2014)

TKNA13. Topić, G., Kristianto, G.Y., Nghiem, M.-Q., Aizawa, A.: The MCAT math retrieval system for NTCIR-10 math track. In: Proceedings of 10th NTCIR Conference, Tokyo, Japan, pp. 680–685 (2013)

WG10. Wolska, M., Grigore, M.: Symbol declarations in mathematical writing. In: Proceedings of Towards Digital Mathematics Library, DML 2010, Paris, France, 7–8 July, pp. 119–127. Masaryk University, Brno (2010)

YA07. Youssef, A.S., Altamimi, M.E.: An extensive math query language. In: Proceedings of the 16th International Conference on Software Engineering and Data Engineering (SEDE-2007), 9–11 July, Imperial Palace Hotel Las Vegas, Las Vegas, Nevada, USA, pp. 57–63 (2007)

YA09. Yokoi, K., Aizawa, A.: An approach to similarity search for mathematical expressions using MathML. In: Proceedings of Towards Digital Mathematics Library, DML 2009, Grand Bend, Ontario, Canada, 8–9 July, pp. 27–35. Masaryk University, Brno (2009)

You05. Youssef, A.: Search of mathematical contents: issues and methods. In: Proceedings of the ISCA 14th International Conference on Intelligent and Adaptive Systems and Software Engineering, 20–22 July, Novotel Toronto Centre, Toronto, Canada, pp. 100–105 (2005)

You06. Youssef, A.M.: Roles of math search in mathematics. In: Borwein, J.M., Farmer, W.M. (eds.) MKM 2006. LNCS (LNAI), vol. 4108, pp. 2–16. Springer, Heidelberg (2006)

You07. Youssef, A.S.: Methods of relevance ranking and hit-content generation in math search. In: Kauers, M., Kerber, M., Miner, R., Windsteiger, W. (eds.) MKM/CALCULEMUS 2007. LNCS (LNAI), vol. 4573, pp. 393–406. Springer, Heidelberg (2007)

You08. Youssef, A.S.: Relevance ranking and hit description in math search. Math. Comput. Sci. 2(2), 333–353 (2008)

YS06. Youssef, A., Shatnawi, M.: Math search with equivalence detection using parse-tree normalization. In: The 4th International Conference on Computer Science and Information Technology (2006)

ZB12. Zanibbi, R., Blostein, D.: Recognition and retrieval of mathematical expressions. Int. J. Doc. Anal. Recogn. (IJDAR) 15(4), 331–357 (2012)

ZKT08. Zhao, J., Kan, M.-Y., Theng, Y.L.: Math information retrieval: user requirements and prototype implementation. In: Proceedings of the 8th ACM/IEEE-CS Joint Conference on Digital Libraries, pp. 187–196. ACM (2008)

ZY14. Zhang, Q., Youssef, A.: An approach to math-similarity search. In: Watt, S.M., Davenport, J.H., Sexton, A.P., Sojka, P., Urban, J. (eds.) CICM 2014. LNCS, vol. 8543, pp. 404–418. Springer, Heidelberg (2014)

Towards the Formalization of Fractional Calculus in Higher-Order Logic

Umair Siddique[1]([⊠]), Osman Hasan[2], and Sofiène Tahar[1]

[1] Department of Electrical and Computer Engineering,
Concordia University, Montreal, QC, Canada
{muh_sidd,tahar}@ece.concordia.ca
[2] School of Electrical Engineering and Computer Science,
National University of Sciences and Technology, Islamabad, Pakistan
osman.hasan@seecs.nust.edu.pk
http://save.seecs.nust.edu.pk/projects/fc.html

Abstract. Fractional calculus is a generalization of classical theories of integration and differentiation to arbitrary order (i.e., real or complex numbers). In the last two decades, this new mathematical modeling approach has been widely used toanalyze a wide class of physical systems in various fields of science and engineering. In this paper, we describe an ongoing project which aims at formalizing the basic theories of fractional calculus in the HOL Light theorem prover. Mainly, we present the motivation and application of such formalization efforts, a roadmap to achieve our goals, current status of the project and future milestones.

Keywords: Fractional calculus · Higher-Order logic · Theorem proving

1 Motivation and Background

Physical and engineering systems are classified as continuous, discrete or hybrid depending upon the nature of underlying system parameters. The rich theories of mathematics provide the necessary tools to study the behaviour of such systems ranging from very small biological organisms to the modern Quantum mechanical phenomenons. Generally, differential equations [39] and difference equations [12] are used to characterize the dynamics of these systems. Consequently, the concept of higher-order differentiation and integration are widely studied in diverse disciplines of science and engineering. For example, it is well understood that the first derivative ($\frac{d}{dt}f(t)$) and second derivative ($\frac{d^2}{dt^2}f(t)$) of a function describe the rate of change and measure of concavity, respectively. However, we rarely think what if the order (n) of higher-order derivative ($\frac{d^n}{dt^n}$) becomes a real, complex or an irrational number? One immediate question arises in our minds is the existence or possibility of such a concept in mathematics. Interestingly, this seemingly new concept dates back to 1695 when L'Hôpital asked Leibniz regarding his notation $\frac{d^n y}{dx^n}$: "*what if n is $\frac{1}{2}$*". In reply, Leibniz [20] prophesied in his letter, "*... Thus it follows that $d^{\frac{1}{2}}x$ will be equal to $x\sqrt{dx : x}$.*

© Springer International Publishing Switzerland 2015
M. Kerber et al. (Eds.): CICM 2015, LNAI 9150, pp. 316–324, 2015.
DOI: 10.1007/978-3-319-20615-8_21

This is an apparent paradox from which, one day, useful consequences can be drawn ...". Leibniz's initial work on the problem of defining the derivative of arbitrary order gave birth to a new field of research in mathematics (called *fractional calculus*) and attracted the attention of many physicists, engineers and geometers. Some of the great mathematicians and physicists who touched the field of fractional calculus are Riemann, Liouville, Laurent, Heaviside and Riesz [24].

The concept of fractional calculus has great potential to change the way we model and analyze the systems. It provides good opportunity to scientists and engineers for revisiting the origins. We briefly outline some of the the main applications of fractional calculus in Table 1. The importance of fractional calculus can be realized by the following quote from Miller and Ross [24]. They stated:

> *"... The fractional calculus finds use in many fields of science and engineering, including fluid flow, rheology, diffusive transport akin to diffusion, electrical networks, electromagnetic theory, and probability.... It seems that hardly a field of science or engineering has remained untouched by this topic ..."*

Nowadays engineering systems exhibiting fractional order dynamics are increasingly used in some safety-critical applications such as control systems,

Table 1. Applications of fractional calculus

Field	Applications
Control engineering	- System identification [17]
	- Biomimetic (bionics) control [7]
	- Trajectory control [11]
	- Temperature control [30]
	- Fractional PI^{α} controller [23]
Signal processing	- Fractional order integrator [19]
	- Fractional order FIR differentiator [37]
	- IIR-type fractional order differentiator [38]
	- Modeling of speech signals [18]
Image processing	- Image restoration and edge detection [29]
	- Satellite image classification [6]
Electromagnetics	- Fractional curl operators [13, 25]
	- Fractional rectangular waveguides [14]
Communication	- Secure chaotic communication [1]
	- Informational network traffic modeling [40]
Biology	- Neuron modeling [5]
	- Biophysical processes [8]
	- Modeling of complex dynamics of tissues [22]

signal processing, electromagnetics and electrical networks (as listed in Table 1). For example, fractional meta-materials based devices are used to build sensitive military and defence equipments and electromagnetic stealth technology [21]. Considering these facts, it is quite interesting and important to build a logical reasoning framework which can be used to formally verify such sophisticated applications within the sound core of a proof assistant. In fact, proof assistants have been successfully used to formalize and verify some challenging and paradoxical mathematical results, e.g., the formal proofs of the Kepler Conjecture (Flyspeck project) [16] and the Odd Order Theorem [15].

In this paper, we present details of an ongoing project[1] to develop a formal reasoning support for fractional calculus in higher-order-logic theorem prover. This project was originally started at the System Analysis and Verification (SAVe) lab[2] in 2010. Earlier formalization was done in the HOL4 theorem prover with the main focus on fractional operators for real-valued functions and the verification of fractional order electrical components. Later on, the scope of the project was expanded to formalize fractional calculus involving complex-valued functions due to its various engineering applications (as listed in Table 1). Currently, we are using the HOL Light theorem prover due to the availability of rich multivariate analysis libraries including Harrison's recent formalization of complex-valued Gamma function[3] as well as the interesting related projects like Flyspeck [16] and the formalization of optics theories (i.e., ray, wave, electromagnetic and quantum) [2].

The rest of the paper is organized as follows: In Sect. 2, we briefly review some commonly used notations and definitions of fractional order operators. We provide an outline of the proposed formalization framework in Sect. 3. Consequently, the current status of the formalization and future milestones are discussed in Sect. 4. Finally, we conclude the paper in Sect. 5.

2   Mathematical Framework of Fractional Calculus

There are different notations available for fractional derivatives and integrals. We use $J_a^v f(x)$ and $D^v f(x)$ for fractional integral and fractional derivative, respectively. In these notations, v is the order of integration or differentiation and a is the lower limit of integration.

For every function $(f : \mathbb{C} \to \mathbb{C})$; and for every number $v \in \mathbb{R}$ or $\mathbb{C}$, J_a^v and D^v should be related to f by the following criteria [10].

1. If $f(x)$ is an analytic function, then $J_a^v f(x)$ and $D^v f(x)$ must also be an analytic function of the variable x and of the order v of integration or differentiation.
2. The operations $J_a^v f(x)$ and $D^v f(x)$ must produce the same result as ordinary integration/differentiation when v is a positive integer.

---

[1] http://save.seecs.nust.edu.pk/projects/fc.html.

[2] http://save.seecs.nust.edu.pk.

[3] https://code.google.com/p/hol-light/source/browse/trunk/Multivariate/gamma.ml.

3. The fractional operators must be linear.

$$J_a^v[\alpha f(x) + \beta g(x)] = \alpha J_a^v f(x) + \beta J_a^v g(x) \tag{1}$$

$$D^v[\alpha f(x) + \beta g(x)] = \alpha D^v f(x) + \beta D^v g(x) \tag{2}$$

4. The operation of order zero must leave the function unchanged.

$$J_a^0 f = f \quad \text{and} \quad D^0 f = f \tag{3}$$

5. The law of exponents must hold for integration and differentiation of arbitrary order under sufficient conditions on function f.

$$J_a^u(J_a^v f) = J_a^{u+v} f \quad \text{and} \quad D^u(D^v f) = D^{u+v} f \tag{4}$$

Fractional integrals and fractional derivatives are also referred to as Differintegrals [27] and there are more than ten well-known definitions for Differintegrals [9]. We describe here two of them, which are most widely used in analyzing real-world problems. These are the Riemann-Liouville and Grünwald-Letnikov definitions, which are also equivalent for a wide class of functions [31].

Riemann-Liouville (RL) Definition:

$$J_a^v f(x) = \frac{1}{\Gamma(v)} \int_a^x (x-t)^{v-1} f(t) dt \tag{5}$$

where $J_a^v f(x)$ represents fractional integration with order v and lower integration limit a. The parameter $a = 0$ gives the Riemann definition and $a = -\infty$ gives the Liouville definition of fractional integration. Indeed Eq. (5) is the generalization of Cauchy's repeated integration formula to non-integer v [32]. Where Γ (.) in the above definition denotes the Gamma function which is defined using the well-known improper integral as follows:

$$\Gamma(z) = \int_0^\infty t^{z-1} e^{-t} dt \tag{6}$$

for $Re(z) > 0$.

The fractional differentiation is given as follows:

$$D^v f(x) = (\frac{d}{dx})^m J_a^{m-v} f(x) \tag{7}$$

where m represents the ceiling of v, i.e., $\lceil v \rceil$.

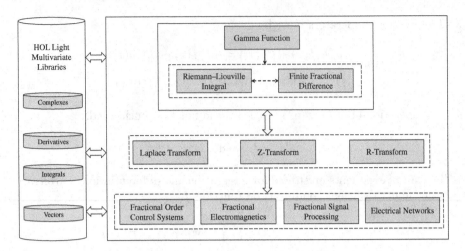

Fig. 1. Formalization framework for fractional calculus

Grünwald-Letnikov (GL) Definition:

$$_cD_x^v f(x) = \lim_{h \to 0} h^{-v} \sum_{k=0}^{[\frac{x-c}{h}]} (-1)^k \binom{v}{k} f(x - kh) \tag{8}$$

Grünwald-Letnikov definition caters for both fractional differentiation and integration, as positive values of v give fractional differentiation and negative values of v give fractional integration. Here, $\binom{v}{k}$ represents the binomial coefficients, which are described in terms of the Gamma function.

3   Formal Analysis Framework

The proposed framework, given in Fig. 1, outlines the main ideas and roadmap to formalize the basic theory behind fractional calculus. The whole framework can be decomposed mainly into three major parts which are the formalization of the core definitions of fractional order operators, formalization of supporting transformations (i.e., Laplace transform [26], Z-Transform [28] and R-Transform [3]) and engineering applications. The first part heavily relies upon the Gamma function as mentioned in Sect. 2. So the core step is to formalize the Gamma function in higher-order logic (HOL) and verify its important properties. Consequently, any definition of fractional order operators can be formalized in HOL. However, our focus is two main definitions, i.e., Riemann-Liouville (RL) and fractional difference which indeed represent continuous and discrete versions of fractional order operators, respectively. This step also involves the validation of all the properties mentioned in Sect. 2. This requires some important results of multivariate calculus such as the notion of Lebesgue measurability and Fubini's theorem which provides the reasoning support for iterated and double integrals. Interestingly, both of these requirements are available in the multivariate analysis libraries of HOL Light. The second step is the formalization of important integral transforms which are necessary to analytically solve linear fractional differential and difference equations. We mainly focus on three transforms, namely the Laplace transform, the Z-Transform, and the recently introduced R-Transform [4]. All of these transformations are used to transform complicated fractional differential (or difference) equations to algebraic equations which are easier to manipulate and to deduce interesting properties. Building upon these fundamentals, our ultimate goal is to formally verify a variety of engineering systems including control systems, signal processing, electromagnetics and electrical networks. All formalization steps make use of different multivariate theories of HOL Light, e.g., derivatives, integrals, complex vectors and measure spaces. Finally, the developed libraries of this project will become part of the existing HOL Light libraries.

4   Current Status and Future Milestones

As mentioned earlier, the project initially considered only real-order fractional operators in the HOL4 theorem prover. The main difficulty was the little support to handle improper integrals which was required to formalize the real-valued Gamma function. Therefore, we extended the integration theory of HOL4 by formalizing a variant of improper integrals using sequential limits. This was then used to formalize the Gamma function and verify some of its main properties, such as the pseudo-recurrence relation ($\Gamma(z+1) = z\Gamma(z)$), the functional equation ($\Gamma(1) = 1$) and the factorial generalization ($\Gamma(k+1) = k!$) [34]. We utilized these foundations to formalize Differintegrals, given in Eqs. (5) and (7), which in turn can be used to represent the dynamics of fractional order systems in higher-order logic. We also verified theorems corresponding to some commonly used properties of Differintegrals namely Identity and Linearity. Consequently, we conducted the formal analysis of a fractional order electrical component namely resistoductor, a fractional integrator and a fractional differentiator circuit [33]. Later on, the scope of the project was revised to include complex-valued functions and complex order fractional operators in HOL Light. The main requirement was to formalize the complex-valued Gamma function, Laplace and Z-Transforms. However, the Gamma function[4] was formalized by Harrison in early 2014. In the meantime, we formalized the basic theories of the Laplace transform [36] and the Z-Transform [35]. Currently, we are working on three main topics which include: (1) formal proofs of the uniqueness of Laplace and Z-Transforms which are required to formally verify the inverses of these transforms; (2) vectorial Z-Transform, which extend the simple Z-Transform over complex vectors; and (3) fractional difference equations, which are mainly based on Gamma function,

---

[4] https://code.google.com/p/hol-light/source/browse/trunk/Multivariate/gamma.ml.

infinite summations and products over complex functions. Finally, we outline the major tasks to achieve the future milestones as follows:

- Formalization of R-Transform.
- Formalization of Differintegrals for complex-valued functions. This is mainly the generalization of the formalization which was developed in HOL4.
- Formalization of linear fractional differential and difference equations with support to analytical solutions using the transform methods.

During the course of this project, two master and two Ph.D. students have contributed to the formalization. Interestingly, all of them are mainly electrical engineers without prior background of formal methods and higher-order-logic theorem proving. Given the complexity and interdisciplinarity of this research project, it is quite encouraging to see people with an engineering (or physics) background to use proof assistants as a complementary tool. The formalization of the fractional calculus is quite challenging as it requires advanced mathematical concepts of vector integration and Lebesgue measurable functions, etc. So expertise in formal reasoning about these complex mathematical phenomena is required for this formalization, which is quite unique compared to reasoning about software and digital hardware systems. The learning curve of HOL Light varies from student to student. Generally, students start proving basic math equations after a couple of months and the pace of formalization increases over time. Learning HOL Light libraries is not difficult once the basic concepts have been grasped by the user. The formalization of the improper integrals, the Gamma function, the fractional calculus, the Z-Transform and the Laplace transform is approximately 15,000 lines of HOL Light code. One of the major obstacles in the formalization was the identification of suitable mathematical definitions and models. Sometimes, textbook proofs do not follow due to various reasons (corner cases, or the proof steps are too abstract, etc.) and they needed to be re-proved on paper with subtle details. Consequently, we have to modify the definitions and thus change the proofs. But now the current formalization seems quite stable as most of the classical properties have been formally verified for our definitions. We believe that future developments can be built on the foundations that have been formalized as most of the work is for general systems. Finally, another important aspect of this project is the potential to apply developed theories to various applications other than fractional calculus. For example, we demonstrated the use of the Gamma function in probability theory [34], the Z-Transform in signal processing [35], and the Laplace transform in power electronics [36].

5   Conclusion

In this paper, we mainly presented the motivation and ongoing activities of our long term project about the formalization of fractional calculus in the HOL Light theorem prover. The main contribution of this project is a comprehensive framework of formal definitions and theorems about fractional calculus which can be used to verify modern control, signal processing and electromagnetic systems. Some future directions and recommendations for HOL Light are the improvements in the visualization of proofs, better automation and more accessible tutorials with examples from different engineering/physics topics.

References

1. Pariz, N., Kiani-B, A., Fallahi, K., Leung, H.: A Chaotic secure communication scheme using fractional chaotic systems based on an extended fractional kalman filter. Commun. Nonlinear Sci. Numer. Simul. **14**, 863–879 (2009)
2. Afshar, S.K., Siddique, U., Mahmoud, M.Y., Aravantinos, V., Seddiki, O., Hasan, O., Tahar, S.: Formal analysis of optical systems. Math. Comput. Sci. **8**(1), 39–70 (2014)
3. Atici, F.M.: A transform method in discrete fractional calculus. Int. J. Diff. Equ. **2**(2), 165–176 (2007)
4. Atici, F.M., Eloe, P.W.: Initial value problems in discrete fractional calculus. Proc. Am. Math. Soc. **137**(3), 981–989 (2009)
5. Auastasio, T.J.: The fractional-order dynamics of brainstem vestibulo-oculomotor neurons. Biol. Cybern. **72**(1), 69–79 (1994)
6. Cuestab, E., Quintanoa, C.: Improving satellite image classification by using fractional type convolution filtering. Int. J. Appl. Earth Obs. Geoinf. **12**(4), 298–301 (2010)
7. Chen, Y.Q., Xue, D., Dou, H.: Fractional calculus and biomimetic control. In: Robotics and Biomimetics, pp. 901–906. IEEE (2004)
8. Marín, M., Domínguez, D.M., Camacho, M.: Macrophage ion currents are fit by a fractional model and therefore are a time series with memory. Eur. Biophys. J. **38**(4), 457–464 (2009)
9. Dalir, M., Bashour, M.: Application of fractional calculus. Appl. Frac. Calc. Phys. **4**(21), 12 (2010)
10. Das, S.: Functional Fractional Calculus for System Identification and Controls. Springer, Heidelberg (2007)
11. Duarte, F.B.M., Machado, J.A.T.: Pseudoinverse trajectory control of redundant manipulators: a fractional calculus perspective. In: International Conference on Robotics and Automation, pp. 2406–2411. IEEE (2002)
12. Elaydi, S.: An Introduction to Difference Equations. Springer, New York (2005)
13. Engheta, N.: Fractional curl operator in electromagnetics. Microw. Opt. Technol. Lett. **17**(2), 86–91 (1998)
14. Faryad, M., Naqvi, Q.A.: Fractional rectangular waveguide. Progr. Electromagn. Res. PIER **75**, 383–396 (2007)
15. Gonthier, G., Asperti, A., Avigad, J., Bertot, Y., Cohen, C., Garillot, F., Le Roux, S., Mahboubi, A., O'Connor, R., Ould Biha, S., Pasca, I., Rideau, L., Solovyev, A., Tassi, E., Théry, L.: A machine-checked proof of the odd order theorem. In: Blazy, S., Paulin-Mohring, C., Pichardie, D. (eds.) ITP 2013. LNCS, vol. 7998, pp. 163–179. Springer, Heidelberg (2013)
16. Hales, T.C.: Introduction to the Flyspeck project. In: Mathematics, Algorithms, Proofs, volume 05021 of Dagstuhl Seminar Proceedings, pp. 1–11 (2005)
17. Hartley, T.T., Lorenzo, C.F.: Fractional system identification: an approach using continuous order distributions. Technical report, National Aeronautics and Space Administration, Glenn Research Cente NASA TM (1999)

18. Ahmad, W.M., Assaleh, K.: Modeling of speech signals using fractional calculus. In: International Symposium on Signal Processing and Its Applications, pp. 1–4. IEEE (2007)
19. Krishna, B.T., Reddy, K.V.V.S.: Design of digital differentiators and integrators of order $\frac{1}{2}$. World J. Model. Simul. **4**, 182–187 (2008)
20. Leibnitz, G.W.: Leibnitzens Mathematische Schriften. SIGDA News Lett. **2**, 301–302 (1962)
21. Lurie, K.A.: An Introduction to the Mathematical Theory of Dynamic Materials. Springer, US (2007)
22. Magin, R.L.: Fractional calculus models of complex dynamics in biological tissues. Comput. Math. Appl. **59**, 1586–1593 (2010)
23. Maione, G., Lino, P.: New tuning rules for fractional pi^{α} controllers. Nonlinear Dyn. **49**(1–2), 251–257 (2007)
24. Miller, K.S., Ross, B.: An Introduction to Fractional Calculus and Fractional Differential Equations. Willey, New York (1993)
25. Naqvi, Q.A., Abbas, M.: Complex and higher order fractional curl operator in electromagnetics. Opt. Commun. **241**, 349–355 (2004)
26. Ogata, K.: Modern Control Engineering. Prentice Hall, Boston (2010)
27. Oldham, K.B., Spanier, J.: The Fractional Calculus. Academic Press, New York (1974)
28. Oppenheim, A.V., Schafer, R.W., Buck, J.R.: Discrete-Time Signal Processing. Prentice Hall, Upper Saddle River (1999)
29. Mathieu, B., Melchior, P., Oustaloup, A., Ceyral, C.: Fractional differentiation for edge detection. Signal Process. **83**, 2421–2432 (2003)
30. Petrás, I., Vinagre, B.M.: Practical application of digital fractional-order controller to temperature control. Acta Montan. Slovaca **7**(2), 131–137 (2002)
31. Podlubny, I.: Fractional Differential Equations. Academic Press, New York (1999)
32. Ross, B.: A brief history and exposition of the fundamental theory of fractional calculus. In: Ross, B. (ed.) Fractional Calculus and Its Applications. Lecture Notes in Mathematics, vol. 457, pp. 1–36. Springer, Heidelberg (1975)
33. Siddique, U., Hasan, O.: Formal analysis of fractional order systems in HOL. In: Formal Methods in Computer Aided Design, pp. 163–170. IEEE (2011)
34. Siddique, U., Hasan, O.: On the formalization of gamma function in HOL. J. Autom. Reason. **53**(4), 407–429 (2014)
35. Siddique, U., Mahmoud, M.Y., Tahar, S.: On the formalization of Z-Transform in HOL. In: Klein, G., Gamboa, R. (eds.) ITP 2014. LNCS, vol. 8558, pp. 483–498. Springer, Heidelberg (2014)
36. Taqdees, S.H., Hasan, O.: Formalization of laplace transform using the multivariable calculus theory of HOL-light. In: McMillan, K., Middeldorp, A., Voronkov, A. (eds.) LPAR-19 2013. LNCS, vol. 8312, pp. 744–758. Springer, Heidelberg (2013)
37. Tseng, C.C.: Design of fractional order digital FIR differentiators. IEEE Signal Process. Lett. **8**(3), 77–79 (2001)
38. Vinagreb, B.M., Chena, Y.Q.: Fractional differentiation for edge detection. Signal Process. **83**, 2359–2365 (2003)
39. Yang, X.S.: Mathematical Modeling with Multidisciplinary Applications. Wiley, New Jersey (2013)
40. Zaborovsky, V., Meylanov, R.: Informational network traffic model based on fractional calculus. In: Proceedings of the International Conference Info-tech and Info-net, pp. 58–63. IEEE (2001)

LeoPARD — A Generic Platform for the Implementation of Higher-Order Reasoners

Max Wisniewski$^{(\boxtimes)}$, Alexander Steen, and Christoph Benzmüller

Department of Mathematics and Computer Science,
Freie Universität Berlin, Berlin, Germany
{max.wisniewski,a.steen,c.benzmueller}@fu-berlin.de

Abstract. LEOPARD supports the implementation of knowledge representation and reasoning tools for higher-order logic(s). It combines a sophisticated data structure layer (polymorphically typed λ-calculus with nameless spine notation, explicit substitutions, and perfect term sharing) with an ambitious multi-agent blackboard architecture (supporting prover parallelism at the term, clause, and search level). Further features of LEOPARD include a parser for all TPTP dialects, a command line interpreter, and generic means for the integration of external reasoners.

1 Introduction

LEOPARD (**Leo's P**arallel **AR**chitecture and **D**atastructures) is designed as a generic system platform for implementing higher-order (HO) logic based knowledge representation, and reasoning tools. In particular, LEOPARD provides the base layer of the new HO automated theorem prover (ATP) LEO-III, the successor of the well known provers LEO-I [4] and LEO-II [7].

Previous experiments with LEO-I and the OANTS mechanism [5] indicate a flexible, multi-agent blackboard architecture is well-suited for automating HO logic [6]. However, (due to project constraints) such an approach has not been realized in LEO-II. Instead, the focus has been on the proof search layer in combination with a simple, sequential collaboration with an external first-order (FO) ATP. LEO-II also provides improved term data structures, term indexing, and term sharing mechanisms, which unfortunately have not been optimally exploited at the clause and the proof search layer. For the development of LEO-III the philosophy therefore has been to allocate sufficient resources for the initial development of a flexible and reusable system platform. The goal has been to bundle, improve, and extend the features with the highest potential of the predecessor systems LEO-I, LEO-II and OANTS.

The result of this initiative is LEOPARD[1], which is written in Scala and currently consists of approx. 13000 lines of code. LEOPARD combines a sophisticated data structure layer [21] (polymorphically typed λ-calculus with nameless

---

This work has been supported by the DFG under grant BE 2501/11-1 (LEO-III).

[1] LEOPARD can be download at: https://github.com/cbenzmueller/LeoPARD.git.

M. Kerber et al. (Eds.): CICM 2015, LNAI 9150, pp. 325–330, 2015.
DOI: 10.1007/978-3-319-20615-8_22

spine notation, explicit substitutions, and perfect term sharing), with a multi-agent blackboard architecture [25] (supporting prover parallelism at the term, clause, and search level) and further tools including a parser for all TPTP [22, 23] syntax dialects, generic support for interfacing with external reasoners, and a command line interpreter. Such a combination of features and support tools is, up to the authors knowledge, not matched in related HO reasoning frameworks.

The intended users of the LEOPARD package are implementors of HO knowledge representation and reasoning systems, including novel ATPs and model finders. In addition, we advocate the system as a platform for the integration and coordination of heterogeneous (external) reasoning tools.

2   Term Data Structure

Data structure choices are a critical part of a theorem prover and permit reliable increases of overall performance when implemented and exploited properly. Key aspects for efficient theorem proving have been an intensive research topic and have reached maturity within FO-ATPs [19, 20]. Naturally, one would expect an even higher impact of the data structure choices in HO-ATPs. However, in the latter context, comparably little effort has been invested yet – probably also because of the inherently more complex nature of HO logic.

Term Language. The LEOPARD term language extends the simply typed λ-calculus with parametric polymorphism, yielding the second-order polymorphically typed λ-calculus (corresponding to $\lambda 2$ in Barendregt's λ-cube [3]). In particular, the system under consideration was independently developed by Reynolds [16] and Girard [14] and is commonly called System F today. Further extensions, for example to admit dependent types, are future work.

Thus, LEOPARD supports the following type and term language:

$$\tau, \nu ::= t \in T \quad \text{(Base type)}$$
$$\mid \alpha \qquad \text{(Type variable)}$$
$$\mid \tau \to \nu \quad \text{(Abstraction type)}$$
$$\mid \forall \alpha.\ \tau \quad \text{(Polymorphic type)}$$

$$s, t ::= X_\tau \in \mathcal{V}_\tau \quad \mid c_\tau \in \Sigma \qquad \text{(Variable / Constant)}$$
$$\mid (\lambda x_\tau\ s_\nu)_{\tau \to \nu} \mid (s_{\tau \to \nu}\ t_\tau)_\nu \qquad \text{(Term abstr. / appl.)}$$
$$\mid (\Lambda \alpha\ s_\tau)_{\forall \alpha\ \tau} \mid (s_{\forall \alpha\ \tau}\ \nu)_{\tau[\alpha/\nu]} \quad \text{(Type abstr. / appl.)}$$

An example term of this language is:

$$\Lambda \alpha \lambda P_{\alpha \to o}\ ((f_{\forall \beta\ (\beta \to o) \to o \to o}\ \alpha)\ (\lambda Y_\alpha\ P\ Y))\ T_o.$$

Nameless Representation. Internally, LEOPARD employs a locally nameless representation (both at the type and term level), that extends de-Bruijn indices to (bound) type variables [15]. The definition of de-Bruijn indices [11] for type variables is analogous to the one for term variables. Thus, the above example term is represented namelessly as

$$\left(\Lambda\lambda_{\underline{1}\to o} \left((f_{\forall(\underline{1}\to o)\to o\to o} \ \underline{1}) \ (\lambda_{\underline{1}} \ 2 \ 1) \right) T_o \right)$$

where de-Bruijn indices for type variables are underlined.

Spine Notation and Explicit Substitutions. On top of nameless terms, LEOPARD employs spine notation [12] and explicit substitutions [1]. The first technique allows quick head symbol queries, and efficient left-to-right traversal, e.g. for unification algorithms. The latter augments the calculus with substitution closures that admit efficient (partial) β-normalization runs. Internally, the above example reads

$$\Lambda\lambda_{\underline{1}\to o} \ f_{\forall(\underline{1}\to o)\to o\to o} \cdot (\underline{1}; \lambda_{\underline{1}} \ 2 \cdot (1); T)$$

where $\cdot$ combines function *heads* to argument lists (*spines*) in which ; denotes concatenation of arguments.

Term Sharing/Indexing. Terms are perfectly shared within LEOPARD, meaning that each term is only constructed once and then reused between different occurrences. This does not only reduce memory consumption in large knowledge bases, but also allows constant-time term comparison for syntactic equality using the term's pointer to its unique physical representation. For fast (sub-)term retrieval based on syntactical criteria (e.g. head symbols, subterm occurrences, etc.) from the term indexing mechanism, terms are kept in β-normal η-long form.

Suite of Normalization Strategies. LEOPARD comes with a number of different (heuristic) β-normalization strategies that adjust the standard leftmost-outermost strategy with different combinations of strict and lazy substitution composition resp. normalization and closure construction. η-normalization is invariant wrt. β-normalization of spine terms and hence η-normalization (to long form) is applied only once for each freshly created term.

Evaluation and Findings. A recent empirical evaluation [21] has shown that there is *no single best reduction strategy* for HO-ATPs. More precisely, for different TPTP problem categories this study identified different best reduction strategies. This motivates future work in which machine learning techniques could be used to suggest suitable strategies.

3   Multi-agent Blackboard Architecture

In addition to supporting classical, sequential theorem proving procedures, LEOPARD offers means for breaking the global ATP loop down into a set of subtasks that can be computed in parallel. This also includes support for subprover parallelism as successfully employed, for example, in Isabelle/HOL's Sledgehammer tool [8]. More generally, LEOPARD is construed to enable parallelism at various levels inside an ATP, including the term, clause, and search level [9]. For this, LEOPARD provides a flexible multi-agent blackboard architecture.

Blackboard Architecture. Process communication in LEOPARD is realized indirectly via a blackboard architecture [24]. The LEOPARD blackboard [25] is a collection of globally shared and accessible data structures which any process, i.e. agent, can query and manipulate at any time in parallel. From the blackboard's perspective each process is a specialist responsible for exactly one kind of problem. The blackboard is generic in the data structures, i.e. it allows the programmer to add various kinds data structures for any kind of data. Insertion into the data structures is handled by the blackboard. Hence, each specialist can indeed by specialized on a single data structure.

The LEoPARD blackboard mechanism and associated data structures provide specific support for nested and-or search trees, meaning that sets of formulae can be split into (nested) and-or contexts. Moreover, for each supercontext respective TPTP SZS status [22] information is automatically inferred from the statuses of its subcontexts.

Agents. In LEoPARD specialist processes can be modeled as agents [25]. Classically, agents are composed of three components: environment perception, decision making, and action execution [24].

The perception of LEoPARD agents is trigger-based, meaning that each agent is notified by a change in the blackboard. LEoPARD agents are to be seen as homomorphisms on the blackboard data together with a filter when to apply an action. Depending on the perceived change of the resp. state of the blackboard an agent decides on an action it wants to execute.

Auction Scheduler. Action execution in LEoPARD is coordinated by an auction based scheduler, which implements an own approximation algorithm [25] for combinatorical auctions [2]. More precisely, each LEoPARD agent computes and places a bid for the execution of its action(s). The auction based scheduler then tries to maximize the global benefit of the particular set of actions to choose.

This selection mechanism works uniformly for all agents that can be implemented in LEoPARD. Balancing the value of the actions is therefore crucial for the performance and the termination of the overall system. A possible generic solution for the agents bidding is to apply machine learning techniques to optimize the bids for the best overall performance. This is future work.

Note that the use of advanced agent technology in LEoPARD is optional. A traditional ATP can still be implemented, for example, as a single, sequential reasoner instantiating exactly one agent in the LEoPARD framework.

Agent Implementation Examples. For illustration purposes, some agent implementations have been exemplarily included in the LEoPARD package. For example, simple agents for *simplification, skolemization, prenex-form, negation-normal-form* and *paramodulation* are provided. Moreover, the agent-based integration of external ATPs is demonstrated and their parallelization is enabled by the LEoPARD agent framework. This includes agents embodying LEO-II and Satallax [10] running remotely on the SystemOnTPTP [22] servers in Miami. These example agents can be easily adapted for other TPTP compliant ATPs.

Each example agent comes with an applicability filter, an action definition and an auction value computation. The provided agents suffice to illustrate the working principles of the LEoPARD multi-agent blackboard architecture to interested implementors. After the official release of LEO-III, further, more sophisticated agents will be included and offered for academic reuse.

4 Other Components

The LEOPARD framework provides useful further components. For example, a generic parser is provided that supports all TPTP syntax dialects. Moreover, a command line interpreter supports fine grained interaction with the system. This is useful not only for debugging but also for training and demonstration purposes. As pointed at above, useful support is also provided for the integration of external reasoners based on the TPTP infrastructure. This also includes comprehensive support for the TPTP SZS result ontology. Moreover, ongoing and future work aims at generic means for the transformation and integration of (external) proof protocols, ideally by exploiting results of projects such as ProofCert[2].

5 Related Work

There is comparably little related work to LEOPARD, since higher-order theorem provers typically implement their own data structures. Related systems (mostly concerning term representation) include λProlog and Teyjus [17], the Abella interactive theorem prover [13], and the logical framework Twelf [18].

Acknowledgements.. We thank the reviewers for their valuable feedback. Moreover, we thank Tomer Libal and the students of the LEO-III project for their contributions to LEOPARD.

References

1. Abadi, M., Cardelli, L., Curien, P.-L., Levy, J.-J.: Explicit substitutions. In: Proceedings of the 17th ACM SIGPLAN-SIGACT Symposium on Principles of Programming Languages, POPL 1990, pp. 31–46, ACM, New York (1990)
2. Arrow, K.J.: Social Choice and Individual Values. Wiley, New York (1951)
3. Barendregt, H.P.: Introduction to generalized type systems. J. of Funct. Program. **1**(2), 125–154 (1991)
4. Benzmüller, C.E., Kohlhase, M.: System description: LEO - a higher-order theorem prover. In: Kirchner, C., Kirchner, H. (eds.) CADE 1998. LNCS (LNAI), vol. 1421, p. 139. Springer, Heidelberg (1998)
5. Benzmüller, C., Sorge, V.: OANTS - combining interactive and automated theorem proving. In: Kerber, M., Kohlhase, M. (eds.) Symbolic Computation and Automated Reasoning, pp. 81–97. A.K.Peters, Massachusetts (2001)
6. Benzmüller, C., Sorge, V., Jamnik, M., Kerber, M.: Combined reasoning by automated cooperation. J. Appl. Log. **6**(3), 318–342 (2008)

---

[2] See https://team.inria.fr/parsifal/proofcert/.

7. Benzmüller, C., Paulson, L., Theiss, F., Fietzke, A.: LEO-II - a cooperative automatic theorem prover for classical higher-order logic (System description). In: Armando, A., Baumgartner, P., Dowek, G. (eds.) IJCAR 2008. LNCS (LNAI), vol. 5195, pp. 162–170. Springer, Heidelberg (2008)

8. Blanchette, J., Böhme, S., Paulson, L.: Extending sledgehammer with SMT solvers. J. Autom. Reason. **51**(1), 109–128 (2013)

9. Bonacina, M.P.: A taxonomy of parallel strategies for deduction. Ann. Math. Artif. Intell. **29**(1–4), 223–257 (2000)

10. Brown, C.E.: Satallax: an automatic higher-order prover. In: Gramlich, B., Miller, D., Sattler, U. (eds.) IJCAR 2012. LNCS, vol. 7364, pp. 111–117. Springer, Heidelberg (2012)

11. De Bruijn, N.G.: Lambda calculus notation with nameless dummies, a tool for automatic formula manipulation, with application to the Church-Rosser theorem. INDAG. MATH **34**, 381–392 (1972)

12. Cervesato, I., Pfenning, F.: A linear spine calculus. J. Log. Comput. **13**(5), 639–688 (2003)

13. Gacek, A.: The Abella interactive theorem prover (System description). In: Armando, A., Baumgartner, P., Dowek, G. (eds.) IJCAR 2008. LNCS (LNAI), vol. 5195, pp. 154–161. Springer, Heidelberg (2008)

14. Girard, J.Y.: Interprétation fonctionnelle et élimination des coupures de l'arithmétique d'ordre supérieur. Ph.D. thesis, Paris VII (1972)

15. Kfoury, A.J., Ronchi Della Rocca, S., Tiuryn, J., Urzyczyn, P.: Della Rocca, J. Tiuryn, and P. Urzyczyn. Alpha-conversion and typability. Inf. Comput. **150**(1), 1–21 (1999)

16. Reynolds, J.C.: Towards a theory of type structure. In: Robinet, B. (ed.) Symposium on Programming. LNCS, vol. 19, pp. 408–423. Springer, Heidelberg (1974)

17. Liang, C., Mitchell, D.: System description: Teyjus - a compiler and abstract machine based implementation of λProlog. In: Ganzinger, H. (ed.) CADE 1999. LNCS (LNAI), vol. 1632, pp. 287–291. Springer, Heidelberg (1999)

18. Pfenning, F., Schürmann, C.: System description: Twelf - a meta-logical framework for deductive systems. In: Ganzinger, H. (ed.) CADE 1999. LNCS (LNAI), vol. 1632, pp. 202–206. Springer, Heidelberg (1999)

19. Riazanov, A.: Implementing an efficient theorem prover. Ph.D. thesis, University of Manchester (2003)

20. Sekar, R., Ramakrishnan, I.V., Voronkov, A.: Term indexing. In: Robinson, A., Voronkov, A. (eds.) Handbook of Automated Reasoning, pp. 1853–1964. Elsevier Science Publishers B.V., Amsterdam (2001)

21. Steen, A.: Efficient data structures for automated theorem proving in expressive higher-order logics. Master's thesis, Freie Universität Berlin (2014). http://userpage.fu-berlin.de/~lex/drop/steen_datastructures.pdf

22. Sutcliffe, G.: The TPTP problem library and associated infrastructure. J. Autom. Reason. **43**(4), 337–362 (2009)

23. Sutcliffe, G., Benzmüller, C.: Automated reasoning in higher-order logic using the TPTP THF infrastructure. J. Formaliz. Reason. **3**(1), 1–27 (2010)

24. Weiss, G. (ed.): Multiagent Systems. MIT Press, Cambridge (2013)

25. Wisniewski, M.: Agent-based blackboard architecture for a higher-order theorem prover. Master's thesis, Freie Universität Berlin (2014). http://userpage.fu-berlin.de/~lex/drop/wisniewski_architecture.pdf

Systems and Data

TIP: Tons of Inductive Problems

Koen Claessen, Moa Johansson, Dan Rosén, and Nicholas Smallbone[✉]

Department of Computer Science and Engineering,
Chalmers University of Technology, Gothenburg, Sweden
{koen,jomoa,danr,nicsma}@chalmers.se

Abstract. This paper describes our collection of benchmarks for inductive theorem provers. The recent spur of interest in automated inductive theorem proving has increased the demands for evaluation and comparison between systems. We expect the benchmark suite to continually grow as more problems are submitted by the community. New challenge problems will promote further development of provers which will greatly benefit both developers and users of inductive theorem provers.

1 Introduction

We have recently seen increased interest in inductive theorem proving, both with specialised provers such as IsaPlanner, Zeno and HipSpec [3,5,13], SMT-solvers such as CVC4 [11], the auto-active prover Dafny [10], recent work on the first-order SPASS prover [14], as well as some support in proof assistants [8,9].

To ease evaluation and development, and compare the relative strengths of the different systems, it is important to have good standard benchmarks. The contribution of this paper is an accessible standard benchmark suite for inductive theorem provers which can be extended by users and developers. The benchmarks are publicly available at: https://github.com/tip-org/benchmarks.

We have so far collected 340 problems in our benchmark suite, which we have called "TIP", for Tons of Inductive Problems—so named in the hope of attracting many more problems! We invite the community to submit additional problems and challenges and expect the collection to continuously grow and provide new challenges for developers.

2 The Benchmark Format

The benchmarks are expressed in a variant of SMT-LIB [1], extended with support for algebraic datatypes and higher-order functions. The format is described in detail in [4]. Starting from the basic SMT-LIB format, we import the following features from existing systems:

- Algebraic datatypes, which are declared using the `declare-datatypes` syntax as supported in Z3 and CVC4.
- Recursive function definitions, which use the `define-funs-rec` syntax implemented in CVC4 and proposed for SMT-LIB 2.5 [1].

© Springer International Publishing Switzerland 2015
M. Kerber et al. (Eds.): CICM 2015, LNAI 9150, pp. 333–337, 2015.
DOI: 10.1007/978-3-319-20615-8_23

– Polymorphic functions, using the proposed **par** syntax [2], which is implemented in a version of CVC4.

We then add more features of our own, which are specific to TIP:

– In the standard `declare-datatypes` syntax, functions over algebraic datatypes are defined using projection functions like **head** and **tail**. We add a pattern-matching syntax, which is more convenient for many provers.
– Many of our benchmark problems use higher-order functions, such as **map**. We add syntax for lambda functions and higher-order functions.
– Many inductive provers treat the goal specially, as opposed to SMT-LIB which expresses the goal as a negated axiom. We add a construct (`assert-not p`) which declares p as the goal; it is semantically equivalent to (`assert (not p)`).

We do not necessarily expect every prover to support TIP natively. Instead, we have made a tool which can translate TIP problems to and from a variety of other formats. Currently our tool can translate TIP problems to a CVC4-compatible version of SMT-LIB or to WhyML, and can compile Haskell programs into TIP properties. It can also perform a number of transformations for tools which do not support the full TIP language, such as removing higher-order functions by defunctionalisation [12]. Our aim is to support many different source and target formats in this tool.

2.1 Example

As an example of what the benchmarks look like, we show property 12 from the IsaPlanner benchmark set (see Sect. 3.1 below), which states that the functions **drop** and **map** distribute over one another:

$$\text{drop n (map f xs) = map f (drop n xs)}.$$

We declare two simple algebraic datatypes representing natural numbers and polymorphic lists.

```
(declare-datatypes (a)
  ((list (nil) (cons (head a) (tail (list a))))))
(declare-datatypes () ((Nat (Z) (S (p Nat)))))
```

Next, we declare two recursive functions: **map**, which is a higher-order function applying a unary function f to each element of a list, and **drop**, which recursively drops a given number of elements from the front of the list.

```
(define-funs-rec
  ((par (a b) (map ((f (=> a b)) (xs (list a))) (list b))))
  ((match xs
    (case nil (as nil (list b)))
    (case (cons y ys) (cons (@ f y) (map f ys))))))
```

```
(define-funs-rec
  ((par (a) (drop ((n Nat) (xs (list a))) (list a))))
  ((match n
    (case Z xs)
    (case (S m)
      (match xs
        (case nil xs)
        (case (cons y ys) (drop m ys)))))))
```

These definitions illustrate several features of TIP:

- Polymorphism: **par** introduces type variables.
- Higher-order functions: (=> a b) is the type of functions from a to b, and @ applies first-class functions to their arguments.
- Pattern-matching using **match** and **case**, which binds new variables.

Finally, the benchmark problem itself is declared with the keyword **assert-not**:

```
(assert-not
  (par (a b)
    (forall ((n Nat) (f (=> a b)) (xs (list a)))
      (= (drop n (map f xs)) (map f (drop n xs))))))
(check-sat)
```

Each benchmark file is stand-alone and only contains one property.

3   Sample Benchmarks

In this section we give a short overview of some the benchmark problems currently available in the repository. At the moment, there are three different main sources of problems. We expect more to be added.

3.1   IsaPlanner's Rippling and Case-Analysis Benchmarks

This set of 85 problems comes from the evaluation of IsaPlanner's rippling-heuristic for guiding rewriting in inductive proofs in the context of functions with case- and if-statements [7]. It has been used in the evaluation of many of the recent inductive theorem provers and includes theorems about lists, natural numbers and binary trees. The problems are relatively easy, most of the theorems can be proved by structural induction using only the function definitions and only 15 require auxiliary lemmas to be discovered.

3.2   Productive Use of Failure Benchmarks

This is another benchmark suite which has been used to evaluate several recent provers. It consists of 50 theorems about lists and natural numbers and originates from evaluation of techniques for discovering auxiliary lemmas in the CLAM prover [6]. The original paper did not provide definitions for the functions used in the benchmarks, so the definitions provided here come from the evaluation of the HipSpec system [3]. These proofs are generally a bit harder, and may require additional lemmas to be found and proved (by another induction) or generalisation of the conjecture in order to strengthen the inductive hypothesis.

3.3   New TIP Benchmarks

This set contains 205 new benchmarks including, amongst others, properties of
the Agda standard library[1] implementation of integers on top of natural numbers, problems about natural numbers in binary representation, various sorting
functions with correctness properties expressed in alternative ways, problems
about regular expressions, binary search trees, grammars and skew heaps. The
problems about sorting, regular expressions, grammars and heaps have to our
knowledge not all been fully automated yet and are offered as challenges!

4   Contribute to TIP

We invite the theorem proving community to contribute additional inductive
benchmarks and challenge problems to TIP. Instructions for how to submit problems can be found in the README file for the repository (https://github.com/
tip-org/benchmarks).

We are developing a toolchain for translating to and from our format. The
development is in its own repository (https://github.com/tip-org/tools). The
tool can currently read in problems in TIP format or Haskell, and output TIP,
SMT-LIB and WhyML, with some caveats:

- The generated SMT-LIB uses `declare-datatypes`, `define-funs-rec` and
  polymorphism.
- The generated WhyML makes no special effort to pass Why3's termination
  checker.

We encourage the community to request and contribute additional input and
output formats to our tool chain.

5   Conclusion and Further Work

TIP is intended to be a standard benchmark suite for developers and users of
inductive theorem provers. We hope that this initiative will ease comparison
and evaluation of systems and spur further collaboration and development by
attracting submissions of additional challenge problems from the community.

In addition to serving as a standard benchmark suite for inductive provers,
the TIP benchmarks may also be useful for developers of theory exploration
systems. Theory exploration is a technique for automatically discovering interesting conjectures about a given set of functions and datatypes, and is used in
for example the HipSpec prover to discover lemmas. The TIP benchmarks can
be compared to the output from the theory explorer in precision/recall analysis
to assess the quality and interestingness of the conjectures generated. A good
theory exploration system may also be used to generate new benchmarks for TIP.

Another aim is to also support non-theorems, for evaluation of tools for
counter-example finding. This requires no extension to the format at all, but

---

[1] https://github.com/agda/agda-stdlib.

it requires a standardization on how to annotate problems with their expected answer (theorem or non-theorem), as well as a common solution format.

It is important to have good tool support if TIP is to be used by the community. Our tool is currently in active development, in order to support more input and output formats, as well as various strategies for encoding features of TIP for provers that do not support the full language.

In the future, we may want to extend the language to support a richer variety of problems. For example, we may want to include problems about lazy functions and co-datatypes.

References

1. Barrett, C., Stump, A., Tinelli, C.: The SMT-LIB standard - version 2.0. In: Proceedings of the 8th International Workshop on Satisfiability Modulo Theories (SMT 2010), Edinburgh, Scotland, July 2010
2. Bobot, F.: [RFC] Add adhoc polymorphism. https://github.com/CVC4/CVC4/pull/51
3. Claessen, K., Johansson, M., Rosén, D., Smallbone, N.: Automating inductive proofs using theory exploration. In: Bonacina, M.P. (ed.) CADE 2013. LNCS, vol. 7898, pp. 392–406. Springer, Heidelberg (2013)
4. Claessen, K., Johansson, M., Rosén, D., Smallbone, N.: The TIP language. http://tip-org.github.io/format.html
5. Dixon, L., Johansson, M.: IsaPlanner 2: A proof planner in Isabelle. Technical report, University of Edinburgh (2007)
6. Ireland, A., Bundy, A.: Productive use of failure in inductive proof. J. Automated Reasoning 16, 79–111 (1996)
7. Johansson, M., Dixon, L., Bundy, A.: Case-analysis for rippling and inductive proof. In: Kaufmann, M., Paulson, L.C. (eds.) ITP 2010. LNCS, vol. 6172, pp. 291–306. Springer, Heidelberg (2010)
8. Johansson, M., Rosén, D., Smallbone, N., Claessen, K.: Hipster: integrating theory exploration in a proof assistant. In: Watt, S.M., Davenport, J.H., Sexton, A.P., Sojka, P., Urban, J. (eds.) CICM 2014. LNCS, vol. 8543, pp. 108–122. Springer, Heidelberg (2014)
9. Kaufmann, M., Panagiotis, M., Moore, J.S.: Computer-Aided Reasoning: An Approach. Kluwer Academic Publishers, Norwell (2000)
10. Leino, K.R.M.: Automating induction with an SMT solver. In: Kuncak, V., Rybalchenko, A. (eds.) VMCAI 2012. LNCS, vol. 7148, pp. 315–331. Springer, Heidelberg (2012)
11. Reynolds, A., Kuncak, V.: Induction for SMT solvers. In: D'Souza, D., Lal, A., Larsen, K.G. (eds.) VMCAI 2015. LNCS, vol. 8931, pp. 80–98. Springer, Heidelberg (2015)
12. Reynolds, J.C.: Definitional interpreters for higher-order programming languages. In: Proceedings of the ACM Annual Conference, ACM 1972, vol. 2, pp. 717–740. ACM, New York (1972)
13. Sonnex, W., Drossopoulou, S., Eisenbach, S.: Zeno: an automated prover for properties of recursive data structures. In: Flanagan, C., König, B. (eds.) TACAS 2012. LNCS, vol. 7214, pp. 407–421. Springer, Heidelberg (2012)
14. Wand, D., Weidenbach, C.: Automatic induction inside superposition. https://people.mpi-inf.mpg.de/dwand/datasup/draft.pdf

Semantic Enrichment of Mathematics via 'tooltips'

Ross Moore[(✉)]

Macquarie University, Sydney, Australia
ross.moore@mq.edu.au

Abstract. A package mathsem for pdf-LATEX implements a way to provide semantic meaning to symbols, without adding a large syntactical burden to the specification of a mathematical expression. It uses a concept of 'active comment', allowing the '%' character at the beginning of a new line to become an active token under highly-controlled circumstances. With a strictly defined syntax, words to express the semantic meaning of a variable ('x' say) can be associated with each occurrence of 'x' in the expression following. The words become content of a tooltip, that 'pops-up' by the symbol in a PDF document. The idea extends to:

1. allow multiple instances of the same symbol have distinct meanings;
2. attach semantics to macro-names as well as character symbols;
3. allow nested tooltip rectangles, for sub-expressions;
4. assign defaults to be attached to symbols and macros, at either global or local levels, to maintain consistency of meaning within extended portions of a document.

It is planned to use the same syntactical constructions to provide words for spoken 'alternative text', in the context of fully-tagged, accessible, mathematical content within PDF documents

Thanks to Michael Kohlhase for ideas suggesting such a package.

1 Introduction

Mathematical notation has been refined over centuries into a very succinct syntax, particularly for input to computer programs, where input such as \(a + b\) using LATEX, produces the typeset visual form: $a + b$. — Call this *Example A*. However, such an expression contains no real *semantics* describing what the mathematics actually represents. How can an author provide such information conveniently, without compromising the high-quality visual representation?

To address this we take a lead from computer software interfaces, where use of 'tooltips', small windows containing a short textual description that pop-up, has become common-place. An extra benefit of tooltips is that their content is, according to the PDF Reference specifications [1,3], intended to be vocalised (e.g., by Adobe Reader's 'Read Out Loud') as a replacement for the content which lies under the tooltip's Button annotation, or by other means. Thus throughout this paper the term 'semantics' could be taken to mean a 'well-structured, meaningful, vocal or extra rendition' of mathematical content, understandable without the need to see the visual form of the expression.

© Springer International Publishing Switzerland 2015
M. Kerber et al. (Eds.): CICM 2015, LNAI 9150, pp. 338–342, 2015.
DOI: 10.1007/978-3-319-20615-8_24

Several LATEX packages[1] support tool-tips, with all employing rather verbose coding. For example, in a linguistics context, one might want something like $S + O$ using coding[2] as follows — call this *Example B*: (Test it[3] yourself.)

```
\pdftooltip{S}{S,Subject}\pdftooltip{+}{plus,Verb}\pdftooltip{O}{O,Object}
```

This puts quite a heavy burden on a document's author to keep both the visual form of the mathematics, and associated semantics, fully correct during editing.

This kind of 'low-level' semantic enrichment of mathematical content is different to, but need not be incompatible with, using LATEX to semantically enrich documents for a 'Semantic Web' as described in [2]. These use the sTEX collection of LATEX macro packages [6]; see *Example F* below for relevant remarks.

2   Adding Semantics to Math Environments

To help alleviate the burden, the mathsem package[4] employs a concept of 'active comment' (or 'semantic comment'). This allows extra words to be introduced into the TEX-based processing and stored for later use. These can appear as text in a tooltip, or be used in other ways such as 'alternative text' for vocalisation purposes. We use the term 'tagging' to refer to addition of such information, with symbols and expressions becoming 'tagged'. Without the package, the comments are simply ignored, thus giving the correct visual form with no edits required.

A series of examples follow, in sequence with A and B above. Each introduces concepts and explains syntax used in the package. *Example D* shows how user-defined macros overcome an inherent difficulty. This opens up great flexibility allowing semantics to be applied to not just individual symbols, but subexpressions and the environment as a whole; see *Examples E* and *F*.

Example C. At the simplest level, meanings of symbols are passed using TEX comments, without interrupting the coding of the mathematical expression.

```
\(
%$semantics
% a $ the dog
% b $ the bone; another bone
% + $ eats
% 2 $ many bones
%$endsemantics
a + b + b^{b^{2}}
% other comments ignored
\)
```

Spacing and vertical positioning of symbols should be unaffected.

Non-tagged output: $a + b + b^{b^2}$

Tagged output: $a + b + b^{b^2}$

---

[1] pdfcomment, fancytooltips, cooltooltips. Click for links to CTAN.
[2] This uses the \pdftooltip command from the pdfcomment package [7].
[3] Not all PDF readers support tooltips; Adobe Reader is recommended.
[4] ... downloadable from https://rutherglen.science.mq.edu.au/~maths/CICM/.

TEXnical Note. This works using TEX's *category-code* mechanism [8] to adjust the roles of '%' and end-of-line, so that lines beginning with '%' no longer need be ignored. Such 'semantic comment' lines are parsed to identify characters to which semantics can be attached. Valid syntax is lines of the following forms:

% ⟨token⟩ $ ⟨semantics⟩ ⟨line-end⟩
% $⟨keyword⟩ ⟨line-end⟩

with spaces allowed either side of the ⟨token⟩ and '$' delimiter, chosen since '$' can never occur within a math-environment. The ⟨semantics⟩ can specify a list, delimited with ';' (semi-colon), of text snippets to be used with successive instances of the same character with the final one persisting for continued use. (See the line for 'b' in *Example C*). The $semantics keyword is optional, reminding an author of the purpose of semantic comments which follow.

After reading a semantic comment and storing its ⟨semantics⟩, TEX 'looks ahead'[5] to see whether the next line also starts with '%'. If so, this is scanned for a semantic comment; otherwise the scope terminates. Any line beginning with an active '%' followed by nothing, '$endsemantics', an unrecognised keyword or other violation of the strict syntax, is treated as an ordinary comment; the scope for active '%' terminates. Next, any found ⟨token⟩ characters are 'activated', by having their \catcode set to 13, before actual mathematical content starts. Such 'active' characters behave as a single-letter name for a macro, which expands to produce the tooltip before placing an (unactive) character into the math-list for normal processing. A single-digit exponent must be enclosed within braces (e.g., ..^{b^{2}} in *Example C*), since the '2' now expands into more than just a single character being superscripted. This active nature of characters is carefully controlled; in particular, being restricted to just a single math-environment.

Example D. Use of active characters, as outlined in the *TEXnical Note*, has one drawback; but there is an easy work-around. When a letter variable, '*a*' say, is activated for semantic enrichment, this letter cannot be used in any macro names employed directly within the same math-environment.

```
\(
%$semantics
% + $ added to
% a$ complex number a; a; a conjugate
%$endsemantics
\Re a = \tfrac12(a + \bar{a})
\)
```

The result can be surprising, producing either wrong output or a TEX error, or both: $\Re a = ac12(a + \underline{a}ra)$

```
! Undefined control sequence.
1.331 \Re a = \tfr
                  ac12(a + \bar{a})
```

A warning message also is issued; *viz.*

```
LaTeX Warning: Command \b invalid in math mode on input line 331.
```

The reason is that the letter '*a*' cannot be used within macro names, as it no longer has the category code of a letter when it is later used as part of the mathematics. User-defined macros provide the solution as in the next example.

---

[5] Using TEX's \futurelet primitive command.

Example E.

```
\DeclareRobustCommand{\HALFOF}%
   {\tfrac12}%
\DeclareRobustCommand
   {\ACONJUGATE}{\bar{a}}%
\(
%$semantics
%\Re $ real part of
% a $ a complex number a ; a
% = $ equals
%\HALFOF $ half of
% ( $ open bracket
% + $ added to
%\ACONJUGATE$ its complex conjugate
% ) $ close bracket
%$endsemantics
  \Re a = \HALFOF (a + \ACONJUGATE)
\)
```

Existing macro names can also be enriched with semantics, provided the macro-name does not use any letters that are to be made active; in this example \Re is 'safe', but \tfrac and \bar are not, because of the active letter 'a'. An 'unsafe' macro is protected within the expansion of a previously defined 'safe' macro. Since most usual TeX and LaTeX commands use only lowercase letters to form the name, we employ just uppercase letters in the new names used here.

Non-tagged: $\Re a = \frac{1}{2}(a + \bar{a})$

Tagged: $\Re a = \frac{1}{2}(a + \bar{a})$

Being already active, the existing macro expansion is first saved under an internal name. Just as with active characters, a new expansion creates the tooltip before placing the actual mathematics using this saved expansion. Macro names themselves are best chosen to correspond to the semantic meaning, as in [2,5,6]. \DeclareRobustCommand allows for better log-messages and safe captions, etc.

Example F. This final example is based upon Example 1 in the STeX documentation [6], using macros \CsumLimits and \Cpower (incorrectly stated as \Cexp in [6]), as defined for STeX.

```
{\DeclareRobustCommand{\InfinSumXexpK}{\CsumLimits{k}1\infty{\XtotheK}}%
\DeclareRobustCommand{\XtotheK}{\Cpower{\varx}{\vark}}%
\DeclareRobustCommand{\vark}{k}%
\DeclareRobustCommand{\varx}{x}%
\(
%$semantics
% \vark $ bound variable k
% \varx $ free variable x
% \infty $ infinity
%\XtotheK $ variable x raised to the power k
%\InfinSumXexpK $ sum with k from 1 to infinity, of x to the power k
\InfinSumXexpK
\)}%
```

Non-tagged: $\sum_{k=1}^{\infty} x^k$

Tagged: $\sum_{k=1}^{\infty} x^k$

To indicate the available flexibility, this example uses the character 'k' both tagged, via \vark, and untagged in the lower limit of the sum. The upper limit of '∞' is tagged, whereas the lower limit '1' is not — but could have been. Being provided internally by the macro \CSumLimits, it is not possible to tag the '=' (equals sign). This would be possible if STeX had used a macro, \EqualsSign say, in its definition for \CSumLimits.

3 Possible Future Developments

With 'active comments' being a viable way to include extra (semantic) information into the processing of math-environments, one can envisage other places to use this; e.g., chemical formulæ, chess layouts, contexts where single letters represent complicated objects. More keywords can be implemented; e.g., to add meaningful /Alt text for PDF tagging (using /Formula<</Alt(...)...>> BDC) of complete mathematical formulae, in documents conforming to the PDF/UA [4] standard. All formulæ must have alternative text for the benefit of screen-readers and other assistive technology. Code for *Example B* might then start as follows.

```
\(
% $all
% "S + O" representing "Subject, Verb, Object"
% $semantics
```

Such tagging of structure with /Alt text applies to "Tagged PDF" generally, as in the author's earlier work [9–11]. Those methods need an interface to allow a document's author to supply words to override built-in defaults, where the meaning of a mathematical symbol can be dependent on context.

References

1. Adobe Systems Inc.; PDF Reference 1.7, November 2006. Also available as [3]. http://www.adobe.com/devnet/pdf/pdf_reference.html
2. Groza, Tudor, Handschuh, Siegfried, Möller, Knud, Decker, Stefan: SALT - Semantically Annotated LaTeX for Scientific Publications. In: Franconi, Enrico, Kifer, Michael, May, Wolfgang (eds.) ESWC 2007. LNCS, vol. 4519, pp. 518–532. Springer, Heidelberg (2007)
3. ISO 32000–1:2008; Document management – Portable document format (PDF 1.7); Technical Committee ISO/TC 171/SC 2, July 2008
4. ISO 14289–1:2012; Document management applications - Electronic document file format enhancement for accessibility - Part 1: Use of ISO 32000–1 (PDF/UA-1); Technical Committee ISO/TC 171/SC 2. Corrected version, December 2014
5. Kohlhase, M.: Using LaTeX as a Semantic Markup Format, Mathematics in Computer Science (2:2), pp. 279–304. Birkhäuser (2008). https://svn.kwarc.info/repos/stex/doc/mcs08/stex.pdf
6. Kohlhase, M.: STEX: Semantic Markup in STEX. Documentation of the LaTeX packages. https://trac.kwarc.info/sTeX/export/2430/trunk/sty/stex.pdf
7. Kleber, J.: pdfcomment – A user-friendly interface to pdf annotations. LaTeX package; available from CTAN; version 2.3a, September 2012
8. Knuth, D.E.: The TeXbook. Addison Wesley (1984, 1986 and later revisions)
9. Moore, R.R.: Ongoing efforts to generate "tagged PDF" using pdfTeX, Muni Press, 2009. Reprinted as: TUGboat, Vol. 30, No. 2, pp. 170–175 (2009)
10. Moore, R.R.: Tagged mathematics in PDFs for accessibility and other purposes. In: CICM-WS-WiP 2013. CEUR Workshops Proceedings, vol. 1010, paper-01.pdf (2013)
11. Moore, Ross: PDF/A-3u as an Archival Format for Accessible Mathematics. In: Watt, Stephen M., Davenport, James H., Sexton, Alan P., Sojka, Petr, Urban, Josef (eds.) CICM 2014. LNCS, vol. 8543, pp. 184–199. Springer, Heidelberg (2014)

Documentation Generator Focusing on Symbols for the HTML-ized Mizar Library

Kazuhisa Nakasho[✉] and Yasunari Shidama

Shinshu University, Matsumoto, Japan
13st205f@shinshu-u.ac.jp, shidama@cs.shinshu-u.ac.jp

Abstract. The purpose of this project is to collect symbol information in the Mizar Mathematical Library and manipulate it into practical and organized documentation. Inspired by the MathWiki project and API reference systems for computer programs, we developed a documentation generator focusing on symbols for the HTML-ized Mizar library. The system has several helpful features, including a symbol list, incremental search, and a referrer list. It targets those who use proof assistance systems, the volume of whose libraries has been rapidly increasing year by year.

Keywords: Mizar · Mathematical knowledge management · Search system · Documentation generator

1 Motivation

In mathematical knowledge management (MKM), expanding of the fields covered by formal methods has led to the rapid growth of formal mathematical libraries. For instance, the Mizar Mathematical Library (MML) [5–7] has grown to more than 2.7 million lines in 2015, and it has been increasing by approximately 0.1 million lines per year.

The development of formal mathematical libraries facilitates the reuse of mathematical symbols and theorems, thereby improving the efficiency of writing formal proofs. However, the increased volume of the libraries makes it difficult for users to grasp what and where symbols and theorems are defined. In recent years, developers of formal proofs have spent considerable time on search tasks in large-scale libraries, thereby decreasing the productivity of formal verification. Therefore, searching and browsing efficiency in large-scale libraries has been a crucial issue in MKM.

2 Survey and Design Decision

We analyze some existing tools for searching and browsing the Mizar library. The HTML-ized Mizar library [1,12] is one of the most successful documentation tools for formal mathematical libraries. The HTML-linked MML was first developed

© Springer International Publishing Switzerland 2015
M. Kerber et al. (Eds.): CICM 2015, LNAI 9150, pp. 343–347, 2015.
DOI: 10.1007/978-3-319-20615-8_25

in the late 1990 s for the former "Journal of Formalized Mathematics"[1] and then re-implemented by Dr. Josef Urban using the XML/XSLT technology. This new HTML-ization was also used in the MathWiki Project.[2] The system is capable of intuitive and rapid browsing as a result of hyperlinks being embedded into symbols, enabling users to jump from symbol occurrences to their definitions by clicking them. The system has been widely used by Mizar users because of its effectiveness and user-friendly design. However, because this system does not have retrieval functions, users are frequently obliged to grep symbols in the MML using text editors. Moreover, although the hyperlinks allow users to jump to their definitions, it is still difficult to, inversely, enumerate the symbols that include a particular symbol in their definitions.

MML Query [3,4] is the most flexible and sophisticated search system for the MML. This system has its own query language, and users can input more detailed information regarding search objects than is possible using grep. However, users must learn and master the query language, thus this is a burden for beginners.

Conversely, in software development, most widely used programming languages have several types of API documentation generators, and almost all of the widely used libraries have their own online API documentation systems. Those API reference systems have common features, such as incremental search and a list of symbols that is automatically created by API documentation generators during library updates. Many documentation generators, such as Doxygen[3] and RDoc[4], have contributed to the acceleration of software development.

We apply the software development approach to developing a documentation generator that works on the MML in order to overcome the drawback of existing search and browsing systems.

3   Application

Using the programming language Python,[5] we developed a documentation generator[6] that comprises the following three steps:

1. Parse the HTML-ized MML and collect symbols and their mutual relationships.
2. Clean and arrange those data.
3. Output reference documents in HTML format. Each file corresponds to one symbol.

These steps take only a few minutes in total.[7]

---

[1] http://mizar.org/JFM.

[2] http://www.ru.nl/foundations/research/projects/mathwiki/.

[3] http://www.doxygen.org/.

[4] https://github.com/rdoc/rdoc.

[5] https://www.python.org/.

[6] https://github.com/aabaa/mmlfrontend.

[7] Windows 7, CPU: AMD A10-5800K 3.8 GHz (4-core), Memory: 16.0 GB.

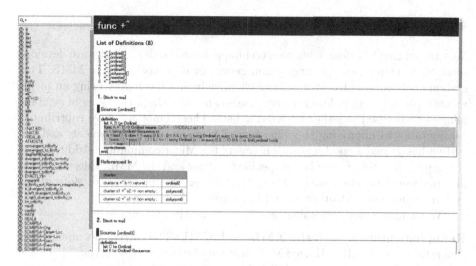

Fig. 1. Screenshot of the reference system.

The latest reference system produced by the generator is available at a website.[8] Figure 1 shows a screenshot of the system.

The reference system offers the following helpful features:

Symbol List: There are nearly 9,000 symbols (predicate, mode, structure, functor, and attribute) in the MML, all of which are listed in the left pane of the system. The type of each symbol can be distinguished by the icon next to the symbol. Clicking a symbol in the list causes the corresponding page to be loaded into the main frame in left pane.

Incremental Search: An incremental search function is located at the top of the left pane. When several search words separated by blanks are input, the system combines the symbol list into symbols that contain all of the indicated words. As the system has an original search table, the function returns search results immediately. Users can quickly look up symbols defined in the MML, even without knowing the correct spelling.

Source Code: The symbol definition source code is imported from the HTML-ized MML. Symbols in bold font are hyperlinked to their definitions. Internal links pointing to their definitions in this reference system are in blue. External links pointing to their definitions in the original HTML-ized MML are in red.

Referrer List: Although the HTML-ized MML enables users to jump from symbol occurrences to their definitions by clicking them, it does not have a function to enumerate symbols that are used in the definitions of particular symbols. The new system organizes the list of referrers for each symbol, and users can check them easily.

---

[8] http://webmizar.cs.shinshu-u.ac.jp/mmlfe/current/.

4   Conclusion and Future Work

We utilized the API documentation technique from the field of software development to develop a new documentation generator that works on the MML. This system enables users to retrieve symbols quickly and intuitively using an incremental search function. Furthermore, users can easily check the types of symbols allowed to be used together by referrer lists. These functions have contributed considerably to improving the efficiency of formal proof development, and the system has gained a good reputation among the Mizar community. Additionally, the approach of the system is not specific to Mizar and the MML, thus all formal libraries would benefit from such a system. Therefore, the future versions of the system should support other formal languages and libraries.

We mention three remaining issues regarding the system:

Reimplementation with the XML-ized MML: The current documentation generator parses the HTML-ized MML instead of the XML-ized Mizar [9]. This is because the former represents relationships between symbols and their definitions as embedded hyperlinks, whereas it is difficult to collect these relationships from the latter. However, the extra process required to generate the HTML-ized MML takes considerable time. Therefore, we would like to change the system to work with the XML-ized MML in the future.

Theorem Search: A theorem search system requires semantic analysis, and machine learning would be a promising approach. Because this research is underway for automated reasoning [2,11], we would like to apply the technique to an interactive search engine.

Tagged Comments: In software development, most documentation generators collect tagged comments, such as authors, purposes, and usages, and reflect them in API documents, whereas the current Mizar library does not have any tagged comments. Although Mizar is a comparatively readable formal language, it is sometimes difficult to discern a writer's intention from a source code. Consequently, such a function would work beneficially, if it were implemented. Furthermore, there is no standard for tagged comments in formal libraries, such a format should be developed in future work and then adopted by all formal libraries.

We also suggest a possible application of the system:

Code Completion: Other major proof assistants have developed graphical interfaces, such as the jEdit plugins for Coq and Isabelle [8,13,14]. Although the Mizar system provides an Emacs plugin [10], some users hope that a newer one will be offered on a modern integrated development environment (IDE). The incremental function of the system would assist in implementation of code completion for those IDE systems.

Acknowledgment. The authors wish to thank the members of the MathWiki Project for their preceding work. Our research is deeply dependent on their product. We especially express our gratitude to Dr. Josef Urban, who is well known as a member of the

MathWiki Project, for giving us some beneficial advice for the study. We would also like to thank to Dr. Adam Naumowicz for helping us improve our paper.

References

1. Alama, J., Brink, K., Mamane, L., Urban, J.: Large formal wikis: issues and solutions. In: Davenport, J.H., Farmer, W.M., Urban, J., Rabe, F. (eds.) Calculemus/MKM 2011. LNCS, vol. 6824, pp. 133–148. Springer, Heidelberg (2011)
2. Alama, J., Heskes, T., Kuhlwein, D., Tsivtsivadze, E., Urban, J.: Premise selection for mathematics by corpus analysis and kernel methods. J. Autom. Reasoning **52**(2), 191–213 (2014)
3. Bancerek, G., Rudnicki, P.: Information retrieval in MML. In: Asperti, Andrea, Buchberger, Bruno, Davenport, James H. (eds.) MKM 2003. LNCS, vol. 2594, pp. 119–132. Springer, Heidelberg (2003)
4. Bancerek, G., Urban, J.: Integrated semantic browsing of the mizar mathematical library for authoring mizar articles. In: Asperti, A., Bancerek, G., Trybulec, A. (eds.) MKM 2004. LNCS, vol. 3119, pp. 44–57. Springer, Heidelberg (2004)
5. Grabowski, A., Kornilowicz, A., Naumowicz, A.: Mizar in a nutshell. J. Formalized Reasoning **3**(2), 153–245 (2010)
6. Matuszewski, R., Rudnicki, P.: Mizar: the first 30 years. Mechanized Math. Appl. **4**(1), 3–24 (2005)
7. Naumowicz, A., Korniłowicz, A.: A brief overview of Mizar. In: Berghofer, S., Nipkow, T., Urban, C., Wenzel, M. (eds.) TPHOLs 2009. LNCS, vol. 5674, pp. 67–72. Springer, Heidelberg (2009)
8. Carst Tankink: PIDE for Asynchronous Interaction with Coq. UITP 2014: 73–83
9. Urban, J.: XML-izing mizar: making semantic processing and presentation of MML easy. In: Kohlhase, M. (ed.) MKM 2005. LNCS (LNAI), vol. 3863, pp. 346–360. Springer, Heidelberg (2006)
10. Urban, J.: MizarMode - an integrated proof assistance tool for the Mizar way of formalizing mathematics. J. Appl. Logic **4**(4), 414–427 (2006)
11. Urban, J.: Momm - fast interreduction and retrieval in large libraries of formalized mathematics. Int. J. Artif. Intell. Tools **15**(1), 109–130 (2006)
12. Urban, J., Alama, J., Rudnicki, P., Geuvers, H.: A wiki for Mizar: motivation, considerations, and initial prototype. In: Autexier, S., Calmet, J., Delahaye, D., Ion, P.D.F., Rideau, L., Rioboo, R., Sexton, A.P. (eds.) AISC 2010. LNCS, vol. 6167, pp. 455–469. Springer, Heidelberg (2010)
13. Wenzel, M.: Isabelle/jEdit – a prover IDE within the PIDE framework. In: Campbell, J.A., Jeuring, J., Carette, J., Dos Reis, G., Sojka, P., Wenzel, M., Sorge, V. (eds.) CICM 2012. LNCS, vol. 7362, pp. 468–471. Springer, Heidelberg (2012)
14. Wenzel, M.: PIDE as front-end technology for Coq. CoRR abs/1304.6626 (2013)

Tools for MML Environment Analysis

Adam Naumowicz[✉]

Institute of Informatics, University of Białystok, Białystok, Poland
adamn@mizar.org

Abstract. In this paper we describe a collection of tools used to support the management of the MIZAR Mathematical Library (MML). The tools handle the dependencies between the texts collected in MML based on the information stored in every article's environment declaration. The application of these tools helps reduce redundant information in the library and speed up its regeneration after revisions.

1 Introduction

MIZAR is a proof assistant system accompanied by a large database of interrelated mathematical developments (articles) that contain formalizations of theorems and their proofs [4,11]. The meaning of all notions used in a MIZAR proof script can be traced back to the few primitive notions of the set-theoretic axiom system (TG), which forms the basis of the MIZAR Mathematical Library (MML). The information about all the facts a user wants to import from the available database goes into an article's environment part. Therefore, the actual semantics of each article depends on the content of the imported ones.

The contents of the MML is available for experimentation as open source documents [2]. The current MIZAR licensing policy is that the proof checking system is freely distributed in pre-compiled binary form for most major platforms, but only the members of the Association of MIZAR Users can actually have access to the system's source code repository if they want to develop their own MIZAR tools using the shared code base. That is why some functionality of the system has been developed independently from that code base to allow wider access. The task of analyzing the dependencies between MML articles indeed does not require too much special MIZAR functionality and can be accomplished by means of popular open source tools [3]. In this paper we briefly describe a collection of easily adaptable and self-documenting Perl scripts sharing a common code that can be applied to perform several tasks connected with analyzing the environment of MIZAR articles. The tools should work with the current official MIZAR version (8.1.02[1]). All the tools and accompanying scripts can be downloaded from the author's web site: http://mizar.uwb.edu.pl/~softadm/envtools.

---

[1] ftp://mizar.uwb.edu.pl/pub/system/i386-linux/mizar-8.1.02_5.22.1191-i386-linux.tar.

© Springer International Publishing Switzerland 2015
M. Kerber et al. (Eds.): CICM 2015, LNAI 9150, pp. 348–352, 2015.
DOI: 10.1007/978-3-319-20615-8_26

2   The Toolbox

The environment of a typical MIZAR article contains a series of directives importing various sorts of notions from articles previously available in the MML (or a local data base). More detailed information about each directive can be found in the user's manual [4]. Let us only recall that the directives are vocabularies, notations, constructors, registrations, requirements, definitions, equalities, expansions, theorems and schemes. Each directive contains a list of (usually a couple dozen) article names whose data should be imported to give a certain meaning to the newly developed formalization.

2.1   The clearenv.pl Script

A typical MIZAR development usually draws information from many articles available in the mathematical knowledge base, so it is a common practice that authors simply try to merge the environment of the articles they want to use. Although the MIZAR verifier points out all repetitions that may be encountered, but that is about all it offers as far as the optimization of the environment is concerned. Therefore the first tool developed to help users and MML maintainers generate relatively simple environments was an environment cleaning utility. In this Perl-based toolset, the program is called clearenv.pl. Given an article, the tool tries to remove one by one all imports in a round-robin manner as far as the article verifies without error (using the -s command line option to stop immediately on first error encountered), so it is necessary for the cleared article to be free of any errors (any still missing gaps in the proof steps should be commented out or marked with appropriate pragmas to suppress error reporting). This simple tool can therefore eliminate all unnecessary imported articles from the environment, replacing the functionality of dedicated MIZAR parser's-based tools like irrvoc and irrths, see [4]. The resulting article's environment is then simplified, easier to expand, but also in many cases makes the overall verification time of the article significantly shorter. Please note that some of the directives are connected with each other, so the order in which we try to eliminate them matters. The script must obviously take this into account. For example, notations depend on vocabularies and constructors. Moreover, because of constructors's recursive nature, it is possible that the needed constructor is already imported by another article's constructors even if it does not explicitly appear in the environment.

2.2   The sortenv.pl Script

The next script in this collection, sortenv.pl, helps to keep the article's environments in sync with the natural ordering of how the MML has been built from the axiomatic notions. In every MML version, such a reference list is available in the mml.lar file, which lists all the processable MML articles, starting from the axiomatic tarski. Preserving the ordering is useful to avoid errors caused by the overloading of popular symbols heavily used in the library. However, there might be cases of overly complicated environments, that the current state of the library may not allow to apply this normalization. So, users are generally encouraged to construct their environments in accordance with this ordering, if it can be done without difficulty. It should be noted that for some environment directives the order is irrelevant, e.g. `vocabularies`, or `requirements` [10], so there is little sense in sorting them (they are left out in the script); for others it may be relevant because of the concrete implementation limitations, e.g. `registrations` of adjectives [9] and reductions [7] or `equalities` [8]; and for `notations` and `definitions` the ordering is meaningful by design.

2.3 The `lastenv.pl` Script

The script `lastenv.pl` can be used to display the last article (according to an order provided in a given list, e.g. `mml.lar`) that the article depends on. Similarly to the above, the users can use the tool to sort out any overloading related problems more easily. But another important usage is for the library developers to check whether some library refactoring (revision) involving environment changes [5] does not disturb the ordering, e.g. we try to make an article precede another one it depends on according to current environment. An example of such a situation when the check is always needed are revisions over the division of MML into its classical (not requiring structures) and abstract part (based on some structural types, e.g. topology, geometry, algebra etc.). The tool can also be helpful for automatic categorizing the MML knowledge [6].

2.4 The `makeenv.pl` Script

The `makeenv.pl`[2] tool can be used to generate a GNU `make` rules out of the dependence structure of MIZAR articles. Another utility shell script, `makemake.sh`, can then be applied to create a complete working `Makefile` prepending to it a stub file `Makefile0` containing a customized environment and general rules. Once the `Makefile` is created, one can use it to process the contents of the MML in the order which is not linear, but reflects the net of MML dependencies. This allows parallelizing some routine tasks that are usually done in a serial manner, such as a complete regeneration of the whole database e.g. as a consequence of a thorough refactoring. Most time-consuming tasks of MIZAR processing are connected with CHECKER verification and this task has been successfully parallelized [12]. The library generation is less time consuming, but on the other hand it has to be performed repeatedly many times during a library revision, so parallelizing this process can also be very useful. For instance, the complete run of three MIZAR environment utilities[3]: `accom`, `exporter` and

---

[2] Should not be confused with the makeenv binary from the MIZAR distribution [4] that serves a slightly different purpose (checking whether the environment of a currently verified document has changed since last accommodation).

[3] To try it with the exemplary customized Makefile from the author's website, one should invoke the process like this: `make iniprel; make -j all`.

`transfer` needed to locally re-create the `prel` directory on a commodity Linux machine with Intel(R) Core(TM)2 Quad CPU at 2.83 GHz takes about 29 min. With 2 CPU cores, the runtime reduces to 15 min., and with 4 cores it needs 9 min. to complete the task. Obviously, this scaling stops when we reach the maximal number of processes which can be simultaneously run because of the dependencies based on the net of MIZAR articles in the current MML. The processes can also be slowed down by too many simultaneous I/O operations. These factors are no longer an issue if we make the test on a system with a big enough number of CPU cores available and a high-throughput storage. In such an environment we can observe performance close to the theoretical ideal, where each processing operation can be done as soon as its dependencies have been fulfilled. So, we can also experimentally evaluate the best runtime that can be achieved in practice and show what number of CPU cores are needed for it. An experiment on a cluster machine running single image Kerrighed SMP kernel on 16 nodes with double Intel(R) E5-2650v2 CPU at 2.60 GHz (16 physical cores each, which accounts to 256 virtual cores) allowed to check that the number of MIZAR processes did not exceed 42 (14.699 on average) and in consequence the runtime was only reduced to about 5 min. But let us note that during MML revisions, a complete rebuild from scratch is rarely required. Much more often the refactoring is applied only to a part of the net of articles, so the parallelized library regeneration in such cases still can be close to interactive.

3   Conclusions

We have presented examples of MIZAR tools implemented in the Perl scripting language for better portability, flexibility and adaptability. Their functionality is useful for end users as well as maintainers of the centralized MIZAR Mathematical Library. Such tools can be further developed to support more involved tasks related to the dependency of formalized documents, for example to the exploration and reverse mathematics on top of MML [1] or in projects like the Wiki for Mizar [13] where interactive regeneration of some parts of the library is among the basic requirements.

References

1. Alama, J.: mizar-items: exploring fine-grained dependencies in the Mizar mathematical library. In: Davenport, J.H., Farmer, W.M., Urban, J., Rabe, F. (eds.) MKM 2011 and Calculemus 2011. LNCS, vol. 6824, pp. 276–277. Springer, Heidelberg (2011). http://dx.doi.org/10.1007/978-3-642-22673-1_19
2. Alama, J., Kohlhase, M., Mamane, L., Naumowicz, A., Rudnicki, P., Urban, J.: Licensing the Mizar mathematical library. In: Davenport, J.H., Farmer, W.M., Urban, J., Rabe, F. (eds.) MKM 2011 and Calculemus 2011. LNCS, vol. 6824, pp. 149–163. Springer, Heidelberg (2011). http://dl.acm.org/citation.cfm?id=2032713. 2032726

3. Alama, J., Mamane, L., Urban, J.: Dependencies in formal mathematics: applications and extraction for Coq and Mizar. In: Campbell, J.A., Jeuring, J., Carette, J., Dos Reis, G., Sojka, P., Wenzel, M., Sorge, V. (eds.) CICM 2012. LNCS, vol. 7362, pp. 1–16. Springer, Heidelberg (2012)

4. Grabowski, A., Korniłowicz, A., Naumowicz, A.: Mizar in a nutshell. J. Formalized Reasoning **3**(2), 153–245 (2010). Special Issue: User Tutorials I

5. Grabowski, A., Schwarzweller, C.: Revisions as an essential tool to maintain mathematical repositories. In: Kauers, M., Kerber, M., Miner, R., Windsteiger, W. (eds.) MKM/CALCULEMUS 2007. LNCS (LNAI), vol. 4573, pp. 235–249. Springer, Heidelberg (2007). http://dx.doi.org/10.1007/978-3-540-73086-6_20

6. Grabowski, A., Schwarzweller, C.: Towards automatically categorizing mathematical knowledge. In: Ganzha, M., Maciaszek, L.A., Paprzycki, M. (eds.) Proceedings of the Federated Conference on Computer Science and Information Systems - FedCSIS 2012, Wroclaw, Poland, 9–12 September 2012, pp. 63–68 (2012)

7. Korniłowicz, A.: On rewriting rules in Mizar. J. Automated Reasoning **50**(2), 203–210 (2013). http://dx.doi.org/10.1007/s10817-012-9261-6

8. Korniłowicz, A.: Equalities in Mizar. In: Gomolińska, A., Grabowski, A., Hryniewicka, M., Kacprzyk, M., Schmeidel, E. (eds.) Trends in Contemporary Computer Science, Podlasie 2014, pp. 59–69 (2014)

9. Naumowicz, A.: Enhanced processing of adjectives in Mizar. In: Grabowski, A., Naumowicz, A. (eds.) Computer Reconstruction of the Body of Mathematics, Studies in Logic, Grammar and Rhetoric, vol. 18(31), pp. 89–101. University of Białystok (2009)

10. Naumowicz, A., Byliński, C.: Improving Mizar texts with *Properties* and *Requirements*. In: Asperti, A., Bancerek, G., Trybulec, A. (eds.) MKM 2004. LNCS, vol. 3119, pp. 290–301. Springer, Heidelberg (2004)

11. Naumowicz, A., Korniłowicz, A.: A brief overview of Mizar. In: Berghofer, S., Nipkow, T., Urban, C., Wenzel, M. (eds.) TPHOLs 2009. LNCS, vol. 5674, pp. 67–72. Springer, Heidelberg (2009). http://dx.doi.org/10.1007/978-3-642-03359-9_5

12. Urban, J.: Parallelizing Mizar. CoRR abs/1206.0141 (2012)

13. Urban, J., Alama, J., Rudnicki, P., Geuvers, H.: A wiki for Mizar: motivation, considerations, and initial prototype. In: Autexier, S., Calmet, J., Delahaye, D., Ion, P.D.F., Rideau, L., Rioboo, R., Sexton, A.P. (eds.) AISC 2010. LNCS, vol. 6167, pp. 455–469. Springer, Heidelberg (2010)

Enabling Symbolic and Numerical Computations in HOL Light

Ons Seddiki$^{(\boxtimes)}$, Cvetan Dunchev, Sanaz Khan-Afshar, and Sofiène Tahar

Department of Electrical and Computer Engineering, Concordia University,
1455 de Maisonneuve W., Montreal, QC H3G 1M8, Canada
{o_sed,dunchev,s_khanaf,tahar}@ece.concordia.ca

Abstract. Verifying mathematical statements by interactive theorem provers often requires algebraic computation. Since many Mechanized Mathematical Systems (MMS) support the OpenMath standard, we propose to link the HOL Light theorem prover to other MMSs via OpenMath. In particular, we present an interface between HOL Light and Mathematica enabling HOL Light users to evaluate arithmetic, transcendental and linear algebraic expressions, using Mathematica.

1 Introduction

Theorem proving is a technique which proves or checks the validity of logical statements. It is based on sequential applications of sound inference rules to a given axiomatic system. The statements proved by the theorem prover, accepting that its core is sound, are absolutely accurate in contrast to paper-and-pencil methods or computer simulations. Often in the process of interactive theorem proving one needs to perform a symbolic computation which might be a tedious task requiring hundreds of inference rules. For example, computing the value of a polynomial over $\mathbb{R}$ needs many inference rules and auxiliary theorems over the theory of real numbers. Furthermore, computing the roots of the same polynomial by the theorem prover is a very hard task. To avoid such limitations, one may make use of a Computer Algebra System (CAS) which has the needed functionality to perform the computation. The result of the CAS is transformed to an axiom which is added to the list of axioms and used by the theorem prover.

Many researchers have addressed the issue of combining symbolic/numeric computation with logical reasoning. One solution is building a CAS inside a theorem prover (e.g., [7]) or building a theorem prover inside a CAS (e.g., [2,10]). The second approach implements a bridge between theorem provers and CAS (e.g., PVS and Maple [1], Isabelle and Maple [3], and HOL and Maple [6]). This connection involves a master-slave relation in which the theorem prover is usually considered as a master and the CAS as a slave, with the assumption that there is no trust in the CAS. The third approach (e.g. Mathscheme[1]) is to build an integrated framework that provides the functionalities of both CAS and theorem proving integrating them into a single tool without sacrificing the soundness

---

[1] http://www.cas.mcmaster.ca/research/mathscheme/.

© Springer International Publishing Switzerland 2015
M. Kerber et al. (Eds.): CICM 2015, LNAI 9150, pp. 353–358, 2015.
DOI: 10.1007/978-3-319-20615-8_27

and without using an intermediate language. Finally, the fourth approach is to define a framework using a standard for mathematical information (such as MathML[2] and OpenMath[3]) that can be exchanged between different Mechanized Mathematical Systems (MMS). For instance, in [4] the authors used OpenMath to develop a Java client-server applet between Maple as a client and the Lego theorem prover as a server.

In this paper, we propose a tool linking HOL Light[4] to Mathematica[5] through the OpenMath standard. In contrast to [4] where the authors present a Java applet which takes a Maple expression as input and returns a Lego expression through the translation to OpenMath, our work is a combination between two external tools where OpenMath is used as a middleware. Another difference between our approach and [4] is that we do not rely on the communication layer established between the client and the server, but on the direct translation of a HOL Light statement to a Mathematica term and vice-versa. Therefore, the performance of the computation is increased.

The proposed tool is part of a general framework providing a heterogeneous problem-solving environment, which connects HOL Light to any MMS. Figure 1 illustrates the general approach of this framework which encompasses a variety of MMSs that support OpenMath such as the theorem provers LEGO[6] and COQ[7], the CASs Maple[8], Gap[9] and Mathematica, or the numerical solver Mupad[10] with the intention of solving and reasoning over larger sets of problems.

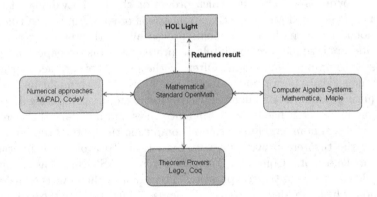

Fig. 1. Connecting Different MMS to HOL Light using OpenMath

---

[2] http://www.w3.org/Math/.

[3] http://www.openmath.org/overview/index.html.

[4] http://www.cl.cam.ac.uk/~jrh13/hol-light/.

[5] http://www.wolfram.com/.

[6] http://www.dcs.ed.ac.uk/home/lego/.

[7] https://coq.inria.fr/.

[8] http://www.maplesoft.com/products/maple/.

[9] http://www.gap-system.org/.

[10] http://de.mathworks.com/discovery/mupad.html.

2   Tool Description

The proposed linkage tool starts by translating the HOL Light statement into an OpenMath object. Then, a Java phrasebook [5], which is a collection of encoding/decoding methods between OpenMath and Mathematica, converts the OpenMath object into an expression, that is passed to Mathematica. The computation of Mathematica is translated back to an OpenMath object using again the Java phrasebook. Finally, the latter is parsed by our tool and converted to a HOL Light axiom. We developed a translator from HOL Light to OpenMath and visa-versa, which enables HOL Light users to access Mathematica's kernel using the Phrasebook OpenMath-Mathematica proposed by Caprotti [5]. After the computation, the returned result from Mathematica is represented as an axiom in HOL Light tagged by `Mathematica` in the form `Mathematica ⊢ Ψ`, where Ψ is the expression performed by Mathematica. Moreover, each theorem derived from this axiom inherits the tag `Mathematica`. This procedure helps to easily trace the axioms created from the interaction with an external tool. After the computation, the returned result can also be represented in another form as a sub-goal and added to the assumption of a main goal. One needs to prove it in order to pursue further proofs. Figure 2 depicts the structure of the tool connecting HOL Light to Mathematica, which is comprised of the following three modules:

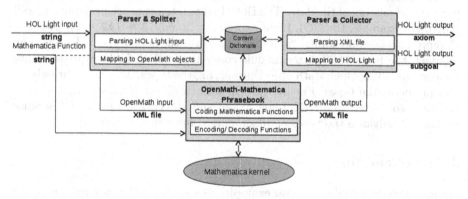

Fig. 2. Tool structure

The *Parser & Splitter* transforms the HOL Light statement into a corresponding OpenMath object as understood by means of the Content Dictionaries (CDs)[11]. First, it parses the HOL Light expression according to a grammar [9] which converts a HOL Light expression to the corresponding OpenMath object. Then, it decomposes the HOL Light input statement into a list of operations and operands. Thereafter, it maps each element of the list with the corresponding OpenMath symbol as understood by means of the related CDs. Finally, it stores the description of the OpenMath object in an XML file.

---

[11] http://www.openmath.org/cd/.

The *OpenMath-Mathematica Phrasebook*[12] defines a collection of Java classes, which provide two sets of methods. The first set represents the encoding and decoding methods between OpenMath and Mathematica based on the declaration of the corresponding CDs. The second one describes the built-in Mathematica call function with the tag already specified by the user. The phrasebook translates the XML file that describes the OpenMath input object into a Mathematica statement, which is then passed to the Mathematica kernel through a Mathematica service. (See footnote 12) This service allows users to remotely call the Mathematica kernel as a computational engine using MathLink.[13] This connection is established via the TCP/IP protocol. Once the result is computed, the Mathematica output statement is translated back to OpenMath and an XML file is generated.

The *Parser & Collector* translates the OpenMath object which encodes the output of Mathematica into the corresponding HOL Light symbols in the relevant CDs. Then, it collects all the HOL Light symbols and returns an axiom tagged by the name of the CAS (i.e., `Mathematica`). In other cases, we can generate the returned result as a sub-goal and prove it in HOL Light. This provides some kind of a determinism to the proof process, because when one knows the result of the computation, finding the proof is more straightforward rather than searching for the result during the whole process of proof derivation.

The above process is sound in the sense that it preserves the types during the parsing and passing of the data. The HOL Light statements are by definition well typed. For example, the HOL Light expression "x pow 3 + (&2 * x) pow 2 + x" represents the polynomial "$x^3 + (2x)^2 + x$" over the field of the reals. Based on this fact, the Parser & Splitter module converts the HOL Light expression into the corresponding OpenMath object preserving the types. The Java Phrasebook also preserves the types. Finally, the OpenMath object obtained from Mathematica is converted to a HOL Light axiom and all types are preserved because we know in advance the types of all supported functions.

3   Applications

We have used our tool on several examples like solving or evaluating non-closed form formulas such as arithmetic or polynomial manipulations or matrix operations, simplifying integrals, derivatives and transcendental functions, computing eigenvectors, checking inequalities, finding roots and factorization of complex polynomials. In the following, we give two simple examples. The input is a string which represents an expression given to HOL Light. The expression consists of two parts. The first part is an ordinary HOL Light expression, whereas the second part represents the built-in Mathematica symbol such as "Factor", "Simplify", etc. The output is an axiom in HOL Light. For example, using the built-in symbol "FullSimplify" we show the computation of the integral $\int_1^{10}(x + 1)\,\mathrm{d}x$:

---

[12] http://mathdox.org/new-web/index.html.
[13] http://reference.wolfram.com/mathematica/tutorial/IntroductionToMathLink. html.

Input:

`#call_mathematica "real_integral (real_interval [&1, &10]) (\x.x + &1)"`
`"FullSimplify";;`

The computed result, $117/2$, is returned to HOL Light as an axiom:

Output:

`val it: thm = Mathematica ⊢ real_integral (real_interval [&1, &10])`
`(\x.x + &1) = &117/&2`

The next example shows the factorization of the polynomial $x^3 + 2.x^2 + x$:

Input:

`#call_mathematica "x pow 3 + &2 * (x pow 2) + x" "Factor";;`

The expected result, $x.(x + 1)^2$, is returned as a HOL Light axiom:

Output:

`val it: thm = Mathematica ⊢ x pow 3+&2*(x pow 2) +x = x*(&1+x) pow 2`

We have conducted several more comprehensive experiments, which can be found in [9]. Moreover, our tool was successfully applied in the formal verification of optical systems [8], where we send from HOL Light the expression of a boundary condition of an optical interface described with electromagnetic fields to Mathematicia in order to be simplified. Details of these experiments can be found in [9]. These examples emphasize not only the benefits of computing such Mathematica expressions within HOL Light but also the efficient performance of our tool in terms of execution time. Our tool, called *HolMatica*, is implemented in a way that we can easily adapt it to any other CAS or theorem provers that support OpenMath. The *HolMatica* tool and running examples can be downloaded from http://hvg.ece.concordia.ca/research/tools/holmatica/

References

1. Adams, A., Dunstan, M.N., Gottliebsen, H., Kelsey, T., Martin, U., Owre, S.: Computer algebra meets automated theorem proving: integrating maple and PVS. In: Boulton, R.J., Jackson, P.B. (eds.) TPHOLs 2001. LNCS, vol. 2152, pp. 27–42. Springer, Heidelberg (2001)
2. Bauer, A., et al.: Analytica - an experiment in combining theorem proving and symbolic computation. JAR **21**(3), 295–325 (1998)
3. Ballarin, C., et al.: Theorems and Algorithms: An Interface between Isabelle and Maple. In: ISSAC, pp. 150–157. ACM (1995)
4. Caprotti, O., Cohen, A.M.: Integrating computational and deduction systems using OpenMath. ENTCS **23**(3), 469–480 (1999)
5. Caprotti, O., Cohen, A.M., Riem, M.: Java phrasebooks for computer algebra and automated deduction. SIGSAM Bulltin **34**, 33–37 (2000)
6. Harrison, J., Théry, L.: A skeptic's approach to combining HOL and maple. JAR **21**, 279–294 (1998)
7. Kaliszyk, C., Wiedijk, F.: Certified computer algebra on top of an interactive theorem prover. In: Kauers, M., Kerber, M., Miner, R., Windsteiger, W. (eds.) MKM/CALCULEMUS 2007. LNCS (LNAI), vol. 4573, pp. 94–105. Springer, Heidelberg (2007)
8. Afshar, S.K., et al.: formal analysis of optical systems. MCS **8**(1), 39–70 (2014)

9. Seddiki, O.: Linking HOL Light to Mathematica using OpenMath. Master's thesis, Concordia University, Montreal, QC, Canada, October 2014
10. Windsteiger, W.: Theorema 2.0: a graphical user interface for a mathematical assistant system. CEUR Workshop Proceedings, vol. 118, pp. 73–81 (2012)

Author Index

Printed in the United States
By Bookmasters